Repetitive DNA Sequences

Repetitive DNA Sequences

Special Issue Editors

Andrew G. Clark
Daniel A. Barbash
Sarah E. Lower
Anne-Marie Dion-Côté

MDPI • Basel • Beijing • Wuhan • Barcelona • Belgrade • Manchester • Tokyo • Cluj • Tianjin

Special Issue Editors

Andrew G. Clark
Department of Molecular
Biology and Genetics,
Cornell University
USA

Daniel A. Barbash
Department of Molecular
Biology and Genetics,
Cornell University
USA

Sarah E. Lower
Department of Biology,
Bucknell University
USA

Anne-Marie Dion-Côté
Département de Biologie,
Université de Moncton
Canada

Editorial Office
MDPI
St. Alban-Anlage 66
4052 Basel, Switzerland

This is a reprint of articles from the Special Issue published online in the open access journal *Genes* (ISSN 2073-4425) (available at: https://www.mdpi.com/journal/genes/special_issues/Repetitive_DNA_Sequences).

For citation purposes, cite each article independently as indicated on the article page online and as indicated below:

LastName, A.A.; LastName, B.B.; LastName, C.C. Article Title. *Journal Name* **Year**, *Article Number*, Page Range.

ISBN 978-3-03928-366-8 (Pbk)
ISBN 978-3-03928-367-5 (PDF)

© 2020 by the authors. Articles in this book are Open Access and distributed under the Creative Commons Attribution (CC BY) license, which allows users to download, copy and build upon published articles, as long as the author and publisher are properly credited, which ensures maximum dissemination and a wider impact of our publications.

The book as a whole is distributed by MDPI under the terms and conditions of the Creative Commons license CC BY-NC-ND.

Contents

About the Special Issue Editors vii

Sarah E. Lower, Anne-Marie Dion-Côté, Andrew G. Clark and Daniel A. Barbash
Special Issue: Repetitive DNA Sequences
Reprinted from: *Genes* **2019**, *10*, 896, doi:10.3390/genes10110896 1

Justin P. Blumenstiel
Birth, School, Work, Death, and Resurrection: The Life Stages and Dynamics of Transposable Element Proliferation
Reprinted from: *Genes* **2019**, *10*, 336, doi:10.3390/genes10050336 5

Yann Bourgeois and Stéphane Boissinot
On the Population Dynamics of Junk: A Review on the Population Genomics of Transposable Elements
Reprinted from: *Genes* **2019**, *10*, 419, doi:10.3390/genes10060419 19

Jesper Boman, Carolina Frankl-Vilches, Michelly da Silva dos Santos, Edivaldo H. C. de Oliveira, Manfred Gahr and Alexander Suh
The Genome of Blue-Capped Cordon-Bleu Uncovers Hidden Diversity of LTR Retrotransposons in Zebra Finch
Reprinted from: *Genes* **2019**, *10*, 301, doi:10.3390/genes10040301 43

Elena Dalla Benetta, Omar S. Akbari and Patrick M. Ferree
Sequence Expression of Supernumerary B Chromosomes: Function or Fluff?
Reprinted from: *Genes* **2019**, *10*, 123, doi:10.3390/genes10020123 61

Gabrielle Hartley and Rachel J. O'Neill
Centromere Repeats: Hidden Gems of the Genome
Reprinted from: *Genes* **2019**, *10*, 223, doi:10.3390/genes10030223 75

Romain Lannes, Carène Rizzon and Emmanuelle Lerat
Does the Presence of Transposable Elements Impact the Epigenetic Environment of Human Duplicated Genes?
Reprinted from: *Genes* **2019**, *10*, 249, doi:10.3390/genes10030249 97

Karen H. Miga
Centromeric Satellite DNAs: Hidden Sequence Variation in the Human Population
Reprinted from: *Genes* **2019**, *10*, 352, doi:10.3390/genes10050352 117

Mats E. Pettersson and Patric Jern
Whole-Genome Analysis of Domestic Chicken Selection Lines Suggests Segregating Variation in ERV Makeups
Reprinted from: *Genes* **2019**, *10*, 162, doi:10.3390/genes10020162 131

Elizaveta Radion, Olesya Sokolova, Sergei Ryazansky, Pavel A. Komarov, Yuri Abramov and Alla Kalmykova
The Integrity of piRNA Clusters is Abolished by Insulators in the *Drosophila* Germline
Reprinted from: *Genes* **2019**, *10*, 209, doi:10.3390/genes10030209 143

Radka Symonová
Integrative rDNAomics—Importance of the Oldest Repetitive Fraction of the Eukaryote Genome
Reprinted from: *Genes* **2019**, *10*, 345, doi:10.3390/genes10050345 . . . . **157**

Changcheng Wu and Jian Lu
Diversification of Transposable Elements in Arthropods and Its Impact on Genome Evolution
Reprinted from: *Genes* **2019**, *10*, 338, doi:10.3390/genes10050338 . . . . **173**

About the Special Issue Editors

Daniel Barbash is a Professor in the Department of Molecular Biology and Genetics at Cornell University. The Barbash lab investigates genome evolution, in order to understand the forces that drive genomic change and how evolution at the DNA level leads to phenotypic divergence and speciation.

Andrew Clark is the Jacob Gould Schurman Professor of Population Genetics and Nancy and Peter Meinig Family Investigator, and Professor in the Department of Molecular Biology and Genetics at Cornell University. The Clark lab is focused on empirical and analytical problems associated with genetic variation in populations.

Sarah Lower is an Assistant Professor in the Department of Biology at Bucknell University. The Lower lab integrates ecological, molecular, and computational approaches to investigate questions about how and why organisms are so diverse. In particular, they are interested in the genetic mechanisms and evolutionary processes underlying the species diversity of fireflies.

Anne-Marie Dion-Côté is an Assistant Professor in the Department of Biology at Université de Moncton. Research in her lab is focused on the role of genome stability in speciation and evolution. They are particularly interested in the molecular basis of reproductive isolation

Editorial

Special Issue: Repetitive DNA Sequences

Sarah E. Lower [1,2], Anne-Marie Dion-Côté [2,3], Andrew G. Clark [2] and Daniel A. Barbash [2,*]

1. Department of Biology, Bucknell University, Lewisburg, PA 17837, USA; s.lower@bucknell.edu
2. Department of Molecular Biology and Genetics, Cornell University, Ithaca, NY 14850, USA; anne-marie.dion-cote@umoncton.ca (A.-M.D.-C.); ac347@cornell.edu (A.G.C.)
3. Biology Department, Université de Moncton, Moncton, NB E1A 3E9, Canada
* Correspondence: barbash@cornell.edu

Received: 2 October 2019; Accepted: 24 October 2019; Published: 6 November 2019

Abstract: Repetitive DNAs are ubiquitous in eukaryotic genomes and, in many species, comprise the bulk of the genome. Repeats include transposable elements that can self-mobilize and disperse around the genome and tandemly-repeated satellite DNAs that increase in copy number due to replication slippage and unequal crossing over. Despite their abundance, repetitive DNAs are often ignored in genomic studies due to technical challenges in identifying, assembling, and quantifying them. New technologies and methods are now allowing unprecedented power to analyze repetitive DNAs across diverse taxa. Repetitive DNAs are of particular interest because they can represent distinct modes of genome evolution. Some repetitive DNAs form essential genome structures, such as telomeres and centromeres, that are required for proper chromosome maintenance and segregation, while others form piRNA clusters that regulate transposable elements; thus, these elements are expected to evolve under purifying selection. In contrast, other repeats evolve selfishly and cause genetic conflicts with their host species that drive adaptive evolution of host defense systems. However, the majority of repeats likely accumulate in eukaryotes in the absence of selection due to mechanisms of transposition and unequal crossing over. However, even these "neutral" repeats may indirectly influence genome evolution as they reach high abundance. In this Special Issue, the contributing authors explore these questions from a range of perspectives.

Keywords: repetitive DNA; transposable element; heterochromatin; genome evolution; genomic conflict

---

Repetitive DNAs include both short and long sequences that repeat in tandem or are interspersed throughout the genome, such as transposable elements (TE), ribosomal rRNA genes (rDNA), and satellite DNA. Repetitive DNA is ubiquitous in eukaryotic genomes, but despite this universality, their possible functions and predictable patterns of evolution remain relatively poorly characterized across taxa. Empirical evidence suggests important roles of repetitive DNA in chromosome stability and segregation, as well as gene regulation. Theory predicts roles of both neutral processes (unequal crossing over, gene conversion) and selection, as well as selfish (non-Mendelian) transmission, in determining patterns of sequence variation in repetitive regions. Despite a wealth of theory, until recently, this fraction of the genome has remained largely overlooked due to technological constraints on sequencing and quantifying repetitive DNA genome-wide. With the advent of high-throughput sequencing technologies, this portion of the genome has become more accessible, though inherent biases due to sequencing chemistry and computational identification pipelines remain challenges.

In this special issue, 11 articles review the evolution and function of the different classes of repetitive DNA and empirically investigate their predicted functions and evolutionary patterns from a variety of perspectives. Two articles approach TE evolution from different angles—Blumenstiel [1] describes the life cycle of a TE and uses this analogy to develop predictions for how TEs evolve, using known examples to describe persisting TEs as quickly proliferating genome invaders, long-lasting residents, and even as "resurrectors" from previously "dead" copies. In another review, Bourgeois and Boissinot [2] synthesize perspectives on the roles of adaptive and non-adaptive processes in TE evolution and offer ways forward to model TE evolution at the population level.

Two studies test predictions about TE evolution using a macroevolution approach. Wu and Lu [3] first develop a new pipeline for identifying transposable elements and then apply it to examine TE proliferation and diversification across 500 million years of arthropod evolution. They introduce the Arthropod TE database as a resource for TE consensus sequences for the community to use and build on. Bohman et al. [4] provide a genome assembly for the Blue-capped Cordon-Bleu, a small East African finch, whose karyotype and annotated transposon content enable new detailed examination of TE evolution in birds, particularly relatives of the model zebra finch. Their results highlight the utility of employing a comparative approach to investigate TE evolution. Together, these papers offer a dynamic view of TE evolution.

Three papers examine the role of adaptive and non-adaptive processes in TE evolution using genomic and functional approaches. Taking a computational approach, Pettersson and Jern [5] find a greater role for neutral evolution rather than selection in endogenous retrovirus (ERV) diversification across domestic chicken lineages. In contrast, Radion et al. [6] use functional and genomic analyses to examine the transcriptional regulation of piRNA clusters and TEs and find evidence for selective constraints. Lannes et al. [7] provide evidence for links between TE presence/absence and regulation of their activity via epigenetic modifications, implicating selection on their regulation. Together, these papers demonstrate the interplay of selection and neutral processes in different groups and emphasize the need for more studies to test broadly applicable "rules" for TE evolution.

Four papers focus on the evolution and function of other less-studied repetitive DNA types. Symonová [8] reviews studies of rDNA, from their function to their use in phylogeny and integrates these perspectives to provide a wider view of rDNA importance and evolution. Benetta et al. [9] synthesize recent work on the non-Mendelian transmission of repetitive facultative (B) chromosomes. Miga [10] reviews recent work on the links between satellite DNA and disease, highlighting the importance of their study to human health. Hartley and O'Neill [11] discuss the evolution and function of satellite DNA and TEs in centromeres. These papers highlight overlooked types of repetitive DNA and identify key challenges to move the field forward.

This special issue demonstrates the benefits of applying multiple perspectives to tackle questions about repetitive DNA evolution, function, and adaptation. They paint a picture of the complex processes involved and reveal the need for additional work. With more affordable sequencing, and a growing arsenal of genetic tools and widely-available annotation databases, it is a promising time to tackle fundamental questions about repetitive DNA with important implications for our understanding of the fundamental rules of chromosome segregation, genome evolution, and human health. We would like to thank all of the authors and reviewers for their contributions to this issue.

Author Contributions: S.E.L., writing—original draft preparation. S.E.L, A.D., A.G.C., and D.A.B., writing—review and editing.

Funding: This editorial was funded by a Ruth L. Kirschstein Postdoctoral Individual National Research Service Award (F32GM126736) to S.E.L.; a post-doctoral scholarship from the Fonds de Recherche de Santé du Québec-Santé (FRQ-S 33616), the Natural Sciences and Engineering Research Council of Canada (NSERC PDF-516851-2018), the Lawski Foundation and a NSERC Discovery grant (RGPIN-2019-05744) to A.D.; and National Institutes of Health R01 GM119125 to D.A.B. and A.G.C. The content is solely the responsibility of the authors and does not necessarily represent the official views of the National Institutes of Health.

Acknowledgments: Many thanks are extended to the authors and reviewers for their contributions to this issue.

Conflicts of Interest: The authors declare no conflict of interest. The funders had no role in the writing of the manuscript.

References

1. Blumenstiel, J.P. Birth, school, work, death, and resurrection: The life stages and dynamics of transposable element proliferation. *Genes* **2019**, *10*, 336. [CrossRef] [PubMed]
2. Bourgeois, Y.; Boissinot, S. On the population dynamics of junk: A review on the population genomics of transposable elements. *Genes* **2019**, *10*, 419. [CrossRef] [PubMed]
3. Wu, C.; Lu, J. Diversification of transposable elements in arthropods and its impact on genome evolution. *Genes* **2019**, *10*, 338. [CrossRef] [PubMed]
4. Boman, J.; Frankl, V.C.; da Silva dos Santos, M.; de Oliveira, E.H.C.; Gahr, M.; Suh, A. The genome of Blue-capped Cordon-Bleu uncovers hidden diversity of LTR retrotransposons in zebra finch. *Genes* **2019**, *10*, 301. [CrossRef] [PubMed]
5. Petterson, M.E.; Jern, P. Whole-genome analysis of domestic chicken selection lines suggests segregating variation in ERV makeups. *Genes* **2019**, *10*, 162. [CrossRef] [PubMed]
6. Radion, E.; Sokolova, S.; Ryazansky, S.; Komarov, P.A.; Abramov, Y.; Kalmykova, A. The integrity of piRNA clusters is abolished by insulators in the Drosophila germline. *Genes* **2019**, *10*, 209. [CrossRef] [PubMed]
7. Lannes, R.; Rizzon, C.; Lerat, E. Does the presence of transposable elements impact the epigenetic environment of human duplicated genes. *Genes* **2019**, *10*, 249. [CrossRef] [PubMed]
8. Symonová, R. Integrative rDNAomics—Importance of the oldest repetitive fraction of the eukaryote genome. *Genes* **2019**, *10*, 345. [CrossRef] [PubMed]
9. Benetta, E.D.; Akbari, O.S.; Feree, F.M. Sequence expression of supernumerary B chromosomes: Function or fluff. *Genes* **2019**, *10*, 123. [CrossRef] [PubMed]
10. Miga, K.H. Centromeric satellite DNAs: Hidden sequence variation in the human population. *Genes* **2019**, *10*, 352. [CrossRef] [PubMed]
11. Hartley, G.; O'Neill, R.J. Centromere repeats: Hidden gems of the genome. *Genes* **2019**, *10*, 223. [CrossRef] [PubMed]

© 2019 by the authors. Licensee MDPI, Basel, Switzerland. This article is an open access article distributed under the terms and conditions of the Creative Commons Attribution (CC BY) license (http://creativecommons.org/licenses/by/4.0/).

Review

Birth, School, Work, Death, and Resurrection: The Life Stages and Dynamics of Transposable Element Proliferation

Justin P. Blumenstiel

Department of Ecology and Evolutionary Biology, University of Kansas, Lawrence, KS 66049, USA; jblumens@ku.edu

Received: 21 March 2019; Accepted: 23 April 2019; Published: 3 May 2019

Abstract: Transposable elements (TEs) can be maintained in sexually reproducing species even if they are harmful. However, the evolutionary strategies that TEs employ during proliferation can modulate their impact. In this review, I outline the different life stages of a TE lineage, from birth to proliferation to extinction. Through their interactions with the host, TEs can exploit diverse strategies that range from long-term coexistence to recurrent movement across species boundaries by horizontal transfer. TEs can also engage in a poorly understood phenomenon of TE resurrection, where TE lineages can apparently go extinct, only to proliferate again. By determining how this is possible, we may obtain new insights into the evolutionary dynamics of TEs and how they shape the genomes of their hosts.

Keywords: transposable element; horizontal transfer; arms race; LINE-1; *Alu*; *hobo*; *I* element

1. Introduction

"And he that was dead came forth, bound hand and foot with graveclothes." John 11:44.

Transposable elements (TEs) have an intimate relationship with the genomes of their hosts. Like any form of parasite they cause harm but they are also dependent on the host for fitness. However, unlike typical parasites, they are directly embedded in the genomes of their hosts. How can such parasites spread if they are harmful? Alleles that are harmful are expected to be lost, but transposable elements exist in essentially all forms of life. In eukaryotes, the persistence of TEs is explained by the fact that sexual reproduction allows TEs to spread even if their net effect is a reduction in host fitness. Gamete fusion allows TEs to colonize new genomes [1] and recombination breaks up the association between progenitor copies and harmful descendant copies [2,3]. However, if TEs proliferate too rapidly within genomes, the consequences of their harm can indeed become too high and impede their success [4]. Transposable elements must walk a fine line between a sufficient rate of proliferation and one that is not so great that TEs become too burdened by the harmful effects that they impose.

The nature of this tension depends on the degree of intimacy with the host genome and is illuminated by considering the moment when a TE and the host genome first meet. This occurs during horizontal transfer, which is the first stage in the life cycle of a TE (see reviews on TE life cycles [5–9]). When a TE first invades a genome, it is a particularly fragile moment for the TE family because such events are likely to be serendipitous. For an element to be successful during the early stages of invasion, it must exploit these chance moments and avoid being lost from the population by drift. Studies show that the optimal TE strategy during horizontal transfer is to have a very high initial transposition rate [4,10]. This arises from the fact that the probability that a new TE becomes established is similar to the probability of fixation for a new beneficial allele. In the case of a new beneficial allele, the probability of fixation is ~$2s$, where s is the beneficial selection coefficient. For a transposon, the probability of establishment is ~$2(u - s)$, where u is the transposition rate and s is the selection coefficient that measures the average harmful effect of each new single insertion [4,10,11]. Establishment is achieved when, on average, each individual in the population has one copy. Rather than fixation, I consider establishment to be a more appropriate term for TE families because fixation is a term that is more appropriate for alleles. Transposable elements insertions within the population are non-allelic if they reside at different locations in the genome. So, if each individual on average carries one insertion, the TE family can be considered established. A single TE insertion allele can be considered fixed if there are no non-insertion variants segregating in the population at that locus.

For both a new beneficial allele and a new transposable element, the fixation (or establishment) probabilities do not depend much on the population size since the dynamics of stochastic loss by drift when the novel variant first appears are the same whether the population size is one million or one trillion. However, a transposition rate that is too high, while it will increase the probability that a TE becomes established, may also impose such a burden that the host may become extinct if the selection regime fails to limit the ever-increasing copy number. Thus, it has been shown that the optimal strategy for a transposable element is to have a high transposition rate during early invasion, followed afterwards by a period with a lower rate of movement [4]. This lower rate of movement may be enabled by host TE suppression mechanisms such as small RNA silencing.

It is not apparent that selection on TE lineages would be efficient enough to directly select such a tunable strategy. However, this tension reveals that optimal TE strategies will depend on the nature of the relationship with a genome. On one end of the continuum, TEs may be long term residents. On the other end, TEs may adopt a strategy of rapid invasion and movement from species to species. In the first part of this review, I discuss the nature and implications of these two strategies. Then, I consider an interesting phenomenon of TE lineages that appear to reside within genomes, go extinct, and then apparently come "back to life" many generations later. I will argue that TEs that show this pattern—I will designate them Lazarus elements—may highlight interesting aspects of TE biology and host interaction.

2. Long-Lasting Relationships

Some TE lineages are long-lived residents of their host genomes. In some cases, this is because TEs have adopted a cooperative strategy with the host. For example, in *Drosophila*, telomere function has been assumed by TEs [12]. However, for TEs that remain parasitic with respect to the host, there may be no better example of long-term coexistence than the LINE-1 elements of mammalian genomes. LINE-1 elements are a member of the non-LTR retrotransposon class and have been residents of mammalian genomes since early in the radiation of mammals [13–15]. In humans, the LINE-1 element has had a profound role in shaping the genome and there are approximately 500,000 copies of this element [16,17]. The LINE-1 family is shared across most mammals due to continued vertical transmission since early in the mammalian radiation [18,19]. Vertebrates that include reptiles, amphibians and fish also share LINE-1 elements, suggesting that the LINE-1 element may have been present since before the origin of mammals [14]. Alternately, it has been proposed that LINE-1 elements entered the therian mammal ancestor (rather than the ancestor of all mammals) through horizontal transfer. This is suggested by the observation that monotremes lack LINE-1 elements and have no clear signature of their previous activity [20]. In either case, the LINE-1 lineage shows a striking level of persistence and success across mammals through ongoing vertical transmission.

What has enabled this intimate relationship for millions of years within mammals? Phylogenetic analysis of LINE-1 elements within mammals has revealed a particular feature of LINE-1 persistence. In particular, phylogenetic trees of LINE-1 elements within a genome frequently have a "ladder-like" appearance [14,21,22]. This represents a scenario in which, through evolutionary time, there is typically only one or few proliferating lineages. This phylogenetic pattern has been proposed to be driven by an ongoing evolutionary arms-race with the host [14]. In particular, as mechanisms of LINE-1 control evolve on the part of the host, evolutionary innovation on the part of the TE lineage enables escape from host control. Recurrent cycles of adaptation and innovation—in both host and TE—can thus lead to the persistence of a single successful TE lineage [21]. This pattern may also be driven by the smaller effective population sizes that are likely more common in mammals. In very large populations, the fixation of an active and harmful TE insertion allele by drift is unlikely. However, in smaller populations, drift may allow such insertion alleles to fix. When an active copy becomes fixed at a particular locus, only decay into a non-functional state will allow the active copy to be lost from the population. Thus, fixation of an active TE insertion allele represents a critical stage in TE-host dynamics.

Faced with the continued presence of the LINE-1 element over millions of years, specialized modes of host control are proposed to contribute to the evolutionary dynamics that yield the arms-race driven "ladder" phylogeny. In particular, new active LINE-1 lineages may carry key innovations that enable specialized modes of escape from repression [23]. Diverse proteins that restrict LINE-1 transposition include APOBEC3, MOV10, ZAP, SAMHD1 and ZNF93 [24]. Signatures of recurrent LINE-1 adaptation that allow evasion from these restricting factors have also been found. For example, within mammals, the 5' UTR of LINE-1 is highly dynamic [22,25–27]. This has been proposed to be driven by the ongoing evolution of KRAB zinc fingers that can evolve specificity to target particular sequences in LINE-1 for repression. In response to this, it appears that selection on the LINE-1 lineage has driven removal of particular target sequences from the 5' UTR [28].

The ongoing persistence of one or few evolving LINE-1 lineages is likely enforced by within lineage competition. Otherwise, we might expect different modes of adaptation to evolve on distinct TE lineages, followed by successful diversification. Competition for host factors required for transposition has been proposed to contribute to this dynamic [29]. Strikingly, and in contrast to mammalian systems, the proliferation of one or few LINE-1 element lineages does not seem to apply in other vertebrates [14]. Rather, multiple lineages of LINE-1 elements have expanded and proliferated in the genomes of reptiles, amphibians and fish [30–32]. This represents a distinct mode of long-term coexistence within the genomes of non-mammalian species. Differences in demographic history and the strength of selection are likely to contribute to this difference. Compared to mammals, some non-mammalian species with greater LINE-1 diversity also show a stronger signature of selection acting to limit the fixation of TE insertion alleles [33]. This suggests that different selection regimes may contribute to the difference in LINE-1 dynamics between mammalian and non-mammalian species (but see [34]). One difference may arise from differences in the probability of ectopic recombination between dispersed repeats [29]. Selection against ectopic recombination is an important determinant of TE dynamics and a low rate across mammalian genomes may decrease the strength of selection against insertions and allow the accumulation of repetitious sequences [35–40]. In addition, if lower levels of ectopic recombination allow greater TE accumulation, persisting copies that fix by drift may intensify competition for host factors. Thus, as genomic copy number increases due to reduced levels of genome-wide ectopic recombination, the magnitude of competition for host factors may increase among competing copies and lineages. This may lead to a greater tendency for a single lineage to outcompete all other lineages. For these reasons, selection on LINE-1 lineages may not simply be to evade host restriction factors. Selection to increase access to host factors that enable transposition, amidst a genome filled with many other copies, may also be critical.

3. Horizontal Transfer: Fast, Cheap and Out of Control

> "Based on our experience in building ground based mobile robots (legged and wheeled), we argue here for fast, cheap missions using large numbers of mass produced simple autonomous robots..." Brooks and Flynn. 1989. Fast, Cheap and Out of Control: A robot invasion of the solar system.

These contrasting modes of LINE-1 evolution—the proliferation of a few lineages in mammals vs. diversification in reptiles, amphibians and fish—represent two forms of long-term co-existence. As previously indicated, long-term co-existence can also be maintained if TEs adapt a strategy of cooperation, as seen in the case of *Drosophila* telomeres. However, for selfish TEs that display parasitic behavior with respect to the host, another strategy relies on horizontal transfer and recurrent invasion. If TEs have the capacity to invade genomes through horizontal transfer, long-term persistence may be enabled by a 'live fast, die young' strategy [41,42]. If a TE family can invade a species, proliferate, and jump to a new species, it may conceivably persist even if it is unlikely to endure within any single species. Studies of the DNA transposon *mariner* in *Drosophila* illustrate how such a strategy is possible [43–45]. *mariner* was discovered in *D. mauritiana*, a close relative of *D. melanogaster*. However, its presence within the *D. melanogaster* species subgroup is considered "spotty" [46,47]. In particular, it appears in several close relatives of *D. melanogaster* but is absent from *D. melanogaster* itself. It has apparently been lost. Interestingly, an additional *mariner* lineage is also found in the genomes of other members of the melanogaster species subgroup, including *D. erecta*, but was apparently lost from the *D. melanogaster*/*D. simulans* clade [48]. This latter *mariner* family also shares 97% sequence similarity with a *mariner* element found in the cat flea, indicating horizontal transfer several million years ago. Overall, these patterns indicate that *mariner* dynamics can be explained by a dynamic process of recurrent horizontal transfer and extinction [48]. In contrast to mammals, it appears that horizontal transfer is rampant in insect species. In a comprehensive analysis of the genomes of nearly 200 insect species, more than 2000 horizontal transfer events were found to have occurred within a span of about 10 million years [49]. Strikingly, the *Tc1/mariner* class of DNA transposons shows the greatest frequency of horizontal transfer. This high propensity for horizontal transfer has been attributed to a lack of dependence on host factors for transposition [50]. *Tc1/mariner cis* regulatory sequences that drive transcription in diverse genomes may also facilitate efficient movement across species [51]. Within a single species, a TE lineage can proliferate if its transposition rate is sufficiently high so that it can increase at a rate faster than its removal due to negative selection. The same principle should also apply across species. If a TE can invade, by horizontal transfer, the genomes of new species at a rate faster than the within species extinction rate, the lineage will also find success. In this case, since TE success depends on being able to move across species, it is unlikely that natural selection will be sufficient for adaptation, on the part of a TE lineage, to a particular host genome. Rather, natural selection will favor a "generalist" strategy that enables movement in the genomes of many species.

4. Extinction

Whether a TE is adapted for continued vertical transmission (as observed for LINE-1 elements) or ongoing movement across species (as perhaps observed for *mariner* elements), TE lineages are not guaranteed perpetual success. Rather, they can also go extinct within a species. Across mammals, LINE-1 extinction has been observed in the rhinoceros and lineages of rodents, bats, insectivores and Afrotherians [15,52–55]. Several mechanisms have been proposed to contribute to LINE-1 extinction. In one scenario, mechanisms of host suppression may be sufficient. It has been noted that the fate of a TE lineage depends on the balance between transposition and the rate of accumulation for degenerating mutations [56]. If the transposition rate is lower than the rate of mutation that renders an element inactive, then the TE lineage will decay. Thus, host control mechanisms that drive a significantly low transposition rate may also drive extinction by decay.

Other factors are also likely to contribute to extinction. TE families may drive other TE families to extinction through direct competition for host factors. For example, LINE-1 extinction in a group of sigmodontine rodents may have been influenced by competition for host factors with an expanding endogenous retrovirus lineage [57,58]. Extinction may also be driven by other TE lineages through direct sequestration of TE-encoded factors that enable transposition. SINE elements, such as the *Alu* element, hijack LINE-1 encoded factors to favor their own increase [59]. Thus, *Alu* amplification may drive LINE-1 extinction through competitive saturation of LINE-1 encoded factors required for LINE-1 transposition [60]. Finally, extinction may also be enabled by a form of lineage "suicide". In the case of DNA transposons, internally deleted copies may titrate functional transposase from fully functional copies [61]. As internally deleted copies within the genome increase, active DNA transposon lineages may lose sufficient access to their own encoded factors.

Finally, stochastic loss and demographic factors may also contribute to lineage extinction. In populations where an active TE does not fix at any particular location, selection or drift may simply lead to the loss of every active copy in the genome. This will be most likely when the transposition rate is sufficiently low, so it is likely to be enhanced by host suppression mechanisms. The dynamics of stochastic loss, in many ways, are likely to be similar to loss by the mutational degeneration of active copies. How long will it take for a TE family to be lost by this mechanism? Using simulation, I have shown that total copy number within the population—rather than population size or per genome copy number—dominates the dynamics of stochastic loss assuming no individual insertion becomes fixed (Figure 1). Selection also plays a role.

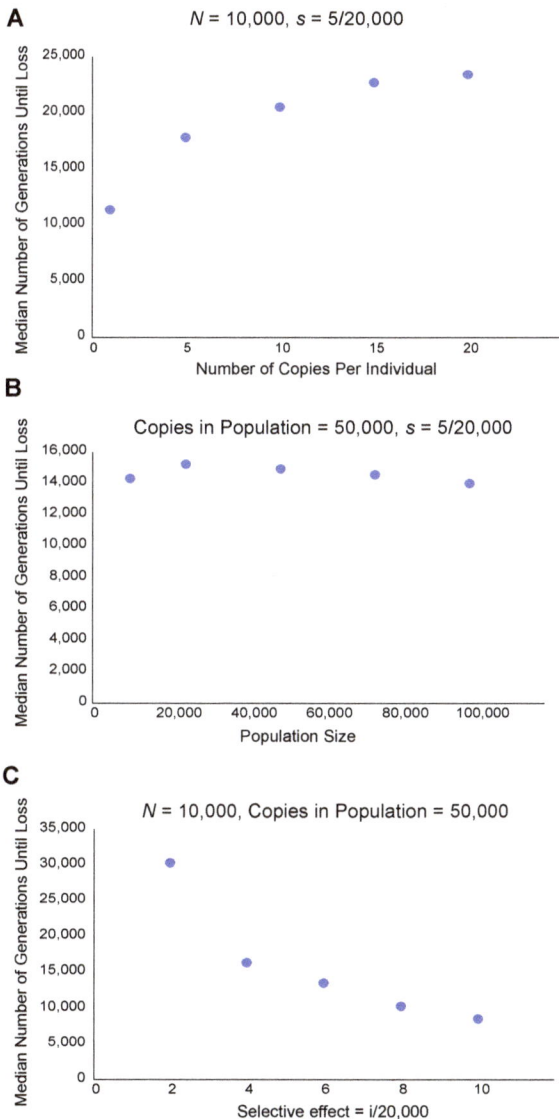

Figure 1. Dynamics of stochastic loss. N is the diploid population size and s is the selection coefficient acting against single insertions. All simulations were performed by simple binomial sampling of insertion

alleles starting at frequency of 1/2N. Sampling was iterated according to frequency in the population for a given number of copies. This procedure implicitly assumes there is no linkage. In addition, by assuming no actual transposition or degradation specifically, it is suitable to a scenario where the rate of transposition is equal to the rate of mutation to a non-functional state. Selection was simulated by adjusting the probability of sampling according to the selection coefficient. (**A**) Fixed population size and negative selection coefficient. The time until loss increases with per individual copy number. Note that the rate of increase declines. (**B**) A fixed number of copies in the population, distributed among individuals of different population sizes. The time until loss is not affected by population size. (**C**) An increasing selection coefficient, as expected, decreases the time until loss.

5. Resurrection

Overall, the canonical life-cycle of a TE family starts with invasion followed by proliferation and eventual extinction. The duration for each of these stages may vary and be influenced by a wide variety of factors, as outlined previously. Extinction is certainly not guaranteed but there are many examples of where this appears to be the case. More striking, however, is that in some cases, extinction seems to be followed by resurrection. This is a mysterious phase of TE dynamics and worthy of investigation because it may shed light on the evolution of TE life-strategies that range between recurrent invasion and long-term coevolution.

Resurrection, also known as re-invasion, occurs when an active TE lineage becomes quiescent and perhaps even extinct, and then later proliferates. Syndromes of hybrid dysgenesis were the first to reveal this phenomenon, in particular the *I-R* syndrome of dysgenesis. Hybrid dysgenesis is a syndrome of intraspecific sterility that occurs when active TE families transmitted paternally are absent or nearly absent from the maternal genome [62–64]. In the absence of abundant maternal copies, a pool of piRNAs that maintain TE repression is not provisioned to the zygote [65,66]. This leads to activation of paternally inherited TEs and sterility. Perhaps the most well understood syndrome of hybrid dysgenesis is the *P-M* system. *P-M* dysgenesis occurs when *P* elements inherited from *P* strain males, mated with *M* strain females, cause germline cell death [67,68] due to excessive transposition in the absence of maternal *P* element piRNAs. In the *P-M* system, the asymmetry in the *P* element abundance between *P* and *M* strains can be explained by recent horizontal transfer rather than resurrection [69]. *M* laboratory strains devoid of *P* elements were established in the early part of the 20th century. *P* element invasion of natural populations via horizontal transfer occurred at a similar time, so natural populations now carry many *P* elements [70,71]. In contrast, *I-R* dysgenesis seems to have arisen from resurrected *I* elements. *I-R* dysgenesis—observed as hatch failure in eggs laid by F1 females—occurs when *I* (inducer) strain males, carrying abundant non-LTR *I* retrotransposons, mate with *R* (reactive) strain females that lack active copies [63]. However, in contrast to the P-M system, the genomes of *R* strains are littered with degraded *I* elements that are the fossils of a previous proliferation event [72–75]. In fact, under certain conditions, the degraded *I* elements can contribute to the piRNA pool and mediate repression of the newer *I* elements [76]. Thus, the genome retains a memory of past invasion and still retains some capacity to restrain new *I* elements.

Two additional cases of hybrid dysgenesis reveal a similar scenario, indicating that resurrection may be a common but poorly appreciated part of the life cycle of TEs. In *Drosophila melanogaster*, a third case of hybrid dysgenesis is driven by the *hobo* element. *hobo* is a DNA transposon that causes hybrid dysgenesis when males carrying multiple active *hobos* are mated with females that lack them [64]. Studies indicate that American populations lacked active *hobo* elements in the 1950s [77]. Strikingly, the genomes of *D. melanogaster* as well as close relatives all carry degraded copies of *hobo*. This suggests that an active version of the *hobo* element was present in an ancestor of all *D. melanogaster*, was lost, but now has proliferated to the extent that it can cause hybrid dysgenesis. This new *hobo* variant also appears among close relatives that include *D. simulans* [78–81]. Finally, a similar scenario is observed in the hybrid dysgenesis syndrome of *Drosophila virilis*. The *Penelope* element likely contributes to this syndrome and it also represents a case of re-invasion. Multiple degraded copies within the genome of *D. virilis* and relatives suggest a scenario of proliferation and extinction, followed by re-invasion [82,83].

As phylogenetic analysis shows (Figures 2 and 3), the mode of past and current proliferation varies across these difference cases. For the *hobo* element, it appears proliferation occurred millions of years ago, prior to the divergence of *D. melanogaster* and the *D. simulans* clade. This corresponds to the upper portion of the *hobo* phylogeny where lineages of *D. melanogaster*, *D. simulans* and *D. sechellia* are intermingled in a complex manner. This was followed by another proliferation, perhaps only in the *D. simulans/D. sechellia* clade. As active and nearly identical copies currently exist in *D. melanogaster*, *D. simulans* and *D. sechellia*, a recent wave of re-activation appears to have occurred across all three species, but especially in *D. melanogaster*. Phylogenetic analysis suggests that the currently active lineage was derived from a clade of proliferating elements derived solely from *D. simulans*, as has been proposed [80]. In contrast to *hobo*, the previous proliferation of the *I* element seems more ongoing, but with one large wave of historical activity in the *D. melanogaster* genome. However, it also appears that the closest relative of the *I* element resides in *D. simulans*. This supports the possibility that both currently active variants of *hobo* and the *I* element were introduced into *D. melanogaster* from *D. simulans*. While these species do not readily produce fertile hybrids, they can in some cases.

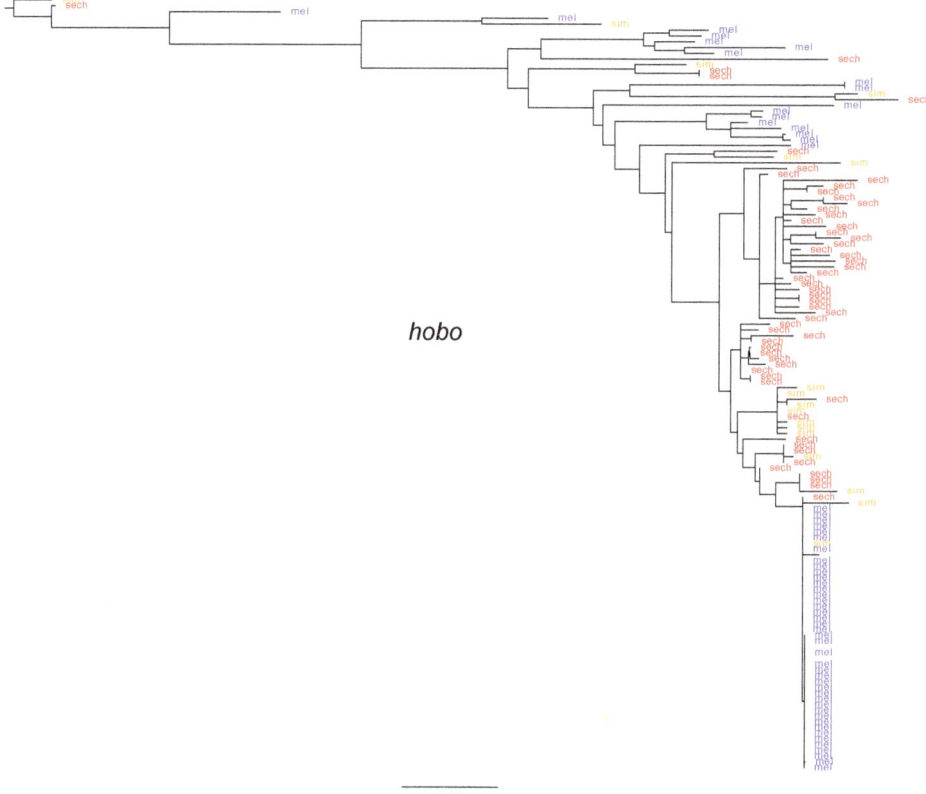

Figure 2. Phylogenetic analysis of *hobo* element fragments extracted from the genomes of *D. melanogaster*, *D. simulans* and *D. sechellia*. Alignments from BLAST output (E cutoff −100) were subjected to phylogenetic analysis with GARLI. Blue indicates *D. melanogaster*, red indicates *D. sechellia* and orange indicates *D. simulans*. The *hobo* phylogeny shows evidence of a previous wave in *D. melanogaster*, as well as new proliferation derived from a lineage residing within *D. simulans* or *D. sechellia*.

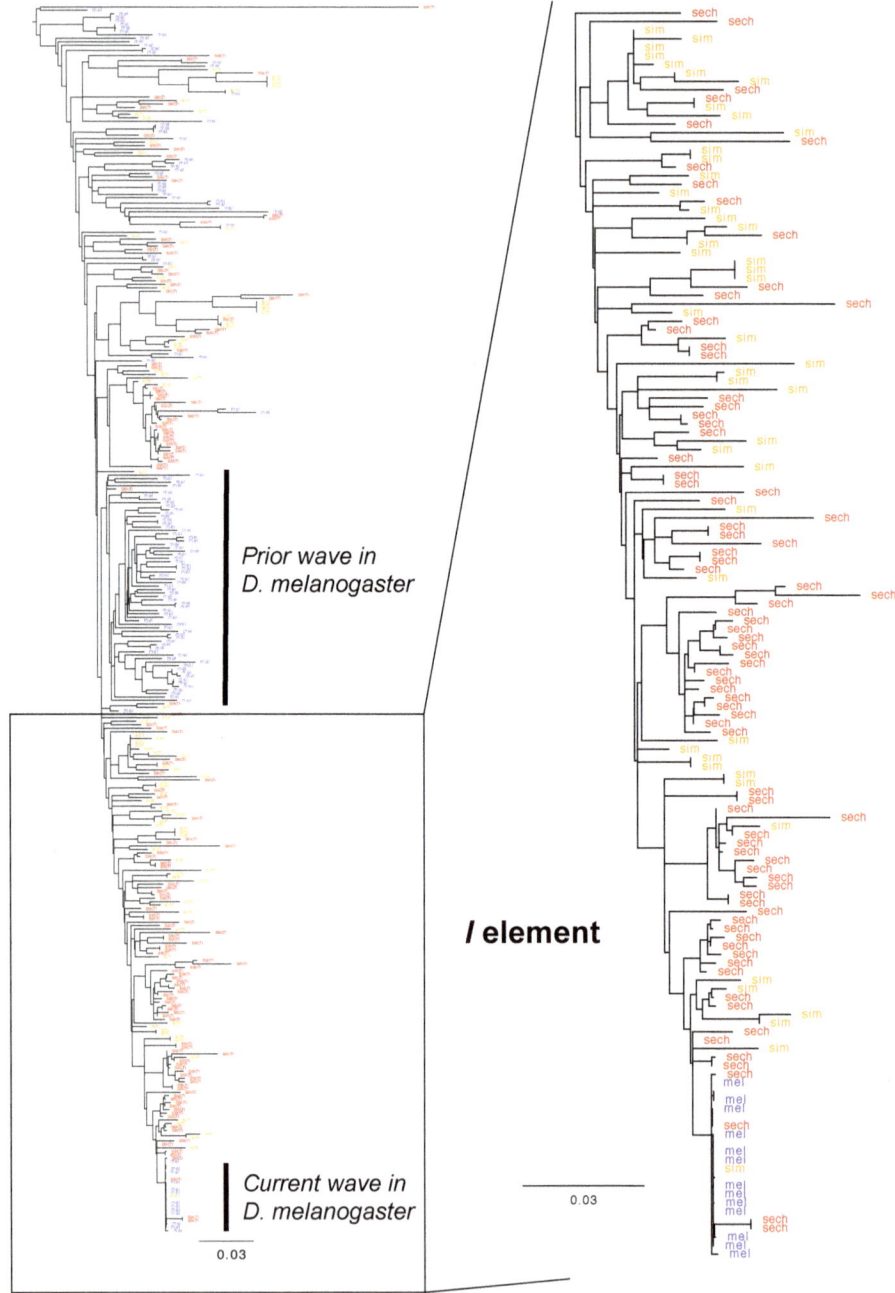

Figure 3. Phylogenetic analysis of *I* element fragments extracted from the genomes of *D. melanogaster*, *D. simulans* and *D. sechellia*. Alignments from BLAST output (E cutoff −100) were subjected to phylogenetic analysis with GARLI. Blue indicates *D. melanogaster*, red indicates *D. sechellia* and orange indicates *D. simulans*. A zoom-in is provided for further detail of recent dynamics. Like *hobo*, the *I* element shows evidence of a previous wave in *D. melanogaster*, as well as new proliferation derived from a lineage residing within *D. simulans* or *D. sechellia*.

How does re-invasion occur? What does the time between waves of proliferation tell us about the likely mechanism of re-invasion? This is a critical question and I consider two possibilities. One possible explanation is that re-invasion arises through iterated rounds of invasion by horizontal transfer (Figure 4A). Specifically, a TE lineage residing in a reservoir species (of any kind, not just a close relative), jumps into the host. After the first round of horizontal transfer, proliferation is followed by extinction. In turn, a second horizontal transfer event is followed by an additional round of proliferation. The reservoir source species of the original horizontal transfer event may not be the same as the second. Under this mechanism, the time between proliferation events may be significant.

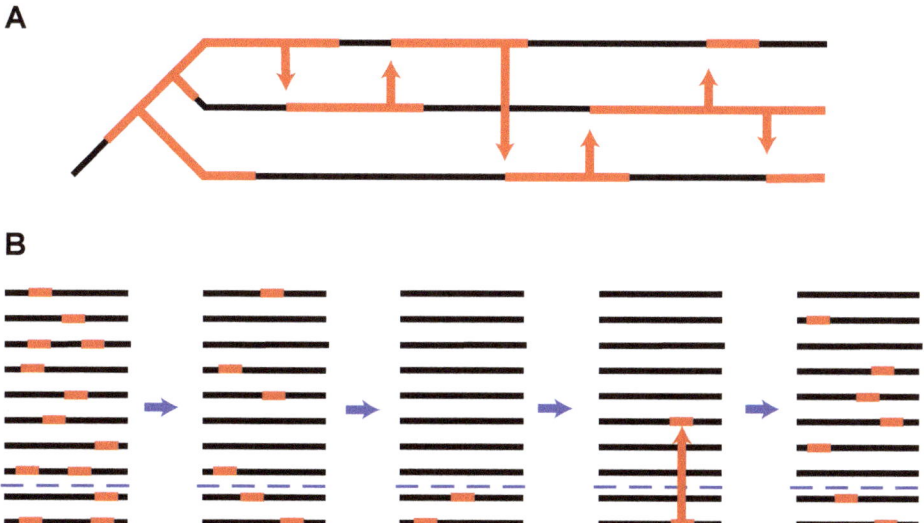

Figure 4. Dynamics of transposable element (TE) resurrection. (**A**) Iterated rounds of horizontal transfer (either through hybridization or other mechanisms) between sympatric close relatives allows extinction events to be followed by new rounds of proliferation. (**B**) Population subdivisions (indicated with the blue dashed line) allow a TE lineage to decrease in abundance, go extinct from one population, but become resurrected through contact with an isolated refuge population.

A second possibility is that a TE lineage becomes quiescent, but active copies linger within the species (Figure 4B). In this case, the TE lineage will have experienced only apparent, but not complete extinction. How could this occur? One scenario is that host suppression mechanisms first become sufficient so as to essentially cease transposition. This may occur through the fixation of host suppressor alleles. For example, a TE copy may land in a piRNA cluster and drive piRNA suppression that is strong enough to drive an extremely low transposition rate. Subsequently, if no TE insertion alleles are fixed within the population (due to selection against insertions), then all functional TE copies will eventually be lost by a combination of drift, selection and degeneration. However, this may take a very long period of time. During this lingering phase, several outcomes are possible. First, selection to retain host suppressor alleles (such as piRNA silencing alleles) may become reduced because there are few TE copies that remain a threat. However, if lingering TE copies take advantage of degraded suppressor alleles that have become frequent within the population, the TE lineage may return to an active state and in turn prevent the fixation of degraded host suppressor alleles. Depending on the timing of these dynamics, this may simply appear to represent a case of continued proliferation. An alternate outcome may give the illusion of resurrection. In this case, host suppression alleles may become degraded and these degraded variants may increase in frequency and perhaps even fix. In the scenario where non-functional host suppressor alleles are neutral, perhaps due to an exceedingly low abundance of lingering TE copies, the time until suppressor decay will be similar to the rate of mutation to the non-functional state. This may be of the order of millions of generations and it may be unlikely that an active lineage could persist for this time. But if lingering TE copies persist, the loss of a host suppressor allele may lead to a new round of TE proliferation. This scenario may be more likely if isolated populations can function as source refugia for new waves of TE proliferation. Depending on the timing of these dynamics, this may give the appearance of TE resurrection. Even if a host suppression allele is not lost, TE resurrection may also be apparent if a novel TE variant arises during the lingering phase. Such a variant may confer resistance to the host suppression allele, thus allowing a new round of TE proliferation. This would be analogous to evolutionary rescue of the TE lineage. Finally, true resurrection may be possible if recombination between non-functional copies leads to restoration of a functional copy.

These two scenarios may explain the appearance of Lazarus elements. In the first case, iterated rounds of horizontal transfer give the appearance of resurrection. In the latter case, an active lineage becomes quiescent for a duration, but lingers. After some period of time, it becomes active again. The key distinction between these two scenarios is the source for the newly proliferating lineage. Is it "from without" or "from within"? The likelihood of these two explanations depends on a large number of unknown parameters. It is apparent that horizontal transfer is quite common for transposable elements and this fact lends support to the "from without" hypothesis. In fact, at least for *Drosophila*, it appears that iterated rounds of horizontal transfer among close relatives might play an important role in the dynamics of re-invasion. For both the *hobo* and the *I* element, copies shared among close relatives are highly similar. Thus, rare hybridization may contribute to continued TE exchange and enable iterated bouts of re-invasion. Horizontal transfer of TEs among close relatives of *D. melanogaster* appears rampant [84]. A similar pattern of ongoing exchange of TEs also appears in close relatives of *D. pseudoobscura* that are in sympatry [85]. Overall, the genomes of closely related species in sympatry may function as an effective higher-level ecosystem for TEs. As modes of TE silencing in one species decay, horizontal transfer mediated by either hybridization or other mechanisms may allow iterated rounds of re-introduction. This dynamic is likely to be shaped by the dynamics of decay of host suppressor alleles. Nonetheless, it is also apparent that systems of host suppression are in rapid flux, and selection for host suppression in any single species may be quite weak. Thus, host alleles that suppress TE movement—such as TE insertions into piRNA clusters—may decay after the threat imposed by a TE family becomes reduced, even if the TE lineage has not yet become completely extinct within the species. In this case, new proliferation events may arise "from within". How might we test between these hypotheses? The key may be to identify the source refugia for resurrected TEs. For the "from without" hypothesis, these may be other species that live in physical proximity, such as close relatives that can hybridize. Alternately, shared parasites that enable horizontal transfer may also function as refugia [86]. However, horizontal transfer can be a very rare event, so proof of source may be extremely challenging. For the "from within" hypothesis, this may require a closer study of TE dynamics in large populations that have geographic structure. Theoretical studies may also examine whether it is plausible for a functional TE lineage to persist at low frequency until a host suppressor allele is lost, followed by re-invasion. Examples of apparent resurrection, where divergence between active and degraded copies are in the 5% to 10% range, suggest that the time would perhaps be too long. However, in the case of the *I* element where degraded copies may be quite young, the "from within" hypothesis might be more plausible. Altogether, distinguishing between these two possibilities will provide insight into the evolutionary strategies that TEs employ to ensure their continued presence across diverse species.

Funding: Funding was provided by NSF Award 1413532 and the University of Kansas.

Acknowledgments: I would like to thank two anonymous reviewers for the very helpful suggestions.

Conflicts of Interest: I declare no conflict of interest.

References

1. Hickey, D.A. Selfish DNA: A sexually-transmitted nuclear parasite. *Genetics* **1982**, *101*, 519–531. [PubMed]
2. Charlesworth, B.; Sniegowski, P.; Stephan, W. The evolutionary dynamics of repetitive DNA in eukaryotes. *Nature* **1994**, *371*, 215–220. [CrossRef]
3. Charlesworth, B.; Charlesworth, D. The population dynamics of transposable elements. *Genet. Res.* **1983**, *42*, 1–27. [CrossRef]
4. Le Rouzic, A.; Capy, P. The first steps of transposable elements invasion: Parasitic strategy vs. genetic drift. *Genetics* **2005**, *169*, 1033–1043. [CrossRef]
5. Song, M.J.; Schaack, S. Evolutionary Conflict between Mobile DNA and Host Genomes. *Am. Nat.* **2018**, *192*, 263–273. [CrossRef] [PubMed]
6. Venner, S.; Miele, V.; Terzian, C.; Biemont, C.; Daubin, V.; Feschotte, C.; Pontier, D. Ecological networks to unravel the routes to horizontal transposon transfers. *PLoS Biol.* **2017**, *15*, e2001536. [CrossRef] [PubMed]
7. Venner, S.; Feschotte, C.; Biemont, C. Dynamics of transposable elements: Towards a community ecology of the genome. *Trends Genet. TIG* **2009**, *25*, 317–323. [CrossRef] [PubMed]
8. Kidwell, M.G.; Lisch, D.R. Perspective: Transposable elements, parasitic DNA, and genome evolution. *Evolution* **2001**, *55*, 1–24. [CrossRef]
9. Hartl, D.L.; Lozovskaya, E.R.; Nurminsky, D.I.; Lohe, A.R. What restricts the activity of mariner-like transposable elements. *Trends Genet. TIG* **1997**, *13*, 197–201. [CrossRef]
10. Kaplan, N.; Darden, T.; Langley, C.H. Evolution and extinction of transposable elements in Mendelian populations. *Genetics* **1985**, *109*, 459–480. [PubMed]
11. Groth, S.B.; Blumenstiel, J.P. Horizontal Transfer Can Drive a Greater Transposable Element Load in Large Populations. *J. Hered.* **2017**, *108*, 36–44. [CrossRef]
12. Casacuberta, E.; Pardue, M.L. Transposon telomeres are widely distributed in the Drosophila genus: TART elements in the virilis group. *Proc. Natl. Acad. Sci. USA* **2003**, *100*, 3363–3368. [CrossRef]
13. Platt, R.N., 2nd; Vandewege, M.W.; Ray, D.A. Mammalian transposable elements and their impacts on genome evolution. *Chromosome Res.* **2018**, *26*, 25–43. [CrossRef] [PubMed]
14. Boissinot, S.; Sookdeo, A. The Evolution of LINE-1 in Vertebrates. *Genome Biol. Evol.* **2016**, *8*, 3485–3507. [CrossRef] [PubMed]
15. Ivancevic, A.M.; Kortschak, R.D.; Bertozzi, T.; Adelson, D.L. LINEs between Species: Evolutionary Dynamics of LINE-1 Retrotransposons across the Eukaryotic Tree of Life. *Genome Biol. Evol.* **2016**, *8*, 3301–3322. [CrossRef] [PubMed]
16. Boissinot, S.; Chevret, P.; Furano, A.V. L1 (LINE-1) retrotransposon evolution and amplification in recent human history. *Mol. Biol. Evol.* **2000**, *17*, 915–928. [CrossRef] [PubMed]
17. Lander, E.S.; Linton, L.M.; Birren, B.; Nusbaum, C.; Zody, M.C.; Baldwin, J.; Devon, K.; Dewar, K.; Doyle, M.; FitzHugh, W.; et al. Initial sequencing and analysis of the human genome. *Nature* **2001**, *409*, 860–921.
18. Burton, F.H.; Loeb, D.D.; Voliva, C.F.; Martin, S.L.; Edgell, M.H.; Hutchison, C.A., 3rd. Conservation throughout mammalia and extensive protein-encoding capacity of the highly repeated DNA long interspersed sequence one. *J. Mol. Biol.* **1986**, *187*, 291–304. [CrossRef]
19. Smit, A.F.; Toth, G.; Riggs, A.D.; Jurka, J. Ancestral, mammalian-wide subfamilies of LINE-1 repetitive sequences. *J. Mol. Biol.* **1995**, *246*, 401–417. [CrossRef]
20. Ivancevic, A.M.; Kortschak, R.D.; Bertozzi, T.; Adelson, D.L. Horizontal transfer of BovB and L1 retrotransposons in eukaryotes. *Genome Biol.* **2018**, *19*, 85. [CrossRef]
21. Furano, A.V. The biological properties and evolutionary dynamics of mammalian LINE-1 retrotransposons. *Prog. Nucleic Acid Res. Mol. Biol.* **2000**, *64*, 255–294. [PubMed]
22. Khan, H.; Smit, A.; Boissinot, S. Molecular evolution and tempo of amplification of human LINE-1 retrotransposons since the origin of primates. *Genome Res.* **2006**, *16*, 78–87. [CrossRef]
23. Boissinot, S.; Furano, A.V. Adaptive evolution in LINE-1 retrotransposons. *Mol. Biol. Evol.* **2001**, *18*, 2186–2194. [CrossRef] [PubMed]
24. Goodier, J.L. Restricting retrotransposons: A review. *Mob. DNA* **2016**, *7*, 16. [CrossRef] [PubMed]
25. Sookdeo, A.; Hepp, C.M.; McClure, M.A.; Boissinot, S. Revisiting the evolution of mouse LINE-1 in the genomic era. *Mob. DNA* **2013**, *4*, 3. [CrossRef] [PubMed]

26. Hayward, B.E.; Zavanelli, M.; Furano, A.V. Recombination creates novel L1 (LINE-1) elements in Rattus norvegicus. *Genetics* **1997**, *146*, 641–654. [PubMed]
27. Wincker, P.; Jubier-Maurin, V.; Roizes, G. Unrelated sequences at the 5' end of mouse LINE-1 repeated elements define two distinct subfamilies. *Nucleic Acids Res.* **1987**, *15*, 8593–8606. [CrossRef]
28. Jacobs, F.M.; Greenberg, D.; Nguyen, N.; Haeussler, M.; Ewing, A.D.; Katzman, S.; Paten, B.; Salama, S.R.; Haussler, D. An evolutionary arms race between KRAB zinc-finger genes ZNF91/93 and SVA/L1 retrotransposons. *Nature* **2014**, *516*, 242–245. [CrossRef]
29. Furano, A.V.; Duvernell, D.D.; Boissinot, S. L1 (LINE-1) retrotransposon diversity differs dramatically between mammals and fish. *Trends Genet. TIG* **2004**, *20*, 9–14. [CrossRef]
30. Novick, P.A.; Basta, H.; Floumanhaft, M.; McClure, M.A.; Boissinot, S. The evolutionary dynamics of autonomous non-LTR retrotransposons in the lizard *Anolis carolinensis* shows more similarity to fish than mammals. *Mol. Biol. Evol.* **2009**, *26*, 1811–1822. [CrossRef]
31. Hellsten, U.; Harland, R.M.; Gilchrist, M.J.; Hendrix, D.; Jurka, J.; Kapitonov, V.; Ovcharenko, I.; Putnam, N.H.; Shu, S.; Taher, L.; et al. The genome of the Western clawed frog Xenopus tropicalis. *Science* **2010**, *328*, 633–636. [CrossRef]
32. Duvernell, D.D.; Pryor, S.R.; Adams, S.M. Teleost fish genomes contain a diverse array of L1 retrotransposon lineages that exhibit a low copy number and high rate of turnover. *J. Mol. Evol.* **2004**, *59*, 298–308. [CrossRef] [PubMed]
33. Tollis, M.; Boissinot, S. Lizards and LINEs: Selection and demography affect the fate of L1 retrotransposons in the genome of the green anole (*Anolis carolinensis*). *Genome Biol. Evol.* **2013**, *5*, 1754–1768. [CrossRef]
34. Xue, A.T.; Ruggiero, R.P.; Hickerson, M.J.; Boissinot, S. Differential Effect of Selection against LINE Retrotransposons among Vertebrates Inferred from Whole-Genome Data and Demographic Modeling. *Genome Biol. Evol.* **2018**, *10*, 1265–1281. [CrossRef]
35. Petrov, D.A.; Aminetzach, Y.T.; Davis, J.C.; Bensasson, D.; Hirsh, A.E. Size matters: Non-LTR retrotransposable elements and ectopic recombination in Drosophila. *Mol. Biol. Evol.* **2003**, *20*, 880–892. [CrossRef] [PubMed]
36. Montgomery, E.A.; Huang, S.M.; Langley, C.H.; Judd, B.H. Chromosome rearrangement by ectopic recombination in *Drosophila melanogaster*: Genome structure and evolution. *Genetics* **1991**, *129*, 1085–1098.
37. Lee, Y.C.; Langley, C.H. Long-term and short-term evolutionary impacts of transposable elements on Drosophila. *Genetics* **2012**, *192*, 1411–1432. [CrossRef] [PubMed]
38. Charlesworth, B.; Langley, C.H.; Sniegowski, P.D. Transposable element distributions in Drosophila. *Genetics* **1997**, *147*, 1993–1995.
39. Charlesworth, B.; Langley, C.H. The population genetics of *Drosophila* transposable elements. *Annu. Rev. Genet.* **1989**, *23*, 251–287. [CrossRef]
40. Langley, C.H.; Montgomery, E.; Hudson, R.; Kaplan, N.; Charlesworth, B. On the role of unequal exchange in the containment of transposable element copy number. *Genet. Res.* **1988**, *52*, 223–235. [CrossRef] [PubMed]
41. Schaack, S.; Gilbert, C.; Feschotte, C. Promiscuous DNA: Horizontal transfer of transposable elements and why it matters for eukaryotic evolution. *Trends Ecol. Evol.* **2010**, *25*, 537–546. [CrossRef] [PubMed]
42. Gilbert, C.; Feschotte, C. Horizontal acquisition of transposable elements and viral sequences: Patterns and consequences. *Curr. Opin. Genet. Dev.* **2018**, *49*, 15–24. [CrossRef]
43. Hartl, D.L.; Lohe, A.R.; Lozovskaya, E.R. Modern thoughts on an ancyent marinere: Function, evolution, regulation. *Annu. Rev. Genet.* **1997**, *31*, 337–358. [CrossRef]
44. Wallau, G.L.; Capy, P.; Loreto, E.; Le Rouzic, A.; Hua-Van, A. VHICA, a New Method to Discriminate between Vertical and Horizontal Transposon Transfer: Application to the Mariner Family within Drosophila. *Mol. Biol. Evol.* **2016**, *33*, 1094–1109. [CrossRef]
45. Wallau, G.L.; Capy, P.; Loreto, E.; Hua-Van, A. Genomic landscape and evolutionary dynamics of mariner transposable elements within the Drosophila genus. *BMC Genom.* **2014**, *15*, 727. [CrossRef]
46. Capy, P.; David, J.R.; Hartl, D.L. Evolution of the transposable element mariner in the *Drosophila melanogaster* species group. *Genetica* **1992**, *86*, 37–46. [CrossRef]
47. Maruyama, K.; Hartl, D.L. Evolution of the transposable element mariner in Drosophila species. *Genetics* **1991**, *128*, 319–329. [PubMed]
48. Lohe, A.R.; Moriyama, E.N.; Lidholm, D.A.; Hartl, D.L. Horizontal transmission, vertical inactivation, and stochastic loss of mariner-like transposable elements. *Mol. Biol. Evol.* **1995**, *12*, 62–72. [CrossRef] [PubMed]

49. Peccoud, J.; Loiseau, V.; Cordaux, R.; Gilbert, C. Massive horizontal transfer of transposable elements in insects. *Proc. Natl. Acad. Sci. USA* **2017**, *114*, 4721–4726. [CrossRef] [PubMed]
50. Lampe, D.J.; Churchill, M.E.; Robertson, H.M. A purified mariner transposase is sufficient to mediate transposition in vitro. *EMBO J.* **1996**, *15*, 5470–5479. [CrossRef]
51. Palazzo, A.; Caizzi, R.; Viggiano, L.; Marsano, R.M. Does the Promoter Constitute a Barrier in the Horizontal Transposon Transfer Process? Insight from Bari Transposons. *Genome Biol. Evol.* **2017**, *9*, 1637–1645. [CrossRef] [PubMed]
52. Sookdeo, A.; Hepp, C.M.; Boissinot, S. Contrasted patterns of evolution of the LINE-1 retrotransposon in perissodactyls: The history of a LINE-1 extinction. *Mob. DNA* **2018**, *9*, 12. [CrossRef]
53. Cantrell, M.A.; Scott, L.; Brown, C.J.; Martinez, A.R.; Wichman, H.A. Loss of LINE-1 activity in the megabats. *Genetics* **2008**, *178*, 393–404. [CrossRef] [PubMed]
54. Platt, R.N., 2nd; Ray, D.A. A non-LTR retroelement extinction in Spermophilus tridecemlineatus. *Gene* **2012**, *500*, 47–53. [CrossRef]
55. Grahn, R.A.; Rinehart, T.A.; Cantrell, M.A.; Wichman, H.A. Extinction of LINE-1 activity coincident with a major mammalian radiation in rodents. *Cytogenet. Genome Res.* **2005**, *110*, 407–415. [CrossRef]
56. Nuzhdin, S.V. Sure facts, speculations, and open questions about the evolution of transposable element copy number. *Genetica* **1999**, *107*, 129–137. [CrossRef] [PubMed]
57. Erickson, I.K.; Cantrell, M.A.; Scott, L.; Wichman, H.A. Retrofitting the genome: L1 extinction follows endogenous retroviral expansion in a group of muroid rodents. *J. Virol.* **2011**, *85*, 12315–12323. [CrossRef]
58. Cantrell, M.A.; Ederer, M.M.; Erickson, I.K.; Swier, V.J.; Baker, R.J.; Wichman, H.A. MysTR: An endogenous retrovirus family in mammals that is undergoing recent amplifications to unprecedented copy numbers. *J. Virol.* **2005**, *79*, 14698–14707. [CrossRef]
59. Deininger, P. Alu elements: Know the SINEs. *Genome Biol.* **2011**, *12*, 236. [CrossRef]
60. Yang, L.; Wichman, H.A. Tracing the history of LINE and SINE extinction in sigmodontine rodents. *bioRxiv* **2018**. [CrossRef]
61. Hartl, D.L.; Lohe, A.R.; Lozovskaya, E.R. Regulation of the transposable element mariner. *Genetica* **1997**, *100*, 177–184. [CrossRef]
62. Bingham, P.M.; Kidwell, M.G.; Rubin, G.M. The molecular basis of P-M dysgenesis—The role of the P-element, a P-Strain-Specific Transposon Family. *Cell* **1982**, *29*, 995–1004. [CrossRef]
63. Bucheton, A.; Paro, R.; Sang, H.M.; Pelisson, A.; Finnegan, D.J. The molecular basis of the I-R hybrid dysgenesis syndrome in *Drosophila melanogaster*: Identification, cloning and properties of the I-Factor. *Cell* **1984**, *38*, 153–163. [CrossRef]
64. Yannopoulos, G.; Stamatis, N.; Monastirioti, M.; Hatzopoulos, P.; Louis, C. hobo is responsible for the induction of hybrid dysgenesis by strains of *Drosophila melanogater* bearing the male recombination factor 23.5MRF. *Cell* **1987**, *49*, 487–495. [CrossRef]
65. Blumenstiel, J.P.; Hartl, D.L. Evidence for maternally transmitted small interfering RNA in the repression of transposition in Drosophila virilis. *Proc. Natl. Acad. Sci. USA* **2005**, *102*, 15965–15970. [CrossRef]
66. Brennecke, J.; Malone, C.D.; Aravin, A.A.; Sachidanandam, R.; Stark, A.; Hannon, G.J. An Epigenetic Role for Maternally Inherited piRNAs in Transposon Silencing. *Science* **2008**, *322*, 1387–1392. [CrossRef] [PubMed]
67. Dorogova, N.V.; Bolobolova, E.U.; Zakharenko, L.P. Cellular aspects of gonadal atrophy in Drosophila P-M hybrid dysgenesis. *Dev. Biol.* **2017**, *424*, 105–112. [CrossRef] [PubMed]
68. Niki, Y.; Chigusa, S.I. Developmental analysis of gonadal sterility of P-M dysgenesis of *Drosophila melanogaster*. *Jpn. J. Genet.* **1986**, *61*, 147–157. [CrossRef]
69. Daniels, S.B.; Peterson, K.R.; Strausbaugh, L.D.; Kidwell, M.G.; Chovnick, A. Evidence for horizontal transmission of the P transposable element between Drosophila species. *Genetics* **1990**, *124*, 339–355.
70. Crow, J.F. The genesis of dysgenesis. *Genetics* **1988**, *120*, 315–318.
71. Engels, W.R. Invasions of P elements. *Genetics* **1997**, *145*, 11–15.
72. Crozatier, M.; Vaury, C.; Busseau, I.; Pelisson, A.; Bucheton, A. Structure and genomic organization of I elements involved in I-R hybrid dysgenesis in *Drosophila melanogaster*. *Nucleic Acids Res.* **1988**, *16*, 9199–9213. [CrossRef]
73. Bucheton, A.; Simonelig, M.; Vaury, C.; Crozatier, M. Sequences similar to the I transposable element involved in I-R hybrid dysgenesis in *D. melanogaster* occur in other *Drosophila* species. *Nature* **1986**, *322*, 650–652. [CrossRef]

74. Bucheton, A.; Vaury, C.; Chaboissier, M.C.; Abad, P.; Pelisson, A.; Simonelig, M. I elements and the Drosophila genome. *Genetica* **1992**, *86*, 175–190. [CrossRef]
75. Vaury, C.; Abad, P.; Pelisson, A.; Lenoir, A.; Bucheton, A. Molecular characteristics of the heterochromatic I elements from a reactive strain of *Drosophila melanogaster*. *J. Mol. Evol.* **1990**, *31*, 424–431. [CrossRef] [PubMed]
76. Grentzinger, T.; Armenise, C.; Brun, C.; Mugat, B.; Serrano, V.; Pelisson, A.; Chambeyron, S. piRNA-mediated transgenerational inheritance of an acquired trait. *Genome Res.* **2012**, *22*, 1877–1888. [CrossRef] [PubMed]
77. Periquet, G.; Hamelin, M.H.; Bigot, Y.; Lepissier, A. Geographical and historical patterns of the distribution of hobo elements in *Drosophila melanogaster* populations. *J. Evol. Biol.* **1989**, *2*, 223–229. [CrossRef]
78. Boussy, I.A.; Itoh, M. Wanderings of hobo: A transposon in *Drosophila melanogaster* and its close relatives. *Genetica* **2004**, *120*, 125–136. [CrossRef]
79. Ragagnin, G.T.; Bernardo, L.P.; Loreto, E.L. Unraveling the evolutionary scenario of the hobo element in populations of *Drosophila melanogaster* and *D. simulans* in South America using the TPE repeats as markers. *Genet. Mol. Biol.* **2016**, *39*, 145–150. [CrossRef]
80. Boussy, I.A.; Daniels, S.B. hobo transposable elements in *Drosophila melanogaster* and *D. simulans*. *Genet. Res.* **1991**, *58*, 27–34. [CrossRef]
81. Simmons, G.M. Horizontal transfer of hobo transposable elements within the *Drosophila melanogaster* species complex: Evidence from DNA sequencing. *Mol. Biol. Evol.* **1992**, *9*, 1050–1060. [CrossRef] [PubMed]
82. Lyozin, G.T.; Makarova, K.S.; Velikodvorskaja, V.V.; Zelentsova, H.S.; Khechumian, R.R.; Kidwell, M.G.; Koonin, E.V.; Evgen'ev, M.B. The structure and evolution of Penelope in the virilis species group of Drosophila: An ancient lineage of retroelements. *J. Mol. Evol.* **2001**, *52*, 445–456. [CrossRef]
83. Evgen'ev, M.B. What happens when Penelope comes?: An unusual retroelement invades a host species genome exploring different strategies. *Mob. Genet. Elem.* **2013**, *3*, 397–408. [CrossRef] [PubMed]
84. Bartolome, C.; Bello, X.; Maside, X. Widespread evidence for horizontal transfer of transposable elements across Drosophila genomes. *Genome Biol.* **2009**, *10*, R22. [CrossRef] [PubMed]
85. Hill, T.; Betancourt, A.J. Extensive exchange of transposable elements in the *Drosophila pseudoobscura* group. *Mob. DNA* **2018**, *9*, 20. [CrossRef] [PubMed]
86. Gilbert, C.; Schaack, S.; Pace, J.K., 2nd; Brindley, P.J.; Feschotte, C. A role for host-parasite interactions in the horizontal transfer of transposons across phyla. *Nature* **2010**, *464*, 1347–1350. [CrossRef]

© 2019 by the author. Licensee MDPI, Basel, Switzerland. This article is an open access article distributed under the terms and conditions of the Creative Commons Attribution (CC BY) license (http://creativecommons.org/licenses/by/4.0/).

Review

On the Population Dynamics of Junk: A Review on the Population Genomics of Transposable Elements

Yann Bourgeois and Stéphane Boissinot *

New York University Abu Dhabi, P.O. 129188 Saadiyat Island, Abu Dhabi, UAE; yb24@nyu.edu
* Correspondence: sb5272@nyu.edu

Received: 4 April 2019; Accepted: 21 May 2019; Published: 30 May 2019

Abstract: Transposable elements (TEs) play an important role in shaping genomic organization and structure, and may cause dramatic changes in phenotypes. Despite the genetic load they may impose on their host and their importance in microevolutionary processes such as adaptation and speciation, the number of population genetics studies focused on TEs has been rather limited so far compared to single nucleotide polymorphisms (SNPs). Here, we review the current knowledge about the dynamics of transposable elements at recent evolutionary time scales, and discuss the mechanisms that condition their abundance and frequency. We first discuss non-adaptive mechanisms such as purifying selection and the variable rates of transposition and elimination, and then focus on positive and balancing selection, to finally conclude on the potential role of TEs in causing genomic incompatibilities and eventually speciation. We also suggest possible ways to better model TEs dynamics in a population genomics context by incorporating recent advances in TEs into the rich information provided by SNPs about the demography, selection, and intrinsic properties of genomes.

Keywords: transposable elements; population genetics; selection; drift; coevolution

1. Introduction

Transposable elements (TEs) are repetitive DNA sequences that are ubiquitous in the living world and have the ability to replicate and multiply within genomes. Since their discovery, TEs have proven to be of paramount importance in the evolution of genomes, shaping their architecture, diversity, and regulation [1–4]. Given their abundance, the precise quantification of the evolutionary forces and mechanisms that condition their polymorphism and eventual fixation or loss in natural populations is needed.

The theoretical and practical tools provided by population genetics have been crucial to better understand how stochasticity and selection shape TEs dynamics (e.g., [2,5–7]). The first demographic models specifically designed for the analysis of TE polymorphisms were already developed in the 1980s, incorporating transposition and excision rates, effective population size, and purifying selection [4]. Despite this early interest, the investigation of TEs' dynamics in natural populations faded between 1990–2000 [8]. While the precise mechanisms underlying the activity and copy number of TEs have been the topic of many early studies, relatively little attention has been paid to their microevolutionary dynamics in the genomic era, when the focus has been on comparative genomics and on analyses at deeper evolutionary scales. This is mostly explained by the sequencing technologies that have, until recently, produced rather short sequencing reads, which prevent the accurate identification of TE insertions. Instead, most population genomics studies have focused on variation regarding single nucleotide polymorphisms (SNPs). The growing availability of whole-genome resequencing data, as well as the development of new computational tools, has revived the interest of the evolutionary genomics community for the analysis of TE polymorphisms [9,10].

Early reports on the propagation of TEs demonstrated a deleterious effect of their activity. This work, which was mostly based on the investigation of TE polymorphisms in *Drosophila* populations, presented this type of variation as neutral or deleterious [11], and subsequent studies have tried to explain the allele frequency spectrum of TEs within this framework [5,12]. However, TEs can dramatically modify phenotypes, for example by triggering epigenetic mechanisms, by modifying gene expression, or by being a source of ready-to-use functional motifs [13,14]. Thus, TEs can potentially be recruited in adaptive processes and rise in frequency due to positive selection. It remains unclear how the abundance and frequency of TEs are controlled by the host, and to what extent they can become the target of positive selection [9]. In addition, understanding the dynamics of TEs requires jointly studying the host demography, adaptation, and mechanistic views of genome architecture, regulation, and coevolution. This will be crucial if we want to quantify the importance of TEs in adaptive processes and the evolution of species. Here, we summarize the current state of the literature on TEs' evolution at microevolutionary scales, but we also propose possible methodologies to jointly study TEs and traditional markers such as SNPs.

2. Transposable Elements: Classification and Mechanisms of Transposition

"Transposable elements" is an umbrella term that covers a wide diversity of DNA sequences that have the ability to move from one location of a genome to another location. Besides being mobile, these sequences don't have much in common, and they differ considerably in sequence, structure, length, base composition, and mode of transposition. A number of excellent reviews are available on TE diversity (among those, we refer the reader to [15–17]), and we provide here a short synthesis of what is known. TEs are broadly classified into two classes: class I elements (or retrotransposons), which are mobilized by the reverse-transcription of an RNA intermediate, and class II elements (DNA transposons), which use a DNA intermediate. Retrotransposons are further divided into long terminal repeats (LTR) and non-LTR retrotransposons, based on the presence of long terminal repeats (LTR). LTR retrotransposons, which include the *copia* and *gypsy* elements, are mobilized by a process similar to retroviruses. The RNA is reverse-transcribed in the cytoplasm into a double-strand cDNA, which is inserted back into the genome by an integrase. Non-LTR retrotransposons, which include the Long Interspersed Nuclear Elements (LINEs) and *Penelope* elements, are mobilized by a mechanism termed target-primed reverse transcription, where the RNA is reverse-transcribed at the site of insertion [18]. The reverse transcriptase of non-LTR retrotransposons can also act on other transcripts and is responsible for the amplification of non-autonomous elements (also called Short INterspersed Elements, or SINEs), which can considerably outnumber their autonomous counterparts [19]. Class II elements include elements that use a cut-and-paste transposition, such as the *hAT* and *mariner* elements, or elements that have a circular DNA intermediate (*Helitrons*). Class II elements can also mediate the transposition of non-autonomous copies, which, similar to SINEs, can amplify to extremely high copy numbers.

Since TEs are part of the genome of their hosts, they are transmitted vertically from parents to offspring. However, many elements have the ability to invade genomes horizontally, and the recent sequencing of a large number of eukaryotic genomes revealed that this process is not as uncommon as previously thought. Some elements seem to be more prone to horizontal transfer than others. Non-LTR retrotransposons are transmitted mostly vertically [20–22], but some families, such as *RTE*, have been shown to readily transfer across highly divergent taxa, for instance from reptiles to cows [23,24]. The horizontal transfer of LTR retrotransposons is more frequent and seems particularly common in plants and insects [25,26]. Similarly, the horizontal transmission of DNA transposons has been widely documented, and for some unknown reason, some organisms, such as butterflies, bats, and squamate reptiles, seem much more prone to horizontal transfer than others [27–31]. Another case of horizontal transfer occurs when the germline is invaded by retroviruses, which can become stable residents of genomes, keeping the ability to multiply in the genome while lacking infectivity [32,33].

The abundance and diversity of TEs differ considerably among organisms, and the evolutionary mechanisms responsible for these differences remain unclear. The number of TE copies is highly correlated with genome size and can show large variation, even within the same eukaryotic lineage. For instance, among parasitic unicellular eukaryotes, TEs are absent from the genome of *Plasmodium falciparum* [34], while the genome of *Trichomonas vaginalis* is composed of 40% TEs [35]. In plants, ~85% of the maize genome is composed of TEs [36], whereas this number is only ~10% in *Arabidopsis thaliana* [37]. Among vertebrates, the abundance in TEs range from ~6% in the pufferfish to more than 50% in zebrafish and some mammals [1,38]. The diversity of TEs also differs considerably among organisms. For instance, the genome of non-mammalian vertebrates (fish, amphibian, reptiles) typically contains a large diversity of active TEs represented by many families of class I and class II elements, whereas the genome of placental mammals generally harbors a single type of autonomous TE: the LINE-1 (L1) element [1,38–40].

3. How Population Dynamics and Intrinsic Properties of Genomes Shape TEs Polymorphisms

3.1. The Role of Purifying Selection and Demography

As for SNPs, the frequency of TE insertions in natural populations is conditioned by the balance among the drift, selection, and migration between demes (Figure 1A). TEs can disrupt genes and regulatory sequences, and thus can negatively affect the fitness of their host. For instance, in humans, several genetic diseases are caused by TE insertions, such as hereditary cancer [41] or haemophilia [42] (for a more exhaustive review, see [43]). This is also exemplified by the extreme rarity of insertions within exons (e.g., in *Drosophila* [44,45] or *Brachypodium distachyon* [46]), compared to intergenic and intronic regions. Thus, it is expected that purifying selection (i.e., selection against deleterious alleles) against TE insertions plays a major role in shaping their frequency in populations. A consequence of purifying selection is that it prevents or delays the fixation of mutations that reduce fitness in a population. This leads to shifts in the derived allele frequency spectrum (AFS), with an excess of derived variants at low frequencies. Many studies have highlighted this effect, using different approaches. Using a diffusion approximation similar to early models of TE evolution [4], Hazzouri et al. estimated the selective coefficient ($N_e s$) against an *Ac*-like transposon to range between −50 and −10 in *Arabidopsis arenosa* [47]. In *Drosophila melanogaster*, the selective coefficient against insertions from the *BS* family in an African population was estimated at $N_e s \approx -4$ [48], and was as low as −100 for some TE families [45]. In humans, this coefficient was estimated at $N_e s = -1.9$ against L1 retrotransposons [49]. Comparisons of TEs' frequencies with estimates obtained from coalescent simulations often reveal deviations from purely neutral expectations. This is observed in green anoles [50,51], mice [50], or *Arabidopsis* [7,47], for which TEs display an excess of singletons compared to SNPs, which is consistent with purifying selection. A common point between those studies is that they take into account the demographic history of investigated populations to properly estimate the significance of deviation from neutrality, revealing substantial differences with estimates of $N_e s$ obtained assuming stable demography [48].

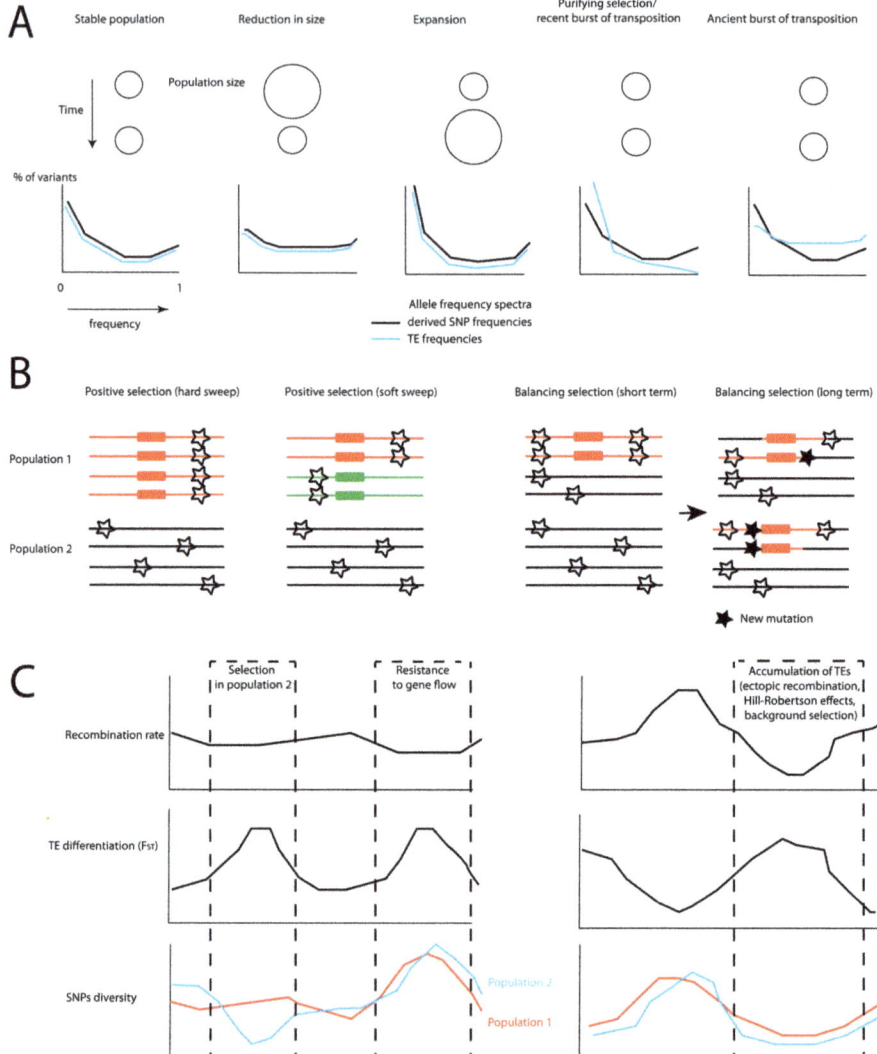

Figure 1. Summary of mechanisms impacting the diversity and frequency of transposable elements (TEs), and their impact on flanking sequences. (**A**) Demographic changes affect the frequency spectra of both TEs and single nucleotide polymorphisms (SNPs) in a similar way, assuming neutrality and a constant rate of transposition. Reductions in effective population sizes should lead to an excess of alleles at intermediate frequencies, while population expansions may lead to an excess of singletons. On the other hand, purifying selection on TEs should lead to an excess of singletons compared to SNPs. Variable rates of transposition may also lead to discrepancies in the spectra between SNPs and TEs. (**B**) TEs involved in adaptation may be detected through their changes in frequencies, but also through the signature left in flanking regions. In the case of positive selection, longer, younger haplotypes should be found nearby positively selected insertions. The similarity of selected haplotypes may be very high in the case of a recent hard sweep, where the insertion is immediately selected and rises in frequency. It may be lower in the case of a so-called soft sweep, where selection either acts after the insertion has already reached an appreciable frequency in the population, or when two insertions with a similar effect on fitness appear at the same time. Positive selection should also result in higher

differentiation at the selected locus compared to populations where selection is not acting. On the other hand, balancing selection may lead to signatures of partial selective sweep when it is recent. Since the selected alleles may be maintained through long periods of time, they have more time to recombine and accumulate new mutations than neutral haplotypes, leading to a narrow signature of high diversity. Since alleles under balancing selection tend to introgress into new populations, and have high diversity, low differentiation is expected at these sites. (**C**) Left panel: Given a constant recombination rate, positive and linked selection in a given population (here, a population of two) may increase differentiation and reduce diversity at selected TEs and flanking regions compared to the rest of the genome. On the other hand, if TEs play a role in incompatibilities after secondary contact, a signature of both elevated differentiation and diversity may be expected. Right panel: However, an excess of TEs in regions of reduced polymorphism, higher differentiation, and lower recombination may be caused by different mechanisms such as purifying selection. This can be due to a reduced effective rate of transposition in regions of high recombination due to deleterious ectopic exchanges, and/or because of the larger-scale effect of selection that accelerates lineage sorting and the differentiation of TEs in regions of low recombination.

The deleterious effect of TEs can have three causes. First, a cost related to where the element inserts (insertional mutagenesis) can affect the host; the number of disease-causing insertions in humans and other organisms constitute prime examples of this [41–43,52]. Second, TEs can produce RNAs or proteins that could be deleterious to the host. For instance, damages induced by the endonuclease encoded by retrotransposons on DNA [53] or the competition of TEs with hosts' genes for transcription factors [54] may lead to a loss in fitness. Third, ectopic recombination between non-allelic copies can lead to deleterious chromosomal rearrangements. Since the 1980s, the relative importance of each of these three mechanisms has been a matter of debate [4,55–57]. However, it has been shown in humans [49], *Drosophila* [57], mouse [50], and anoles [51] that long elements are found at lower frequency in populations than short elements. This suggests that purifying selection acts more strongly against longer copies of elements, and it was shown, in humans, that short elements behave similarly to neutral alleles [49,58]. This pattern could be explained by selection against intact progenitors—which are the longest elements, and the only ones that are capable of producing the RNA and proteins necessary for transposition—or by the ectopic exchange model, since longer elements are more likely to mediate ectopic recombination than shorter ones [50,57,59]. However, selection seems to act against long elements that are not full-length and thus not active, which suggests that the ectopic exchange model plays a preponderant role [50,59]. This model is also supported by the genomic distribution of elements of different length. Long elements tend to be absent from highly recombining regions of genomes [44,60] and accumulate in non-recombining regions such as the human Y chromosome [61,62]. The effect of ectopic recombination will depend on the abundance of elements and the frequency of the insertions. For ectopic recombination to have a substantial effect requires the elements to have reached a copy number threshold so that large families of TEs are more likely to be deleterious than smaller ones [45,57,63]. In addition, heterozygous insertions are more likely to be involved in ectopic recombination because of the lack of an allelic copy on the other chromosome [64]. Thus, elements at low frequency in populations are more likely to be deleterious, since insertions are more likely to be present in the heterozygous state. This suggests that selection against TE insertions may be frequency-dependent, so that the selection coefficient against a specific insertion will decrease when the insertion increases in frequency. Thus, it is expected that rapidly expanding TE families, which are characterized by a high copy number and a majority of insertions in the heterozygous state, are more deleterious than smaller families, where elements are found at high frequency (for instance, after a strong bottleneck effect). These predictions still need to be tested, and this aspect will need to be incorporated in future models of TE evolution.

Genetic drift is the stochastic variation of allele frequencies across generations due to the finite size of natural populations. The effect of genetic drift will depend on the effective size of populations and their past demographic history. When an effective population size is small, genetic drift can cause large changes in allelic frequency, and may even counteract the effect of selection, so that insertions that would be eliminated by selection in large populations can reach high frequency or even fixation in small populations. The stochasticity induced by demographic events explains a significant amount of TEs' diversity in natural populations, which is consistent with theoretical models (e.g., [4,65,66]). For example, in *Arabidopsis lyrata*, smaller populations showed an accumulation of TEs at higher frequencies, due to stronger stochasticity and a reduced efficiency of purifying selection in those populations [7,67], and this has been documented across six TE families. In *B. distachyon*, the loss of retrotransposons across genetic clusters is partly explained by recent bottlenecks and demography [46]. In *Drosophila subobscura*, recent bottlenecks explain the high frequencies of the *bilbo* and *gypsy* elements [68]. A recent study demonstrated that TEs' diversity could be explained by variation in effective population sizes in humans and sticklebacks [50,69], while a joint effect of purifying selection and demography was more obvious in anoles and mice [50,70]. Overall, demography may play an important role in the likelihood for TEs to reach fixation and increase genome size, which is in accordance with the hypothesis that genome size may be directly related to demographic history [71].

3.2. Non-Equilibrium between Transposition and Loss

Another important parameter when characterizing TE dynamics is the interplay between the rate of insertion and the rate at which copies are lost from the population. For the sake of simplicity, early models of population genetics applied to TEs have often assumed that these parameters were in equilibrium [66]. However, the frequency of TEs is likely impacted by shifts in this balance. Sudden bursts of transposition can occur, generating a large cohort of insertions with roughly the same age. Such bursts are well-documented in *Drosophila* [72], rice [73], piciformes [74], fish [75], or mammals [28]. On the other hand, hosts defense mechanisms may be triggered by a high level of transposition. This may lead to waves of extinction, with fast drops in the number of functional TE copies in genomes, and ultimately to the complete cessation of transposition. This alteration between periods of proliferation and elimination has sometimes been described as a life cycle [76,77], which results in genealogies between insertions that are quite different from classical turnover expectations [76]. Some stages of this life cycle may be particularly sensitive to high genetic drift, as the stochastic loss of functional copies may lead to the premature loss of transposition compared to large populations [65]. From a population genomics perspective, this non-equilibrium dynamic has a direct impact on the average age of TE insertions in a given population. This affects not only the copy number, but also the frequency spectrum of these insertions. Ultimately, this can generate complications when interpreting discrepancies between the allele frequency spectra obtained from SNPs and TEs, since they may then be explained by a combination of selection and unbalanced ratios between transposition and elimination rates (Figure 1A). For example, an excess of rare insertions may be due to a recent burst of transposition, leading to an excess of low-frequency TEs insertions [78]. Such a signature would be mistakenly attributed to purifying selection in equilibrium models [7,12].

Non-equilibrium explanations for the excess of rare insertions are considered unlikely [5,45] by some authors. Nevertheless, the direct application on TEs of classical population genetics assumptions that rely on constant mutation rates may not be realistic. For example, in *Drosophila*, the frequency spectra of TEs from different families is directly related to each family's age and their time since inactivation [44]. This may be particularly important for models where little is known about the dynamics of the TEs. To take this issue into account, a test that quantify purifying selection on TEs has been developed [12] that is conditional on the age of elements. However, this age is often overestimated for TE sequences, because of non-equilibrium demography and mutations introduced by transposition errors [12]. Recent advances in modeling may facilitate the deployment of methods that jointly estimate selection and transposition [79].

3.3. Transposition and Variable Rates of Recombination

A consequence of selection limiting the proliferation of TEs in genomes is that TEs should be more frequently found in regions of the genome where natural selection and elimination mechanisms are weaker or less efficient. This requires a better quantification of the relationship between the number and the type of TE insertions and genomic features such as recombination, which is often found to be negatively associated with TE content [60,80]. Regions of low recombination tend to be associated with a lower gene content, which reduces the likelihood for an insertion to be strongly deleterious. Selection is more likely to remove TE insertions in regions of high recombination, since more frequent ectopic recombination should increase the likelihood of deleterious chromosomal rearrangements [56]. In addition, TE silencing is often associated with epigenetic modifications that are negatively associated with recombination [81,82]. Another mechanism is Hill–Robertson interference. Competition between haplotypes harboring different deleterious TE insertions may reduce the efficiency of selection, similar to a reduction of local effective population sizes that enhance the impact of genetic drift in regions of low recombination [83,84]. Ultimately, this may lead to the fixation of TEs through the process of Muller's ratchet, where low recombination prevents the persistence of a haplotype without any insertion, increasing mutational load [56]. However, this latter effect is more likely for TEs in regions of extremely low recombination [56]. The position of recombination hotspots varies across species [85], which can be an alternative explanation to divergent selection when interpreting variation in TE frequencies between species and populations.

Recent studies of recombination landscapes have improved our understanding of TEs dynamics. The expected negative correlation between TEs and recombination rates has been observed for LINEs in humans [59,62], mice, and rats [86]. In *Drosophila*, there is evidence that both reduced gene content in regions of low recombination and ectopic recombination shape the frequency of TEs along the genome [87,88]. However, the insertion process itself varies between different TE families, and may be responsible for variation in abundance and frequency along chromosomes. Indeed, a more detailed examination of the correlation between TEs and recombination shows a heterogeneous pattern, with some TE families [89] and endoviruses [90] found more frequently near recombination hotspots. The same pattern is observed near recombination hotspots in *Ficedula*, which is possibly due to the shared preference of recombination and transposition machineries for open chromatin [85]. A preference for high-recombining regions has also been shown for DNA transposons (but not non-LTR elements) in *Caenorhabditis elegans* [91]. This may be due to the cut-and-paste mechanism of transposition that takes advantage of the double-stranded breaks that initiate recombination events. Another possible explanation lies in the negative correlation between the age of TEs and the recombination rate, suggesting that a long-term effect of recombination is needed to remove TEs from genomes. Overall, this suggests that previous demonstrations of a negative correlation between TE content and recombination rate need to take into account the properties and histories that are specific to each TE family [60,91].

Until recently, most theoretical works on TE dynamics have considered constant recombination rates [56]. The emergence of new simulation tools that can simultaneously incorporate the intrinsic properties of the genome and the evolutionary history of populations may be valuable to disentangle the effects of demography, selection, recombination, and the transposition process of TEs (Figure 2). A promising method is SLIM3 [79], which is able to simulate TEs as well as flank genomic fragments under any arbitrary complex demographic scenario, and can also incorporate variations in transposition rates due to thresholds in abundance or any other feature deemed useful by the user. Then, contrast between simulations and observed data may be performed to quantify the dynamics of TEs, for example through approximate Bayesian computation (ABC) [92] approaches (see [50] for an example).

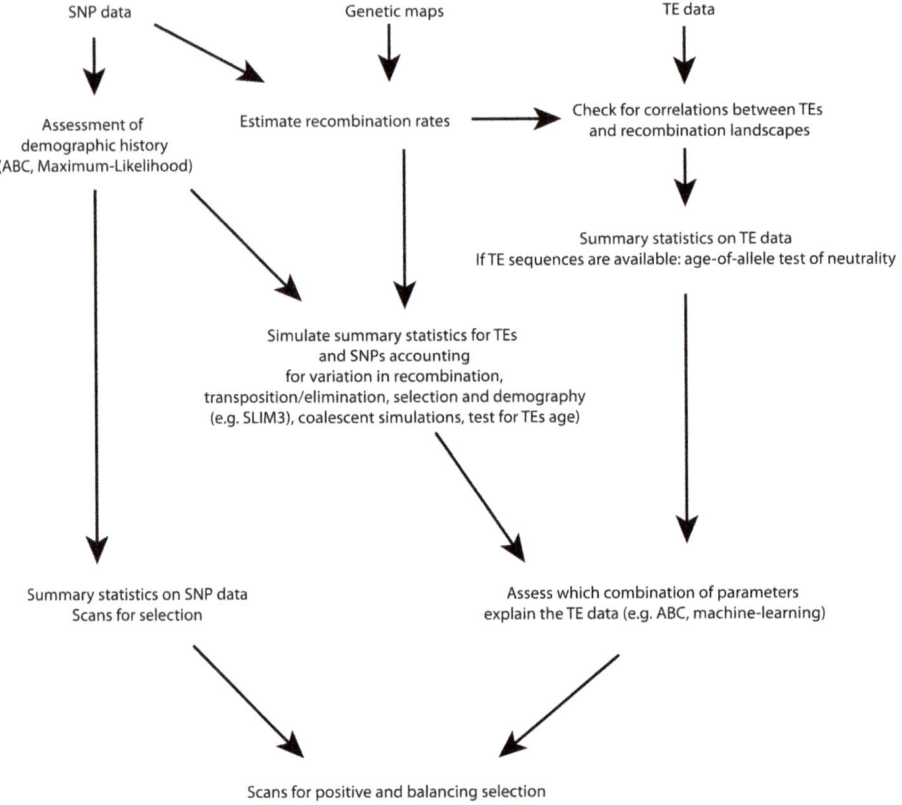

Figure 2. A possible analytical pipeline for population genomics of TEs, highlighting some promising methods. Genetics and genomics may provide information about the intrinsic properties of genomes (e.g., recombination maps) and extrinsic processes such as demographic changes and selection. This information may then be used to build neutral expectations about both TEs and SNPs. Contrasting the observed statistics for TEs (e.g., frequencies, length, properties of flanking regions) with simulations may facilitate the quantification of the mechanisms that act on their diversity.

3.4. Coevolutionary Dynamics

Coevolution between TEs and their hosts is a crucial aspect that shapes TE diversity and impacts the likelihood for insertions to reach high frequencies. Understanding the distribution of TE polymorphisms across genomes and populations requires a better quantification of the mechanisms behind TEs silencing [93]. Refining the timescale of coevolution between TEs and control mechanisms would provide important insights about constraints on the transposition rate. Such knowledge would improve our models of transposition for specific TE families.

Hosts use many mechanisms to control the proliferation of TEs within their genomes (see [94] for an exhaustive review in humans). An important example is the APOBEC enzymes. APOBEC3 proteins inhibit endoviruses by editing dC residues to dU during reverse transcription. This increases the rate of G to A mutation, and ultimately results in the inhibition of transposition. They are also inhibitors of reverse transcription, making them efficient against LINEs and other retrotransposons [95]. Variation in the sequence and structure of APOBEC genes seems to be directly related to their efficiency in controlling TEs [96,97]. There is already evidence that APOBEC proteins act in specific ways on TEs from different families across vertebrates [97]. In vertebrates, epigenetic modifications such as methylation [98] and histone modifications [99] may be responsible for controlling TEs by limiting their expression. In rice, mutants at a chromomethylase, *OsCMT3a*, cannot methylate TEs, and display a burst of transposition [100]. Finally, another control mechanism lies in small RNA pathways, by which TEs RNA is recognized and eliminated. In fruit flies, two main mechanisms regulate TE activity: siRNA/*Dicer* [101] and piRNA [102,103]. Therefore, further refinements of models of TEs' evolution would benefit from the knowledge of the spatial repartition of methylated regions and other control mechanisms that are specific to the host. A promising approach lies in simulations and model-fitting incorporating demography, selection, and control mechanisms to test expectations about TE dynamics. For example, a recent simulation study showed that large, non-recombining clusters of piRNAs are more efficient at trapping TEs and preventing invasions [104]. Transposition rates and population sizes mostly influenced the length during which TEs were active, but not the final amount of TE insertions [104]. Combining experimental evolution with modeling may provide better resolution on the coevolutionary process; an example is provided in [105]. In this work, the authors investigated how synergies between RNAi and methylation pathways effectively controlled TE proliferation, using a set of ordinary differential equations describing transposition, elimination, methylation, and RNA interference. By reanalyzing the expression and transposition of the *Evade* element in two *A. thaliana* inbred lines, they could show that small amounts of RNAi were enough to initiate methylation and silencing. According to the model, the retention of methylated TEs prevented reamplification more efficiently than elimination. Although these models may benefit from further refinements by incorporating unstable demography or linked selection to be broadly applicable, they already provide a solid conceptual and methodological basis.

Importantly, this dynamic implies that there is a coevolution between the different components of the genome, which may have an impact on the diversity of hosts' defense genes. Scanning the genome for loci that display correlation between their diversity and the number of TE families found in the host may be a way to identify which genes in a pathway are of primary functional importance. There are signatures of fast adaptive evolution at genes that are involved in RNA interference in Drosophila [106], with recent selective sweeps encompassing genes from the piRNA pathway [107]. Another compelling example of coevolution is found in primates, where two zinc-finger genes, *ZNF91* and *ZNF93*, evolved rapidly to prevent the expansion of SINE and LINE elements [108]. Besides the need for a more comprehensive understanding of the pathways involved in TEs regulation, there is a need for further investigation in a population genetics context. For example, are demographic fluctuations such as bottlenecks responsible for a relaxation of selective pressures at defense genes that may explain bursts of transposition? Is there a link between diversity at defense genes associated with speciation and environmental adaptation?

4. Transposable Elements as a Source of Adaptation

4.1. Evidence for Positive Selection on TEs and SNPs

Identifying TEs that are under positive selection and therefore rise to high frequency in populations is an exciting alley for research in population genomics. However, detecting positive selection is a challenging task even for traditional markers such as SNPs [109]. TEs idiosyncrasies must also be taken into account, since bursts of transposition or insertion bias due to recombination also shape their diversity. Many TEs have been domesticated by hosts genomes over long evolutionary time scales, leading to the emergence of novel cellular functions through the recruitment of TE-derived coding sections or *cis*-regulatory domains [110]. For example, the RAG genes that are involved in the recombination process of antibodies in jawed vertebrates [111,112] originated from a domesticated *Transib* element [113]. Whole TE families may be domesticated by a host. For example, in *Drosophila*, three non-LTR retrotransposons (TART, TARHE, and HeT-A) preferentially transpose in telomeres and prevent their shortening [114], although their domestication is likely incomplete [115]. TEs are also important for the stability of centromeres during replication [116], and might be involved in speciation.

For example in rice, recent insertions of both class I and class II transposons are responsible for the accelerated differentiation of centromeres between three cultivated species and subspecies [117].

Bursts of transposition are known to occur in organisms put under stressful conditions [118], which may be subsequently recruited by the host for rapid adaptation [2,119]. For example, the increased transposition of BARE-1 may be adaptive and is associated with higher elevation and dryness in natural populations of the wild barley [120]. A burst of transposition is associated with the adaptive radiation of *Anolis* lizards. This has led to an increase in TE insertions within the *HOX* genes clusters compared to other vertebrates, which may be linked to the outstanding morphological diversity in these lizards [121]. In maize, the expansion of Helitrons might have been associated with positive selection over 4% of these elements [122]. Some *Helitrons* subfamilies can capture gene fragments. The survival rate of these elements was correlated with the length of genetic inserts, which might enhance their adaptive potential.

TEs can provide a selective advantage and quickly modify phenotypes, for example by triggering epigenetic mechanisms and enhancing gene expression due to the insertion of a TE promotor [13,123]. A recent example includes the genetic determinism of the industrial melanism trait in peppered moth, which is associated with a TE insertion in the *cortex* gene [124]. In *Drosophila*, there is evidence that TEs may be recruited in adaptation to temperate environment, pesticides [125,126], development [127], or oxidative stress [128,129]. The same insertion may have both positive and negative effects on fitness [127,130], which may prevent fixation due to the associated cost of selection. In humans, analyses based on TE frequencies in 15 populations sampled across Europe, Asia, and Africa highlighted candidate TEs for adaptation that might be responsible for change in gene expression [131]. However, we note that unlike recent studies in *Drosophila* [129], this study focused primarily on TE frequencies, and did not examine signatures of selection in flanking regions, and used a relatively simplistic model of human demography. Importantly, similar to traditional markers such as SNPs, the effects of past demography may mimic expected signatures of selection. For example, in *D. melanogaster*, latitudinal variation in North America and Australia was partly explained by past admixture between African and European populations [6]. Overall, the way that TEs are recruited by the host—either through the recycling of TE-derived coding regions (e.g., *RAG* genes), because of the repeats themselves (e.g., TART) or because of regulatory effects (cortex in peppermoth, [132]—the candidate genes in humans [131]) still need to be quantified.

4.2. Quantifying Positive Selection on TEs

A promising approach consists in the joint analysis of TEs and SNPs to detect candidate insertions for positive selection (Figure 1B,C and Figure 2). SNPs can be used to build neutral demographic models and allele frequency spectra that are expected under neutrality [7,51]. Variation in allele frequencies across populations can be used to detect insertions displaying high differentiation driven by positive selection [10,133]. A common bias in these approaches is that background selection can also lead to unusual allele frequency spectra and patterns of differentiation due to stronger drift in regions of low recombination. A possible way to overcome this issue and identify loci that are truly under positive selection consists of performing genome-wide association with environmental or phenotypic features [109]. Other approaches based on linkage disequilibrium (LD) can help identify insertions that are associated with long haplotypes, and are therefore more likely to be under recent positive selection. The distribution of haplotypes' length may provide useful information to estimate the age of an insertion (see for example [124]). A number of tests, including iHS, XP-EHH, and H2/H1 statistics or nSL [134–137], can be used on datasets combining TE insertions and SNPs.

Other approaches that directly link environmental and phenotypic variation to SNPs may be applied to TEs as well. Methods that track association between allele frequencies and environmental features across populations are increasingly powerful (e.g., BAYPASS [138]). Classical genome-wide association analyses (GWAS) at the scale of individual phenotypes are also a good way to better link TEs variation with relevant ecological mechanisms that may shape diversity. Other potentially fruitful approaches have been developed that facilitate the joint inference of demography and selection and make a better use of whole-genome information. Those include ancestral recombination graphs (ARGs) inference [139], approximate Bayesian computation (ABC) [92], and machine learning [140]. ARGs inference reconstructs coalescent and recombination landscapes along genomic fragments, and is useful to quantitatively estimate the time since selection and completeness of selective sweeps. However, this inference is computationally intensive and unpractical for very large datasets [139]. ABC and machine learning are faster approaches that use summary statistics computed across genomic windows to classify them as selected or not. These approaches allow combining multiple tests for selection such as the ones described above. Then, expectations for these statistics can be obtained by simulations under the hypothesis of selection or neutrality, and algorithms can be trained to classify windows as more or less likely to contain selected sites [141,142]. This type of approach has the advantage of directly including the confounding effects of demography in its implementation, and provides an estimate of false positive and false negative rates.

A general question in the study of adaptation at the genomic level lies in identifying the origin of beneficial alleles. Selected alleles can have independent mutational origins and rise independently in the frequency in each population, as they provide a selective advantage. Selected alleles might originate from novel alleles that quickly reach high frequency due to their benefit (hard sweep) or from pre-existing standing variation (so-called soft sweeps [143]). At last, an allele initially selected in one population can spread through migration to other populations where it provides a selective advantage. These questions are especially interesting for TEs. For example, biases in transposition due to recombination and coevolution with the host may facilitate the repeated emergence of advantageous mutations in the same genomic regions, ultimately promoting convergent evolution. Methods similar to diploS/HIC [144] may be used to disentangle scenarios of neutrality, selection on de novo mutations (hard sweep), or on standing variation (soft sweep). Another recently developed maximum-likelihood approach, dmc [145], aims at distinguishing between different modes of convergent adaptation at candidate sites for selection, and may be useful to use on candidate TEs for adaptation and flanking SNPs.

4.3. Studying Balancing Selection on TEs

Evidence for balancing selection, a type of selection that maintains variation, is still elusive in natural populations, even for SNPs (but see [146] for a discussion of its importance). This type of selection is notoriously difficult to detect due to its very localized effects, especially on long evolutionary time scales. Several recent methods have been specifically developed to detect this type of selection [139,147,148], and may be used on TEs or linked SNPs and haplotypes (Figure 1B). The role of TE insertions in facilitating balancing selection is worth investigating, although neglected [149]. A recent example in a locust is a *Lm1* insertion in the heat-shock protein *Hsp90*, which is found only in the heterozygote state and seems to display latitudinal variation [150]. This insertion is associated with the faster development of embryos, and may control the number of broods that hatch in a year. Instead of directly providing a selective advantage, TEs might facilitate the maintenance of diversity at loci where their expression at the homozygote state would be detrimental, for example at genes of the Major Histocompatibility Complex [151].

4.4. Limitations and Future Improvements

A word of caution is needed, since all those approaches are more likely to identify whole genomic regions than specific TE insertions under selection. Therefore, functional validation remains an essential step to identify TE insertions that have a positive impact on fitness [9]. Moreover, several types of selection remain difficult to detect and quantify, such as multi-locus weak selection or balancing selection [109]. However, it is now possible to address such issues, as recent advances in sequencing will allow for the inclusion of large number of individuals in a dataset, and will thus facilitate the narrowing of candidate regions for selection. Low-depth sequencing becomes an interesting way to obtain genotypic information for many individuals [152], and may be associated with the systematic search for transposable elements using state-of-the-art methods such as MELT, which have been shown to perform well when detecting polymorphic variants, even at relatively low sequencing depths [153]. However, other methods are being developed (Table 1), and may be more suited to a specific design, such as pooled whole-genome resequencing. This may be coupled with recent improvements in GWAS such as mixed linear models that have enhanced power to detect the loci associated with relevant phenotypes and polygenic selection [154] using large sample sizes.

Table 1. Summary of tools commonly used for transposable elements (TE) detection and analysis. Methods that have been compared on human datasets in [155] are highlighted in bold.

Name of the Method	Purpose	Link	Reference
Popoolation_TE2	TE detection in pooled designs	https://sourceforge.net/p/popoolation-te2/wiki/Home/	[156]
T-LEX2	Detection of polymorphic TEs from short reads	http://petrov.stanford.edu/cgi-bin/Tlex.html	[157]
STEAK	Detection of polymorphic TEs from short reads	https://github.com/applevir/STEAK	[158]
TIDAL	Detection of polymorphic TEs from short reads	http://www.bio.brandeis.edu/laulab/Tidal_Fly/Tidal_Fly_Home.html	[159]
MELT	Detection of polymorphic TEs from short reads	http://melt.igs.umaryland.edu/	[153]
LoRTE	Detection of polymorphic TEs from PacBio sequencing	http://www.egce.cnrs-gif.fr/?p=6422	[160]
ITIS	Detection of polymorphic TEs from short reads	https://github.com/Chuan-Jiang/ITIS	[161]
TEMP	Detection of polymorphic TEs from short reads	https://github.com/JialiUMassWengLab/TEMP	[162]
Mobster	Detection of polymorphic TEs from short reads	http://sourceforge.net/projects/mobster/	[163]
Tangram	Detection of polymorphic TEs from short reads	https://github.com/jiantao/Tangram	[164]
RetroSeq	Detection of polymorphic TEs from short reads	https://github.com/tk2/RetroSeq	[165]
RelocaTE2	Detection of polymorphic TEs from short reads	https://github.com/JinfengChen/RelocaTE2	[166]
McClintock	Combination of several methods into a single pipeline	https://github.com/bergmanlab/mcclintock	[167]
Invade	Population genomics modeling (forward-in-time) incorporating coevolution with piRNA clusters	https://sourceforge.net/p/te-tools/code/HEAD/tree/sim3p/	[104]
SLIM3	Population genomics modeling (forward-in-time)	https://messerlab.org/slim/	[79]

5. The Role of Selfish Elements in Genomic Conflicts: Impact in Natural Populations

During speciation, populations may diverge and accumulate private combinations of alleles at multiple loci. The disruption of these allele combinations in hybrids may result in lower fitness, which is a process known as Bateson–Dobzhansky–Muller incompatibilities, and prevents the homogenization of gene pools [168,169]. These incompatibilities can emerge when conflicts between selfish elements and the host lead to different coevolutionary mechanisms in isolated populations [170–173]. Secondary contact between these diverged genomes results in a disruption of the control mechanisms and ultimately the low fitness of hybrids, therefore maintaining differentiated species. TEs may play important roles in these processes (see [174] for a more exhaustive review). A classic example of the hybrid dysgenesis induced by TEs is provided in *D. melanogaster*. In this species, the P-element (a DNA transposon) that expanded recently was probably introduced through horizontal transfer from *D. willistoni* [175,176]. Crosses between females where the P-elements are absent (M females) and P males carrying the element produce progeny exhibiting high mutation rates, chromosomal rearrangements and sterility [177]. This is caused by the deposition of piRNAs in the egg by the females that cannot recognize the P elements provided by the male genome, causing massive expansion. This recent invasion of the P element in *D. melanogaster*, but also in *D. simulans* [178–180], highlights the fast dynamic of coevolutionary mechanisms dealing with genomic conflicts and how they can lead to speciation.

Repeated elements are associated with DNA-binding proteins that shape the chromosome organization. There is evidence for the rapid reorganization of these repeats between closely related species (e.g., in rice, [117]) that shape heterochromatin repartition and ultimately disturb the meiotic process in hybrids. Since TEs are associated with major structural changes and variation in repeat content, they may play an important role in meiotic drive, where driver elements rise in frequency by distorting meiosis [173]. Their abundance and high turnover on sex chromosomes (among other repeats) also suggests that TEs may play an important role in the process of speciation and Haldane's rule, which states that in hybrids between incipient species, the sex that is most likely to display reduced fitness is the heterogametic one [181]. Moreover, TEs can be responsible for gross chromosome rearrangements due to unequal recombination between TE copies [55], which may explain the fast divergence in karyotypes and ultimately speciation (see [182] for a review). TEs may also play a role in dosage compensation between males and females, as demonstrated for a domesticated *Helitron* element in *Drosophila miranda* [183]. In this species, a succession of neo-X chromosomes appeared in the last million years. Gene expression is upregulated by twofold in males by the male specific lethal (MSL) complex that targets an ~21-bp specific sequence harbored by the domesticated element [184]. Domestication of the *Helitron* element occurred each time a new sex chromosome emerged, with a specific motif invading the chromosome and recruiting adjacent genes in dosage compensation.

How can population genomics contribute to the study of TEs involved in incompatibilities and speciation? First, it remains clear that functional assessments and crosses in controlled conditions may be critical to provide definite proof of the role of TEs in maintaining barriers between species [174]. However, cline theory [185] and the information provided by SNPs can be useful to assess which specific elements may be involved in the speciation process. For example, genomes may be scanned for an excess of private TE insertions in regions of low recombination that resist the gene flow between two species. Since Haldane's rule predicts that sex chromosomes should be quicker to accumulate incompatibility loci, contrasting the TE content between sex chromosomes and autosomes may also provide evidence for TE-driven incompatibilities. The analysis of SNP and haplotype diversity in regions flanking TEs may also facilitate the interpretation, for example by estimating the age of haplotypes that contain insertions and whether they display evidence of resisting introgression.

Coevolution between TEs and recombination may be important in maintaining divergence between populations (Figure 1C). TEs may drive variation in recombination rates by inducing changes in chromatin conformation; they may also facilitate the suppression of recombination between diverging lineages through their accumulation in low-recombining regions (see [80] for a discussion). This is why when examining the dynamic of TEs after secondary contact, a careful examination of changes in recombination rates along chromosomes and a comparison of correlation between active and inactive families would be recommended [80]. On a related note, variation at genes that shape the recombination landscape may be relevant to assess in association with TEs dynamics. For example, in mammals, *PRDM9* is involved in the fast-evolving positioning of recombination hotspots [186], but it is also involved in hybrid sterility and speciation [173]. Variation at this gene between incipient species may lead to divergent constraints on transposable elements diversity along genomes, which in turn could facilitate the spread of regions of reduced recombination resisting gene flow.

At last, elements involved in incompatibility may display gradients of association with the environment due to coupling [187], where clines of incompatible alleles drift to match tension zones corresponding to environmental discontinuity. Special care should be taken to identify possible cryptic hybrid zones that can trap incompatible alleles along environmental clines when looking for TEs involved in adaptation to the environment [169,187].

6. Future Directions

Recent methodological progresses should prove useful to obtain a better understanding of the dynamics of TEs in natural populations. It is increasingly acknowledged that local variations in mutation and recombination rate, demography, selective sweeps, and linked and background selection have to be integrated into analyses of genetic variation (e.g., [188,189]). All these factors are also likely to explain local variation in TEs density, forcing us to adopt a more integrative approach when studying TEs' dynamics. Comparisons of simulations-based models are flexible and powerful, and have become increasingly popular in population genomics [92,140]. The challenge with TEs lies in properly simulating the process by which they insert and are removed from genomes, as well as demography and selection. This requires a good preliminary knowledge of the idiosyncrasies of the species and the TEs under investigation. As new methods keep being developed to jointly estimate the effects of demography and selection on genomes, the field of TEs population genomics will move toward more model-based approaches. This will provide quantitative estimates of the forces underlying TEs dynamics.

Another crucial aspect that is still missing for most sequenced species is a high-quality genome assembly. Poor assemblies often omit highly repetitive regions where TEs are more likely to lie. Without proper assembly and annotation, it becomes impossible to perform a near-exhaustive assessment of TE insertions and identification of polymorphisms [9]. This is especially important when investigating the role of repetitive regions in the emergence of incompatibilities. Besides, since the most powerful methods to detect selection use the spatial distribution of allele frequencies and LD, they cannot be used efficiently on highly fragmented genomes. This creates biases; for example, in the Tasmanian devil, poor assembly led to incorrectly assume the inactivation of LINE-1 elements [190]. However, the advent of third-generation sequencing techniques should circumvent this issue and expand the study of TEs to a broader diversity of organisms.

Only a few models are available to study the population genomics of TEs, and drosophilids are clearly over-represented in the field of TE population genetics. This creates a challenge regarding drawing general conclusions about TE dynamics, as well as the relative importance of selection and drift in shaping genomic diversity. The large effective population size of the *Drosophila* species has been hypothesized to facilitate a widespread effect of selection across the genome [189,191], making both demographic inference and the detection of outliers difficult. Besides those on humans, *Drosophila*, and some crops (rice, Arabidopsis, maize), studies remain scarce, with a few studies highlighting the effects of both drift and purifying selection on TE's diversity in green anoles [51] and birds [192]. As whole-genome assembly and resequencing becomes more affordable, there is hope that more general conclusions about the microevolutionary dynamics of TEs may be drawn.

Author Contributions: Y.B. and S.B. contributed to the conceptualization, writing and editing of the review.

Funding: This research was funded by New York University Abu Dhabi (NYUAD) research funds AD180 (to S.B.).

Acknowledgments: The authors thank three anonymous reviewers for their comments on the manuscript.

Conflicts of Interest: The authors declare no conflict of interest.

References

1. Sotero-Caio, C.G.; Platt, R.N.; Suh, A.; Ray, D.A. Evolution and diversity of transposable elements in vertebrate genomes. *Genome Biol. Evol.* **2017**, *9*, 161–177. [CrossRef] [PubMed]
2. Chuong, E.B.; Elde, N.C.; Feschotte, C. Regulatory activities of transposable elements: From conflicts to benefits. *Nat. Rev. Genet.* **2017**, *18*, 71–86. [CrossRef] [PubMed]
3. Song, M.J.; Schaack, S. Evolutionary Conflict between Mobile DNA and Host Genomes. *Am. Nat.* **2018**, *192*, 263–273. [CrossRef] [PubMed]
4. Charlesworth, B.; Charlesworth, D. The Population Genetics of Transposable Elements. *Genet. Res.* **1983**, *42*, 1–27. [CrossRef]

5. Barron, M.G.; Fiston-Lavier, A.-S.; Petrov, D.A.; Gonzalez, J. Population Genomics of Transposable Elements in Drosophila. *Annu. Rev. Genet.* **2014**, *48*, 561–581. [CrossRef] [PubMed]
6. Bergland, A.O.; Tobler, R.; Gonzalez, J.; Schmidt, P.; Petrov, D. Secondary contact and local adaptation contribute to genome-wide patterns of clinal variation in Drosophila melanogaster. *Mol. Ecol.* **2016**, *25*, 1157–1174. [CrossRef] [PubMed]
7. Lockton, S.; Ross-Ibarra, J.; Gaut, B.S. Demography and weak selection drive patterns of transposable element diversity in natural populations of Arabidopsis lyrata. *Proc. Natl. Acad. Sci. USA* **2008**, *105*, 13965–13970. [CrossRef] [PubMed]
8. Biémont, C. A brief history of the status of transposable elements: From junk DNA to major players in evolution. *Genetics* **2010**, *186*, 1085–1093. [CrossRef] [PubMed]
9. Villanueva-Cañas, J.L.; Rech, G.E.; de Cara, M.A.R.; González, J. Beyond SNPs: how to detect selection on transposable element insertions. *Methods Ecol. Evol.* **2017**, *8*, 728–737. [CrossRef]
10. Hoban, S.; Kelley, J.L.; Lotterhos, K.E.; Antolin, M.F.; Bradburd, G.; Lowry, D.B.; Poss, M.L.; Reed, L.K.; Storfer, A.; Whitlock, M.C. Finding the Genomic Basis of Local Adaptation: Pitfalls, Practical Solutions, and Future Directions. *Am. Nat.* **2016**, *188*, 379–397. [CrossRef]
11. Doolittle, W.F.; Sapienza, C. Selfish genes, the phenotype paradigm and genome evolution. *Nature* **1980**, *284*, 601–603. [CrossRef] [PubMed]
12. Blumenstiel, J.P.; Chen, X.; He, M.; Bergman, C.M. An age-of-allele test of neutrality for transposable element insertions. *Genetics* **2014**, *196*, 523–538. [CrossRef] [PubMed]
13. Morgan, H.D.; Sutherland, H.G.; Martin, D.I.; Whitelaw, E. Epigenetic inheritance at the agouti locus in the mouse. *Nat. Genet.* **1999**, *23*, 314–318. [CrossRef] [PubMed]
14. Stuart, T.; Eichten, S.R.; Cahn, J.; Karpievitch, Y.V.; Borevitz, J.O.; Lister, R. Population scale mapping of transposable element diversity reveals links to gene regulation and epigenomic variation. *Elife* **2016**, *5*, 1–27. [CrossRef] [PubMed]
15. Tollis, M.; Boissinot, S. The evolutionary dynamics of transposable elements in eukaryote genomes. In *Genome Dynamics*; MA, G.-R., Ed.; Karger: Basel, Switzerland, 2012; pp. 68–91, ISBN 9783318021509.
16. *Mobile DNA III*; Craig, N.; Chandler, M.; Gellert, M.; Lambowitz, A.; Rice, P.; Sandmeyer, S. (Eds.) American Society for Microbiology (ASM): Washington, DC, USA, 2015.
17. Bourque, G.; Burns, K.H.; Gehring, M.; Gorbunova, V.; Seluanov, A.; Hammell, M.; Imbeault, M.; Izsvák, Z.; Levin, H.L.; Macfarlan, T.S.; et al. Ten things you should know about transposable elements. *Genome Biol.* **2018**, *19*, 199. [CrossRef]
18. Luan, D.D.; Korman, M.H.; Jakubczak, J.L.; Eickbush, T.H. Reverse transcription of R2Bm RNA is primed by a nick at the chromosomal target site: A mechanism for non-LTR retrotransposition. *Cell* **1993**, *72*, 595–605. [CrossRef]
19. Dewannieux, M.; Esnault, C.; Heidmann, T. LINE-mediated retrotransposition of marked Alu sequences. *Nat. Genet.* **2003**, *35*, 41–48. [CrossRef]
20. Malik, H.S.; Burke, W.D.; Eickbush, T.H. The age and evolution of non-LTR retrotransposable elements. *Mol. Biol. Evol.* **1999**, *16*, 793–805. [CrossRef]
21. Kordiš, D.; Lovšin, N.; Gubenšek, F. Phylogenomic analysis of the L1 retrotransposons in Deuterostomia. *Syst. Biol.* **2006**, *55*, 886–901. [CrossRef]
22. Waters, P.D.; Dobigny, G.; Waddell, P.J.; Robinson, T.J. Evolutionary history of LINE-1 in the major clades of placental mammals. *PLoS ONE* **2007**, *2*. [CrossRef]
23. Kordis, D. Unusual horizontal transfer of a long interspersed nuclear element between distant vertebrate classes. *Proc. Natl. Acad. Sci. USA* **1998**, *95*, 10704–10709. [CrossRef] [PubMed]
24. Ivancevic, A.M.; Kortschak, R.D.; Bertozzi, T.; Adelson, D.L. Horizontal transfer of BovB and L1 retrotransposons in eukaryotes. *Genome Biol.* **2018**, *19*, 1–13. [CrossRef] [PubMed]
25. Schaack, S.; Gilbert, C.; Feschotte, C. Promiscuous DNA: Horizontal transfer of transposable elements and why it matters for eukaryotic evolution. *Trends Ecol. Evol.* **2010**, *25*, 537–546. [CrossRef] [PubMed]
26. Bartolomé, C.; Bello, X.; Maside, X. Widespread evidence for horizontal transfer of transposable elements across Drosophila genomes. *Genome Biol.* **2009**, *10*. [CrossRef] [PubMed]
27. Reiss, D.; Mialdea, G.; Miele, V.; de Vienne, D.; Peccoud, J.; Gilbert, C.; Duret, L.; Charlat, S. Global survey of mobile DNA horizontal transfer in arthropods reveals Lepidoptera as a prime hotspot. *PLoS Genet.* **2019**, *15*, e1007965. [CrossRef] [PubMed]

28. Pace, J.K., II; Feschotte, C. The evolutionary history of human DNA transposons: Evidence for intense activity in the primate lineage. *Genome Res.* **2007**, *17*, 422–432. [CrossRef] [PubMed]
29. Thomas, J.; Schaack, S.; Pritham, E.J. Pervasive horizontal transfer of rolling-circle transposons among animals. *Genome Biol. Evol.* **2010**, *2*, 656–664. [CrossRef]
30. Gilbert, C.; Hernandez, S.S.; Flores-Benabib, J.; Smith, E.N.; Feschotte, C. Rampant horizontal transfer of SPIN transposons in squamate reptiles. *Mol. Biol. Evol.* **2012**, *29*, 503–515. [CrossRef]
31. Novick, P.; Smith, J.; Ray, D.; Boissinot, S. Independent and parallel lateral transfer of DNA transposons in tetrapod genomes. *Gene* **2010**, *449*, 85–94. [CrossRef]
32. Ribet, D.; Harper, F.; Dupressoir, A.; Dewannieux, M.; Pierron, G.; Heidmann, T. An infectious progenitor for the murine IAP retrotransposon: Emergence of an intracellular genetic parasite from an ancient retrovirus. *Genome Res.* **2008**, *18*, 597–609. [CrossRef]
33. Gifford, R.; Tristem, M. The evolution, distribution and diversity of endogenous retroviruses. *Virus Genes* **2003**, *26*, 291–316. [CrossRef] [PubMed]
34. Nelson, K.E.; Peterson, J.; Gardner, M.J.; Mungall, C.; White, O.; Angiuoli, S.; Shallom, S.J.; Selengut, J.; Rutherford, K.; Nene, V.; et al. Genome sequence of the human malaria parasite Plasmodium falciparum. *Nature* **2002**, *419*, 498–511.
35. Carlton, J.M.; Hirt, R.P.; Silva, J.C.; Delcher, A.L.; Schatz, M.; Zhao, Q.; Wortman, J.R.; Bidwell, S.L.; Alsmark, U.C.M.; Besteiro, S.; et al. Draft Genome Sequence of the Sexually Transmitted Pathogen Trichomonas vaginalis. *Science* **2007**, *315*, 207–213. [CrossRef] [PubMed]
36. Schnable, P.S.; Page, S.E.E.L.; Pasternak, S.; Liang, C.; Zhang, J.; Fulton, L.; Graves, T.A.; Minx, P.; Reily, A.D.; Courtney, L.; et al. The B73 Maize Genome: Complexity, Diversity, and Dynamics. *Science* **2012**, *326*, 1112–1115. [CrossRef] [PubMed]
37. The Arabidopsis Genome Initiative. Analysis of the genome sequence of the flowering plant Arabidopsis thaliana. *Nature* **2000**, *408*, 796–815. [CrossRef] [PubMed]
38. Chalopin, D.; Naville, M.; Plard, F.; Galiana, D.; Volff, J.N. Comparative analysis of transposable elements highlights mobilome diversity and evolution in vertebrates. *Genome Biol. Evol.* **2015**, *7*, 567–580. [CrossRef]
39. Furano, A.V.; Duvernell, D.D.; Boissinot, S. L1 (LINE-1) retrotransposon diversity differs dramatically between mammals and fish. *Trends Genet.* **2004**, *20*, 9–14. [CrossRef] [PubMed]
40. Boissinot, S.; Sookdeo, A. The Evolution of Line-1 in Vertebrates. *Genome Biol. Evol.* **2016**, *8*, 3485–3507. [CrossRef]
41. Qian, Y.; Mancini-DiNardo, D.; Judkins, T.; Cox, H.C.; Brown, K.; Elias, M.; Singh, N.; Daniels, C.; Holladay, J.; Coffee, B.; et al. Identification of pathogenic retrotransposon insertions in cancer predisposition genes. *Cancer Genet.* **2017**, *216–217*, 159–169. [CrossRef]
42. Green, P.M.; Bagnall, R.D.; Waseem, N.H.; Giannelli, F. Haemophilia A mutations in the UK: Results of screening one-third of the population. *Br. J. Haematol.* **2008**, *143*, 115–128. [CrossRef]
43. Hancks, D.C.; Kazazian, H.H. Roles for retrotransposon insertions in human disease. *Mob. DNA* **2016**, *7*. [CrossRef] [PubMed]
44. Kofler, R.; Betancourt, A.J.; Schlötterer, C. Sequencing of pooled DNA samples (Pool-Seq) uncovers complex dynamics of transposable element insertions in Drosophila melanogaster. *PLoS Genet.* **2012**, *8*. [CrossRef] [PubMed]
45. Petrov, D.A.; Fiston-Lavier, A.-S.; Lipatov, M.; Lenkov, K.; Gonzalez, J. Population Genomics of Transposable Elements in Drosophila melanogaster. *Mol. Biol. Evol.* **2011**, *28*, 1633–1644. [CrossRef] [PubMed]
46. Stritt, C.; Gordon, S.P.; Wicker, T.; Vogel, J.P.; Roulin, A.C. Recent activity in expanding populations and purifying selection have shaped transposable element landscapes across natural accessions of the mediterranean grass Brachypodium distachyon. *Genome Biol. Evol.* **2018**, *10*, 304–318. [CrossRef] [PubMed]
47. Hazzouri, K.M.; Mohajer, A.; Dejak, S.I.; Otto, S.P.; Wright, S.I. Contrasting patterns of transposable-element insertion polymorphism and nucleotide diversity in autotetraploid and allotetraploid Arabidopsis species. *Genetics* **2008**, *179*, 581–592. [CrossRef]
48. González, J.; Macpherson, J.M.; Messer, P.W.; Petrov, D.A. Inferring the strength of selection in Drosophila under complex demographic models. *Mol. Biol. Evol.* **2009**, *26*, 513–526. [CrossRef]
49. Boissinot, S.; Davis, J.; Entezam, A.; Petrov, D.; Furano, A.V. Fitness cost of LINE-1 (L1) activity in humans. *Proc. Natl. Acad. Sci. USA* **2006**, *103*, 9590–9594. [CrossRef]

50. Xue, A.T.; Ruggiero, R.P.; Hickerson, M.J.; Boissinot, S. Differential effect of selection against LINE retrotransposons among vertebrates inferred from whole-genome data and demographic modeling. *Genome Biol. Evol.* **2018**, *10*, 1265–1281. [CrossRef]
51. Ruggiero, R.P.; Bourgeois, Y.; Boissinot, S. LINE Insertion Polymorphisms Are Abundant but at Low Frequencies across Populations of Anolis carolinensis. *Front. Genet.* **2017**, *8*, 1–14. [CrossRef]
52. Quadrana, L.; Silveira, A.B.; Mayhew, G.F.; LeBlanc, C.; Martienssen, R.A.; Jeddeloh, J.A.; Colot, V. The Arabidopsis thaliana mobilome and its impact at the species level. *Elife* **2016**, *5*, 1–25. [CrossRef]
53. Olivares, M.; Alonso, C.; López, M.C. The open reading frame 1 of the L1Tc retrotransposon of Trypanosoma cruzi codes for a protein with apurinic-apyrimidinic nuclease activity. *J. Biol. Chem.* **1997**, *272*, 25224–25228. [CrossRef] [PubMed]
54. Conte, C.; Dastugue, B.; Vaury, C. Promoter competition as a mechanism of transcriptional interference mediated by retrotransposons. *EMBO J.* **2002**, *21*, 3908–3916. [CrossRef] [PubMed]
55. Langley, C.H.; Montgomery, E.A.; Hudson, R.; Kaplan, N.; Charlesworth, B. On the role of unequal exchange in the containment of transposable element copy number. *Genet. Res.* **1988**, *52*, 223–235. [CrossRef] [PubMed]
56. Dolgin, E.S.; Charlesworth, B. The effects of recombination rate on the distribution and abundance of transposable elements. *Genetics* **2008**, *178*, 2169–2177. [CrossRef] [PubMed]
57. Petrov, D.A.; Aminetzach, Y.T.; Davis, J.C.; Bensasson, D.; Hirsh, A.E. Size matters: Non-LTR retrotransposable elements and ectopic recombination in Drosophila. *Mol. Biol. Evol.* **2003**, *20*, 880–892. [CrossRef] [PubMed]
58. Cordaux, R.; Lee, J.; Dinoso, L.; Batzer, M.A. Recently integrated Alu retrotransposons are essentially neutral residents of the human genome. *Gene* **2006**, *373*, 138–144. [CrossRef] [PubMed]
59. Song, M.; Boissinot, S. Selection against LINE-1 retrotransposons results principally from their ability to mediate ectopic recombination. *Gene* **2007**, *390*, 206–213. [CrossRef] [PubMed]
60. Nam, K.; Ellegren, H. Recombination drives vertebrate genome contraction. *PLoS Genet.* **2012**, *8*. [CrossRef]
61. Charlesworth, B. The organization and evolution of the human Y chromosome. *Genome Biol.* **2003**, *4*. [CrossRef] [PubMed]
62. Boissinot, S.; Entezam, A.; Furano, A. V Selection Against Deleterious LINE-1-Containing Loci in the Human Lineage. *Mol. Biol.* **2001**, *18*, 926–935. [CrossRef]
63. Montgomery, E.; Charlesworth, B.; Langley, C. A test for the role of natural selection in the stabilization of transposable element copy number in a population of Drosophila melanogaster. *Genet Res.* **1987**, *49*, 31–41. [CrossRef]
64. Montgomery, E.A.; Huang, S.M.; Langley, C.H.; Judd, B.H. Chromosome rearrangement by ectopic recombination in Drosophila melanogaster: Genome structure and evolution. *Genetics* **1991**, *129*, 1085–1098.
65. Le Rouzic, A.; Boutin, T.S.; Capy, P. Long-term evolution of transposable elements. *Proc. Natl. Acad. Sci. USA* **2007**, *104*, 19375–19380. [CrossRef]
66. Charlesworth, B.; Sniegowski, P.; Stephan, W. The evolutionary dynamics of repetitive DNA in eukaryotes. *Nature* **1994**, *371*, 215–220. [CrossRef] [PubMed]
67. Ross-Ibarra, J.; Wright, S.I.; Foxe, J.P.; Kawabe, A.; DeRose-Wilson, L.; Gos, G.; Charlesworth, D.; Gaut, B.S. Patterns of polymorphism and demographic history in natural populations of Arabidopsis lyrata. *PLoS ONE* **2008**, *3*. [CrossRef]
68. García Guerreiro, M.P.; Chávez-Sandoval, B.E.; Balanyà, J.; Serra, L.; Fontdevila, A. Distribution of the transposable elements bilbo and gypsy in original and colonizing populations of Drosophila subobscura. *BMC Evol. Biol.* **2008**, *8*. [CrossRef] [PubMed]
69. Blass, E.; Bell, M.; Boissinot, S. Accumulation and rapid decay of non-LTR retrotransposons in the genome of the three-spine stickleback. *Genome Biol. Evol.* **2012**, *4*, 687–702. [CrossRef] [PubMed]
70. Tollis, M.; Boissinot, S. Lizards and LINEs: Selection and demography affect the fate of L1 retrotransposons in the genome of the green anole (Anolis carolinensis). *Genome Biol. Evol.* **2013**, *5*, 1754–1768. [CrossRef]
71. Lynch, M.; Conery, J.S. The Origins of Genome Complexity. *Science* **2003**, *302*, 1401–1404. [CrossRef]
72. Vieira, C.; Lepetit, D.; Dumont, S.; Biémont, C. Wake up of transposable elements following Drosophila simulans worldwide colonization. *Mol. Biol. Evol.* **1999**, *16*, 1251–1255. [CrossRef]
73. Piegu, B.; Guyot, R.; Picault, N.; Roulin, A.; Saniyal, A.; Kim, H.; Collura, K.; Brar, D.S.; Jackson, S.; Wing, R.A.; et al. Doubling genome size without polyploidization: Dynamics of retrotransposition-driven genomic expansions in Oryza australiensis, a wild relative of rice. *Genome Res.* **2006**, *16*, 1262–1269. [CrossRef] [PubMed]

74. Manthey, J.D.; Moyle, R.G.; Boissinot, S. Multiple and independent phases of transposable element amplification in the genomes of piciformes (woodpeckers and allies). *Genome Biol. Evol.* **2018**, *10*, 1445–1456. [CrossRef] [PubMed]
75. De Boer, J.G.; Yazawa, R.; Davidson, W.S.; Koop, B.F. Bursts and horizontal evolution of DNA transposons in the speciation of pseudotetraploid salmonids. *BMC Genomics* **2007**, *8*, 1–10. [CrossRef] [PubMed]
76. Hellen, E.H.B.; Brookfield, J.F.Y. The diversity of class II transposable elements in mammalian genomes has arisen from ancestral phylogenetic splits during ancient waves of proliferation through the genome. *Mol. Biol. Evol.* **2013**, *30*, 100–108. [CrossRef] [PubMed]
77. Hellen, E.H.B.; Brookfield, J.F.Y. Transposable element invasions. *Mob. Genet. Elements* **2013**, *3*, e23920. [CrossRef] [PubMed]
78. Bergman, C.M.; Bensasson, D. Recent LTR retrotransposon insertion contrasts with waves of non-LTR insertion since speciation in Drosophila melanogaster. *Proc. Natl. Acad. Sci. USA* **2007**, *104*, 11340–11345. [CrossRef]
79. Haller, B.C.; Messer, P.W. SLiM 3: Forward Genetic Simulations Beyond the Wright-Fisher Model. *Mol. Biol. Evol.* **2019**, *36*, 632–637. [CrossRef] [PubMed]
80. Kent, T.V.; Uzunović, J.; Wright, S.I. Coevolution between transposable elements and recombination. *Philos. Trans. R. Soc. B Biol. Sci.* **2017**, *372*. [CrossRef]
81. Choi, K.; Zhao, X.; Kelly, K.A.; Venn, O.; Higgins, J.D.; Yelina, N.E.; Hardcastle, T.J.; Ziolkowski, P.A.; Copenhaver, G.P.; Franklin, F.C.H.; et al. Arabidopsis meiotic crossover hot spots overlap with H2A.Z nucleosomes at gene promoters. *Nat. Genet.* **2013**, *45*, 1327–1336. [CrossRef]
82. Myers, S.; Bottolo, L.; Freeman, C.; McVean, G.; Donnelly, P. A Fine-Scale Map of Recombination Rates and Hotspots Across the Human Genome. *Science* **2005**, *310*, 321–324. [CrossRef]
83. Hill, W.G.; Robertson, A. Local effects of limited recombination. *Genet. Res.* **1966**, *8*, 269–294. [CrossRef] [PubMed]
84. Felsenstein, J. The evolution advantage of recombination. *Genetics* **1974**, *78*, 737–756. [PubMed]
85. Kawakami, T.; Mugal, C.F.; Suh, A.; Nater, A.; Burri, R.; Smeds, L.; Ellegren, H. Whole-genome patterns of linkage disequilibrium across flycatcher populations clarify the causes and consequences of fine-scale recombination rate variation in birds. *Mol. Ecol.* **2017**, *26*, 4158–4172. [CrossRef] [PubMed]
86. Jensen-Seaman, M.I.; Furey, T.S.; Payseur, B.A.; Lu, Y.; Roskin, K.M.; Chen, C.F.; Thomas, M.A.; Haussler, D.; Jacob, H.J. Comparative recombination rates in the rat, mouse, and human genomes. *Genome Res.* **2004**, *14*, 528–538. [CrossRef] [PubMed]
87. Bartolomé, C.; Maside, X.; Charlesworth, B. On the abundance and distribution of transposable elements in the genome of Drosophila melanogaster. *Mol. Biol. Evol.* **2002**, *19*, 926–937. [CrossRef] [PubMed]
88. Rizzon, C.; Marais, G.; Gouy, M.; Biémont, C. Recombination rate and the distribution of transposable elements in the Drosophila melanogaster genome. *Genome Res.* **2002**, *12*, 400–407. [CrossRef] [PubMed]
89. Myers, S.; Freeman, C.; Auton, A.; Donnelly, P.; McVean, G. A common sequence motif associated with recombination hot spots and genome instability in humans. *Nat. Genet.* **2008**, *40*, 1124–1129. [CrossRef]
90. Campos-Sánchez, R.; Cremona, M.A.; Pini, A.; Chiaromonte, F.; Makova, K.D. Integration and Fixation Preferences of Human and Mouse Endogenous Retroviruses Uncovered with Functional Data Analysis. *PLoS Comput. Biol.* **2016**, *12*, 1–41. [CrossRef]
91. Duret, L.; Marais, G.; Biemont, C. Transposons but not retrotransposons are located preferentially in regions of high recombination rate in Caenorhabditis elegans. *Genetics* **2000**, *156*, 1661–1669.
92. Csilléry, K.; Blum, M.G.B.; Gaggiotti, O.E.; François, O. Approximate Bayesian Computation (ABC) in practice. *Trends Ecol. Evol.* **2010**, *25*, 410–418. [CrossRef]
93. Ågren, J.A.; Wright, S.I. Co-evolution between transposable elements and their hosts: A major factor in genome size evolution? *Chromosom. Res.* **2011**, *19*, 777–786. [CrossRef] [PubMed]
94. Goodier, J.L. Restricting retrotransposons: A review. *Mob. DNA* **2016**, *7*. [CrossRef] [PubMed]
95. Arias, J.F.; Koyama, T.; Kinomoto, M.; Tokunaga, K. Retroelements versus APOBEC3 family members: No great escape from the magnificent seven. *Front. Microbiol.* **2012**, *3*, 1–12. [CrossRef] [PubMed]
96. Koito, A.; Ikeda, T. Intrinsic immunity against retrotransposons by APOBEC cytidine deaminases. *Front. Microbiol.* **2013**, *4*, 1–9. [CrossRef] [PubMed]

97. Lindič, N.; Budič, M.; Petan, T.; Knisbacher, B.A.; Levanon, E.Y.; Lovšin, N. Differential inhibition of LINE1 and LINE2 retrotransposition by vertebrate AID/APOBEC proteins. *Retrovirology* **2013**, *10*, 1–16. [CrossRef] [PubMed]
98. Yoder, J.A.; Walsh, C.P.; Bestor, T.H. Cytosine methylation and the ecology of intragenomic parasites. *Trends Genet.* **1997**, *13*, 335–340. [CrossRef]
99. Huda, A.; Mariño-Ramírez, L.; Jordan, I.K. Epigenetic histone modifications of human transposable elements: Genome defense versus exaptation. *Mob. DNA* **2010**, *1*, 1–12. [CrossRef]
100. Cheng, C.; Tarutani, Y.; Miyao, A.; Ito, T.; Yamazaki, M.; Sakai, H.; Fukai, E.; Hirochika, H. Loss of function mutations in the rice chromomethylase OsCMT3a cause a burst of transposition. *Plant J.* **2015**, *83*, 1069–1081. [CrossRef]
101. Van Rij, R.P.; Berezikov, E. Small RNAs and the control of transposons and viruses in Drosophila. *Trends Microbiol.* **2009**, *17*, 163–171. [CrossRef]
102. Prud'homme, N.; Gans, M.; Masson, M.; Terzian, C.; Bucheton, A. Flamenco, a gene controlling the gypsy retrovirus of Drosophila melanogaster. *Genetics* **1995**, *139*, 697–711.
103. Goriaux, C.; Desset, S.; Renaud, Y.; Vaury, C.; Brasset, E. Transcriptional properties and splicing of the flamenco piRNA cluster. *EMBO Rep.* **2014**, *15*, 411–418. [CrossRef] [PubMed]
104. Kofler, R. Dynamics of transposable element invasions with piRNA clusters. *Mol. Biol. Evol.* **2019**. [CrossRef] [PubMed]
105. Roessler, K.; Bousios, A.; Meca, E.; Gaut, B.S. Modeling Interactions between Transposable Elements and the Plant Epigenetic Response: A Surprising Reliance on Element Retention. *Genome Biol. Evol.* **2018**, *10*, 803–815. [CrossRef] [PubMed]
106. Palmer, W.H.; Hadfield, J.D.; Obbard, D.J. RNA-Interference Pathways Display High Rates of Adaptive Protein Evolution in Multiple Invertebrates. *Genetics* **2018**, *208*, 1585–1599. [CrossRef] [PubMed]
107. Simkin, A.; Wong, A.; Poh, Y.P.; Theurkauf, W.E.; Jensen, J.D. Recurrent and recent selective sweeps in the piRNA pathway. *Evolution* **2013**, *67*, 1081–1090. [CrossRef] [PubMed]
108. Jacobs, F.M.J.; Greenberg, D.; Nguyen, N.; Haeussler, M.; Ewing, A.D.; Katzman, S.; Paten, B.; Salama, S.R.; Haussler, D. An evolutionary arms race between KRAB zinc-finger genes ZNF91/93 and SVA/L1 retrotransposons. *Nature* **2014**, *516*, 242–245. [CrossRef] [PubMed]
109. Haasl, R.J.; Payseur, B.A. Fifteen years of genomewide scans for selection: Trends, lessons and unaddressed genetic sources of complication. *Mol. Ecol.* **2016**, *25*, 5–23. [CrossRef]
110. Miller, W.J.; McDonald, J.F.; Nouaud, D.; Anxolabehere, D. Molecular domestication—More than a sporadic episode in evolution. *Genetica* **1999**, *107*, 197–207. [CrossRef]
111. Jung, D.; Alt, F.W. Unraveling V(D)J Recombination: Insights into Gene Regulation. *Cell* **2004**, *116*, 299–311. [CrossRef]
112. Oettinger, M.A.; Schatz, D.G.; Gorka, C.; Baltimore, D.; Oetringer, M.A. RAG-1 and RAG-2, Adjacent Genes That Synergistically Activate V(D)J Recombination. *Science* **1990**, *248*, 1517–1523. [CrossRef]
113. Kapitonov, V.V.; Koonin, E.V. Evolution of the RAG1-RAG2 locus: Both proteins came from the same transposon. *Biol. Direct* **2015**, *10*, 1–8. [CrossRef] [PubMed]
114. Pardue, M.-L.; DeBaryshe, P.G. Retrotransposons that maintain chromosome ends. *Proc. Natl. Acad. Sci. USA* **2011**, *108*, 20317–20324. [CrossRef] [PubMed]
115. Lee, Y.C.G.; Leek, C.; Levine, M.T. Recurrent Innovation at Genes Required for Telomere Integrity in Drosophila. *Mol. Biol. Evol.* **2017**, *34*, 467–482. [PubMed]
116. Zaratiegui, M.; Vaughn, M.W.; Irvine, D.V.; Goto, D.; Watt, S.; Bähler, J.; Arcangioli, B.; Martienssen, R.A. CENP-B preserves genome integrity at replication forks paused by retrotransposon LTR. *Nature* **2011**, *469*, 112–115. [CrossRef] [PubMed]
117. Gao, D.; Jiang, N.; Wing, R.A.; Jiang, J.; Jackson, S.A. Transposons play an important role in the evolution and diversification of centromeres among closely related species. *Front. Plant Sci.* **2015**, *6*. [CrossRef] [PubMed]
118. Capy, P.; Gasperi, G.; Biémont, C.; Bazin, C. Stress and transposable elements: Co-evolution or useful parasites? *Heredity* **2000**, *85*, 101–106. [CrossRef] [PubMed]
119. Rey, O.; Danchin, E.; Mirouze, M.; Loot, C.; Blanchet, S. Adaptation to Global Change: A Transposable Element-Epigenetics Perspective. *Trends Ecol. Evol.* **2016**, *31*, 514–526. [CrossRef] [PubMed]

120. Kalendar, R.; Tanskanen, J.; Immonen, S.; Nevo, E.; Schulman, A.H. Genome evolution of wild barley (Hordeum spontaneum) by BARE-1 retrotransposon dynamics in response to sharp microclimatic divergence. *Proc. Natl. Acad. Sci. USA* **2000**, *97*, 6603–6607. [CrossRef]
121. Feiner, N. Accumulation of transposable elements in *HOX* gene clusters during adaptive radiation of Anolis lizards. *Proc. Biol. Sci.* **2016**, *283*. [CrossRef]
122. Yang, L.; Bennetzen, J.L. Distribution, diversity, evolution, and survival of Helitrons in the maize genome. *Proc. Natl. Acad. Sci. USA* **2009**, *106*, 19922–19927. [CrossRef]
123. Schrader, L.; Kim, J.W.; Ence, D.; Zimin, A.; Klein, A.; Wyschetzki, K.; Weichselgartner, T.; Kemena, C.; Stökl, J.; Schultner, E.; et al. Transposable element islands facilitate adaptation to novel environments in an invasive species. *Nat. Commun.* **2014**, *5*, 1–10. [CrossRef] [PubMed]
124. Hof, A.E.V.; Campagne, P.; Rigden, D.J.; Yung, C.J.; Lingley, J.; Quail, M.A.; Hall, N.; Darby, A.C.; Saccheri, I.J. The industrial melanism mutation in British peppered moths is a transposable element. *Nature* **2016**, *534*, 102–105. [CrossRef] [PubMed]
125. González, J.; Petrov, D.A. The adaptive role of transposable elements in the Drosophila genome. *Gene* **2009**, *448*, 124–133. [CrossRef] [PubMed]
126. González, J.; Karasov, T.L.; Messer, P.W.; Petrov, D.A. Genome-wide patterns of adaptation to temperate environments associated with transposable elements in Drosophila. *PLoS Genet.* **2010**, *6*, 33–35. [CrossRef] [PubMed]
127. Ullastres, A.; Petit, N.; González, J. Exploring the phenotypic space and the evolutionary history of a natural mutation in drosophila melanogaster. *Mol. Biol. Evol.* **2015**, *32*, 1800–1814. [CrossRef] [PubMed]
128. Guio, L.; Barrón, M.G.; González, J. The transposable element Bari-Jheh mediates oxidative stress response in Drosophila. *Mol. Ecol.* **2014**, *23*, 2020–2030. [CrossRef] [PubMed]
129. Rech, G.E.; Bogaerts-Marquez, M.; Barron, M.G.; Merenciano, M.; Villanueva-Canas, J.L.; Horvath, V.; Fiston-Lavier, A.-S.; Luyten, I.; Venkataram, S.; Quesneville, H.; et al. Stress response, behavior, and development are shaped by transposable element-induced mutations in Drosophila. *PLoS Genet.* **2018**, *15*, e1007900. [CrossRef]
130. González, J.; Macpherson, J.M.; Petrov, D.A. A recent adaptive transposable element insertion near highly conserved developmental loci in Drosophila melanogaster. *Mol. Biol. Evol.* **2009**, *26*, 1949–1961. [CrossRef]
131. Rishishwar, L.; Wang, L.; Wang, J.; Yi, S.V.; Lachance, J.; Jordan, I.K. Evidence for positive selection on recent human transposable element insertions. *Gene* **2018**, *675*, 69–79. [CrossRef]
132. Feschotte, C. Transposable elements and the evolution of regulatory networks. *Nat. Rev. Genet.* **2008**, *9*, 397–405. [CrossRef]
133. Lotterhos, K.E.; Whitlock, M.C. Evaluation of demographic history and neutral parameterization on the performance of FST outlier tests. *Mol. Ecol.* **2014**, *23*, 2178–2192. [CrossRef] [PubMed]
134. Sabeti, P.C.; Schaffner, S.F.; Fry, B.; Lohmueller, J.; Varilly, P.; Shamovsky, O.; Palma, A.; Mikkelsen, T.S.; Altshuler, D.; Lander, E.S. Positive natural selection in the human lineage. *Science* **2006**, *312*, 1614–1620. [CrossRef] [PubMed]
135. Garud, N.R.; Messer, P.W.; Buzbas, E.O.; Petrov, D.A. Recent Selective Sweeps in North American Drosophila melanogaster Show Signatures of Soft Sweeps. *PLoS Genet.* **2015**, *11*, 1–32. [CrossRef] [PubMed]
136. Ferrer-Admetlla, A.; Liang, M.; Korneliussen, T.; Nielsen, R. On detecting incomplete soft or hard selective sweeps using haplotype structure. *Mol. Biol. Evol.* **2014**, *31*, 1275–1291. [CrossRef] [PubMed]
137. McCarroll, S.A.; Sabeti, P.C.; Frazer, K.A.; Varilly, P.; Fry, B.; Ballinger, D.G.; Lohmueller, J.; Cox, D.R.; Hostetter, E.; Hinds, D.A.; et al. Genome-wide detection and characterization of positive selection in human populations. *Nature* **2007**, *449*, 913–918.
138. Gautier, M. Genome-Wide Scan for Adaptive Divergence and Association with Population-Specific Covariates. *Genetics* **2015**, *201*, 1555–1579. [CrossRef] [PubMed]
139. Rasmussen, M.D.; Hubisz, M.J.; Gronau, I.; Siepel, A. Genome-Wide Inference of Ancestral Recombination Graphs. *PLoS Genet.* **2014**, *10*. [CrossRef]
140. Schrider, D.R.; Kern, A.D. Supervised Machine Learning for Population Genetics: A New Paradigm. *Trends Genet.* **2018**, *34*, 301–312. [CrossRef]
141. Schrider, D.R.; Kern, A.D. Machine Learning for Population Genetics: A New Paradigm. *bioRxiv* **2017**, 206482. [CrossRef]

142. Schrider, D.R.; Mendes, F.K.; Hahn, M.W.; Kern, A.D. Soft shoulders ahead: Spurious signatures of soft and partial selective sweeps result from linked hard sweeps. *Genetics* **2015**, *200*, 267–284. [CrossRef]
143. Messer, P.W.; Petrov, D.A. Population genomics of rapid adaptation by soft selective sweeps. *Trends Ecol. Evol.* **2013**, *28*, 659–669. [CrossRef] [PubMed]
144. Kern, A.D.; Schrider, D.R. diploS/HIC: An Updated Approach to Classifying Selective Sweeps. *G3* **2018**, *8*, 1959–1970. [CrossRef] [PubMed]
145. Lee, K.M.; Coop, G. Distinguishing Among Modes of Convergent Adaptation Using Population Genomic Data. *Genetics* **2018**, *207*, 1591–1619. [CrossRef] [PubMed]
146. Sellis, D.; Callahan, B.J.; Petrov, D.A.; Messer, P.W. Heterozygote advantage as a natural consequence of adaptation in diploids. *Proc. Natl. Acad. Sci. USA* **2011**, *108*, 20666–20671. [CrossRef] [PubMed]
147. Siewert, K.M.; Voight, B.F. Detecting Long-Term Balancing Selection Using Allele Frequency Correlation. *Mol. Biol. Evol.* **2017**, *34*, 2996–3005. [CrossRef] [PubMed]
148. DeGiorgio, M.; Lohmueller, K.E.; Nielsen, R. A model-based approach for identifying signatures of ancient balancing selection in genetic data. *PLoS Genet.* **2014**, *10*, e1004561. [CrossRef] [PubMed]
149. Van Oosterhout, C. Transposons in the MHC: The Yin and Yang of the vertebrate immune system. *Heredity* **2009**, *103*, 190–191. [CrossRef] [PubMed]
150. Chen, B.; Zhang, B.; Xu, L.; Li, Q.; Jiang, F.; Yang, P.; Xu, Y.; Kang, L. Transposable Element-Mediated Balancing Selection at Hsp90 Underlies Embryo Developmental Variation. *Mol. Biol. Evol.* **2017**, *34*, 1127–1139. [CrossRef]
151. van Oosterhout, C. A new theory of MHC evolution: Beyond selection on the immune genes. *Proc. Biol. Sci.* **2009**, *276*, 657–665. [CrossRef]
152. Nicod, J.; Davies, R.W.; Cai, N.; Hassett, C.; Goodstadt, L.; Cosgrove, C.; Yee, B.K.; Lionikaite, V.; McIntyre, R.E.; Remme, C.A.; et al. Genome-wide association of multiple complex traits in outbred mice by ultra-low-coverage sequencing. *Nat. Genet.* **2016**, *48*, 912–918. [CrossRef]
153. Gardner, E.J.; Lam, V.K.; Harris, D.N.; Chuang, N.T.; Scott, E.C.; Pittard, W.S.; Mills, R.E.; 1000 Genomes Project Consortium; Devine, S.E. The Mobile Element Locator Tool (MELT): Population-scale mobile element discovery and biology. *Genome Res.* **2017**, *27*, 1916–1929. [CrossRef] [PubMed]
154. Wen, Y.J.; Zhang, H.; Ni, Y.L.; Huang, B.; Zhang, J.; Feng, J.Y.; Wang, S.B.; Dunwell, J.M.; Zhang, Y.M.; Wu, R. Methodological implementation of mixed linear models in multi-locus genome-wide association studies. *Brief. Bioinform.* **2018**, *19*, 700–712. [CrossRef]
155. Rishishwar, L.; Mariño-Ramírez, L.; Jordan, I.K. Benchmarking computational tools for polymorphic transposable element detection. *Brief. Bioinform.* **2017**, *18*, 908–918. [CrossRef] [PubMed]
156. Kofler, R.; Gómez-Sánchez, D.; Schlötterer, C. PoPoolationTE2: Comparative Population Genomics of Transposable Elements Using Pool-Seq. *Mol. Biol. Evol.* **2016**, *33*, 2759–2764. [CrossRef] [PubMed]
157. Fiston-Lavier, A.S.; Barrón, M.G.; Petrov, D.A.; González, J. T-lex2: Genotyping, frequency estimation and re-annotation of transposable elements using single or pooled next-generation sequencing data. *Nucleic Acids Res.* **2015**, *43*. [CrossRef] [PubMed]
158. Santander, C.G.; Gambron, P.; Marchi, E.; Karamitros, T.; Katzourakis, A.; Magiorkinis, G. STEAK: A specific tool for transposable elements and retrovirus detection in high-throughput sequencing data. *Virus Evol.* **2017**, *3*, 1–12. [CrossRef] [PubMed]
159. Rahman, R.; Chirn, G.W.; Kanodia, A.; Sytnikova, Y.A.; Brembs, B.; Bergman, C.M.; Lau, N.C. Unique transposon landscapes are pervasive across Drosophila melanogaster genomes. *Nucleic Acids Res.* **2015**, *43*, 10655–10672. [CrossRef] [PubMed]
160. Disdero, E.; Filée, J. LoRTE: Detecting transposon-induced genomic variants using low coverage PacBio long read sequences. *Mob. DNA* **2017**, *8*, 4–9. [CrossRef]
161. Jiang, C.; Chen, C.; Huang, Z.; Liu, R.; Verdier, J. ITIS, a bioinformatics tool for accurate identification of transposon insertion sites using next-generation sequencing data. *BMC Bioinformatics* **2015**, *16*, 1–8. [CrossRef]
162. Zhuang, J.; Wang, J.; Theurkauf, W.; Weng, Z. TEMP: A computational method for analyzing transposable element polymorphism in populations. *Nucleic Acids Res.* **2014**, *42*, 6826–6838. [CrossRef]
163. Thung, D.T.; de Ligt, J.; Vissers, L.E.M.; Steehouwer, M.; Kroon, M.; de Vries, P.; Slagboom, E.P.; Ye, K.; Veltman, J.A.; Hehir-Kwa, J.Y. Mobster: Accurate detection of mobile element insertions in next generation sequencing data. *Genome Biol.* **2014**, *15*, 488. [CrossRef] [PubMed]

164. Wu, J.; Lee, W.P.; Ward, A.; Walker, J.A.; Konkel, M.K.; Batzer, M.A.; Marth, G.T. Tangram: A comprehensive toolbox for mobile element insertion detection. *BMC Genomics* **2014**, *15*, 1–15. [CrossRef] [PubMed]
165. Keane, T.M.; Wong, K.; Adams, D.J. RetroSeq: Transposable element discovery from next-generation sequencing data. *Bioinformatics* **2013**, *29*, 389–390. [CrossRef] [PubMed]
166. Chen, J.; Wrightsman, T.R.; Wessler, S.R.; Stajich, J.E. RelocaTE2: A high resolution transposable element insertion site mapping tool for population resequencing. *PeerJ* **2017**, *5*, e2942. [CrossRef] [PubMed]
167. Nelson, M.G.; Linheiro, R.S.; Bergman, C.M. McClintock: An Integrated Pipeline for Detecting Transposable Element Insertions in Whole-Genome Shotgun Sequencing Data. *G3* **2017**, *7*, 2763–2778. [CrossRef] [PubMed]
168. Seehausen, O.; Butlin, R.K.; Keller, I.; Wagner, C.E.; Boughman, J.W.; Hohenlohe, P.A.; Peichel, C.L.; Saetre, G.-P.; Bank, C.; Brannstrom, A.; et al. Genomics and the origin of species. *Nat. Rev. Genet.* **2014**, *15*, 176–192. [CrossRef] [PubMed]
169. Butlin, R.K.; Smadja, C.M. Coupling, Reinforcement, and Speciation. *Am. Nat.* **2017**, *191*, 155–172. [CrossRef]
170. Jangam, D.; Feschotte, C.; Betrán, E. Transposable Element Domestication As an Adaptation to Evolutionary Conflicts. *Trends Genet.* **2017**, *33*, 817–831. [CrossRef]
171. Lindholm, A.K.; Dyer, K.A.; Firman, R.C.; Fishman, L.; Forstmeier, W.; Holman, L.; Johannesson, H.; Knief, U.; Kokko, H.; Larracuente, A.M.; et al. The Ecology and Evolutionary Dynamics of Meiotic Drive. *Trends Ecol. Evol.* **2016**, *31*, 315–326. [CrossRef]
172. Gardner, A.; Úbeda, F. The meaning of intragenomic conflict. *Nat. Ecol. Evol.* **2017**, *1*, 1807–1815. [CrossRef]
173. Crespi, B.; Nosil, P. Conflictual speciation: Species formation via genomic conflict. *Trends Ecol. Evol.* **2013**, *28*, 48–57. [CrossRef] [PubMed]
174. Serrato-Capuchina, A.; Matute, D.R. The role of transposable elements in speciation. *Genes* **2018**, *9*, 254. [CrossRef] [PubMed]
175. Daniels, S.B.; Peterson, K.R.; Strausbaugh, L.D.; Kidwell, M.G.; Chovnik, A. Evidence for horizontal transmission of the P transposable element between Drosophila species. *Genetics* **1990**, *124*, 339–355.
176. Kidwell, M.G. Hybrid dysgenesis in Drosophila melanogaster: The relationship between the P–M and I–R interaction systems. *Genet. Res.* **1979**, *33*, 205–217. [CrossRef]
177. Kimura, K.; Kidwell, M.G. Differences in P element population dynamics between the sibling species Drosophila melanogaster and Drosophila simulans. *Genet. Res.* **1994**, *63*, 27–38. [CrossRef] [PubMed]
178. Yoshitake, Y.; Inomata, N.; Sano, M.; Kato, Y.; Itoh, M. The P element invaded rapidly and caused hybrid dysgenesis in natural populations of Drosophila simulans in Japan. *Ecol. Evol.* **2018**, *8*, 9590–9599. [CrossRef] [PubMed]
179. Hill, T.; Schlötterer, C.; Betancourt, A.J. Hybrid Dysgenesis in Drosophila simulans Associated with a Rapid Invasion of the P-Element. *PLoS Genet.* **2016**, *12*, 1–17.
180. Kofler, R.; Hill, T.; Nolte, V.; Betancourt, A.J.; Schlötterer, C. The recent invasion of natural *Drosophila simulans* populations by the P-element. *Proc. Natl. Acad. Sci. USA* **2015**, *112*, 6659–6663. [CrossRef]
181. O'Neill, M.J.; O'Neill, R.J. Sex chromosome repeats tip the balance towards speciation. *Mol. Ecol.* **2018**. [CrossRef]
182. Brown, J.D.; O'Neill, R.J. Chromosomes, Conflict, and Epigenetics: Chromosomal Speciation Revisited. *Annu. Rev. Genom. Hum. Genet.* **2010**, *11*, 291–316. [CrossRef]
183. Ellison, C.; Bachtrog, D. Dosage Compensation via Transposable Element Mediated Rewiring of a Regulatory Network. *Science* **2013**, *342*, 846–850. [CrossRef] [PubMed]
184. Conrad, T.; Akhtar, A. Dosage compensation in Drosophila melanogaster: Epigenetic fine-tuning of chromosome-wide transcription. *Nat. Rev. Genet.* **2012**, *13*, 123–134. [CrossRef] [PubMed]
185. Gay, L.; Crochet, P.-A.; Bell, D.A.; Lenormand, T. Comparing clines on molecular and phenotypic traits in hybrid zones: a window on tension zone models. *Evolution* **2008**, *62*, 2789–2806. [CrossRef] [PubMed]
186. Lesecque, Y.; Glémin, S.; Lartillot, N.; Mouchiroud, D.; Duret, L. The Red Queen Model of Recombination Hotspots Evolution in the Light of Archaic and Modern Human Genomes. *PLoS Genet.* **2014**, *10*, 1–14. [CrossRef] [PubMed]
187. Bierne, N.; Welch, J.; Loire, E.; Bonhomme, F.; David, P. The coupling hypothesis: Why genome scans may fail to map local adaptation genes. *Mol. Ecol.* **2011**, *20*, 2044–2072. [CrossRef] [PubMed]
188. Andrew, R.L.; Bernatchez, L.; Bonin, A.; Buerkle, C.A.; Carstens, B.C.; Emerson, B.C.; Garant, D.; Giraud, T.; Kane, N.C.; Rogers, S.M.; et al. A road map for molecular ecology. *Mol. Ecol.* **2013**, *22*, 2605–2626. [CrossRef] [PubMed]

189. Li, J.; Li, H.; Jakobsson, M.; Li, S.; SjÖdin, P.; Lascoux, M. Joint analysis of demography and selection in population genetics: Where do we stand and where could we go? *Mol. Ecol.* **2012**, *21*, 28–44. [CrossRef] [PubMed]
190. Orozco-terWengel, P. The devil is in the details: The effect of population structure on demographic inference. *Heredity* **2016**, *116*, 349–350. [CrossRef]
191. Sattath, S.; Elyashiv, E.; Kolodny, O.; Rinott, Y.; Sella, G. Pervasive adaptive protein evolution apparent in diversity patterns around amino acid substitutions in drosophila simulans. *PLoS Genet.* **2011**, *7*. [CrossRef]
192. Suh, A.; Smeds, L.; Ellegren, H. Abundant recent activity of retrovirus-like retrotransposons within and among flycatcher species implies a rich source of structural variation in songbird genomes. *Mol. Ecol.* **2018**, *27*, 99–111. [CrossRef]

© 2019 by the authors. Licensee MDPI, Basel, Switzerland. This article is an open access article distributed under the terms and conditions of the Creative Commons Attribution (CC BY) license (http://creativecommons.org/licenses/by/4.0/).

Article

The Genome of Blue-Capped Cordon-Bleu Uncovers Hidden Diversity of LTR Retrotransposons in Zebra Finch

Jesper Boman [1,*], Carolina Frankl-Vilches [2], Michelly da Silva dos Santos [3], Edivaldo H. C. de Oliveira [3], Manfred Gahr [2] and Alexander Suh [1,*]

1. Department of Evolutionary Biology, Evolutionary Biology Centre (EBC), Science for Life Laboratory, Uppsala University, SE-752 36 Uppsala, Sweden
2. Department of Behavioral Neurobiology, Max Planck Institute for Ornithology, 82319 Seewiesen, Germany; frankl@orn.mpg.de (C.F.-V.); gahr@orn.mpg.de (M.G.)
3. Laboratório de Cultura de Tecidos e Citogenética, SAMAM, Instituto Evandro Chagas, Ananindeua, Pará, and Faculdade de Ciências Naturais (ICEN), Universidade Federal do Pará, Belém 66075-110, Brazil; michellyufpa@gmail.com (M.d.S.d.S.); ehco@ufpa.br (E.H.C.d.O.)
* Correspondence: jesper.boman@gmail.com (J.B.); alexander.suh@ebc.uu.se (A.S.)

Received: 6 March 2019; Accepted: 5 April 2019; Published: 13 April 2019

Abstract: Avian genomes have perplexed researchers by being conservative in both size and rearrangements, while simultaneously holding the blueprints for a massive species radiation during the last 65 million years (My). Transposable elements (TEs) in bird genomes are relatively scarce but have been implicated as important hotspots for chromosomal inversions. In zebra finch (*Taeniopygia guttata*), long terminal repeat (LTR) retrotransposons have proliferated and are positively associated with chromosomal breakpoint regions. Here, we present the genome, karyotype and transposons of blue-capped cordon-bleu (*Uraeginthus cyanocephalus*), an African songbird that diverged from zebra finch at the root of estrildid finches 10 million years ago (Mya). This constitutes the third linked-read sequenced genome assembly and fourth in-depth curated TE library of any bird. Exploration of TE diversity on this brief evolutionary timescale constitutes a considerable increase in resolution for avian TE biology and allowed us to uncover 4.5 Mb more LTR retrotransposons in the zebra finch genome. In blue-capped cordon-bleu, we likewise observed a recent LTR accumulation indicating that this is a shared feature of Estrildidae. Curiously, we discovered 25 new endogenous retrovirus-like LTR retrotransposon families of which at least 21 are present in zebra finch but were previously undiscovered. This highlights the importance of studying close relatives of model organisms.

Keywords: transposable elements; transposons; LTR retrotransposons; ERV; genome; genome annotation; karyotype; estrildidae; zebra finch; *Uraeginthus cyanocephalus*

1. Introduction

Birds are remarkable among vertebrates by having small genomes, a low variation (0.91–2.16 pg, 2.4-fold) in genome size and a low density of repetitive elements [1–3]. Small genome sizes of birds are typically explained as an adaption for flight, through association with high metabolic rate which in turn selects for small red blood cells capable of greater gas exchange per unit volume [4–6]. This view is consistent with the observation of smaller genomes in flighted versus flightless birds and more streamlined genomes of bats compared to other eutherians [4,7,8]. However, measurements of insertion and deletion rates suggest that birds with more transposable element (TE) accumulation also have more deletions, resulting in a higher net shrinking and therefore smaller genomes [3]. Larger genome sizes of flightless birds result from low deletion rates and accumulation of TEs, meaning that they have less genomic turnover overall [3]. This might indicate that genome size differences among extant birds do not necessarily reflect adaptation for flight, but instead lineage-specific differences in genome dynamism [3].

Birds are the most species-rich group of land vertebrates as a result of a massive radiation following the demise of other dinosaur fauna at the Cretaceous–Paleogene extinction event 65 Mya [9]. The putative association between TE accumulation and speciation that has been shown in, e.g., mammals [10] is an interesting prospect for avian TE biology. Transposons have for example been implicated as hotspots for chromosomal breakpoint regions [11–13], conceivably associating transposon accumulation with chromosomal inversions. Through recombination suppression, inversions may act as islands of genomic differentiation (e.g., [14]). Research has shown that the genome of the important model organism zebra finch has undergone many inversions on a short evolutionary timescale [15,16]. Zebra finch also has a recent accumulation of endogenous retrovirus (ERV)-like long terminal repeat (LTR) retrotransposons [17], which proliferate through a copy and paste mechanism [18]. Romanov et al. [16] found a positive correlation between LTR retrotransposons and genomic regions especially prone to chromosomal rearrangements, so-called evolutionary breakpoint regions. Moreover, intra-chromosomal rearrangements such as inversions are more frequent in the zebra finch's family Estrildidae, than in other bird lineages [15].

To understand the dynamics of LTR proliferation in Estrildidae, we de-novo sequenced and karyotyped the genome of blue-capped cordon-bleu (*Uraeginthus cyanocephalus*) and performed an in-depth computational prediction and manual curation of TEs. Blue-capped cordon-bleu is an East African estrildid finch and famous for its rapid tap dancing display [19,20]. It belongs to a lineage that split from the Austro-Pacific zebra finch at the root of Estrildidae 10 Mya [15]. In-depth annotations of TEs consisting of both computational prediction and manual curation have so far only been presented for zebra finch, chicken (*Gallus gallus*) and collared flycatcher (*Ficedula albicollis*) [17,21,22]. Each genome curated has revealed a great diversity of new transposon families and subfamilies. Through rigorous manual curation, we discovered 25 new ERV-like retrotransposon families of which 21 are shared with zebra finch. Using repeats from collared flycatcher and blue-capped cordon-bleu, we find an additional 4.5 Mb of LTR elements (i.e., >10% increase in annotated bp) in the zebra finch genome assembly taeGut2, compared with using only previously curated bird repeats from Repbase. Furthermore, we show that blue-capped cordon-bleu has experienced a recent accumulation of LTR retrotransposons, which indicates that this is a shared feature of estrildid finches and likely important in shaping their genomic landscape.

2. Materials and Methods

2.1. Sequencing, Genome Assembly and Karyotyping

We sequenced the genome from heart and testis tissues of a male blue-capped cordon-bleu (*U. cy.*) bred at Max Planck Institute for Ornithology (Germany), Seewiesen animal facility, using the 10X Genomics Chromium linked-read system [23,24] and sequencing of 150-bp paired-end reads on an Illumina HiSeq X instrument, both conducted by SciLifeLab Stockholm (Sweden). Animal handling was carried out in accordance with the European Communities Council Directive 2010/63 EU and the legislation of the state of Upper Bavaria. We used a genome assembly from testis tissue for RepeatModeler prediction (see below), but decided to use an assembly from heart tissue for all analyses, to be more comparable with the somatic repeatomes of zebra finch, collared flycatcher and chicken, due to the recent hypothesis of a germline restricted chromosome being widespread among songbirds [25–27]. Hereafter, "the genome of blue-capped cordon-bleu" refers to the heart assembly. We generated "pseudohaploid" draft genome assemblies using Supernova 2.0 [23,24]. The Chromium system employs a unique barcoding of reads from the same input DNA molecule which potentially allows for the assembly of longer contigs and scaffolds than conventional short-read technologies [24]. We assessed the assembly quality using the assemblathon_stats.pl script [28] and investigated the gene set completeness using the aves_odb9 library in BUSCO2 [29] (Table 1). Karyotyping was performed on fibroblast cells from the embryos of both male and female blue-capped cordon-bleu using established protocols [30,31] with modifications described previously in Santos et al. [32] and Furo et al. [33] (Figure 1).

2.2. Computational and Manual Curation of Transposable Elements

Repetitive element consensus sequences were predicted de novo using RepeatModeler ver. 1.0.8 [34]. The predicted library of consensus sequences was masked with RepeatMasker ver. 4.0.7 using the *Aves* Repbase library [35]. Consensus sequences more than 5% diverged from previously annotated zebra finch repeat consensuses [17] were selected for manual curation. Using a custom script [22], the 20 best BLASTn ver. 2.6.0+ [36] hits of each consensus sequence along with 2-kb flanks were aligned using MAFFT ver. 7.310 [37]. For each repeat predicted by RepeatModeler, a new majority rule consensus sequence was made based on the aligned hits, either manually with an alignment viewer (Aliview [38] or BioEdit [39]) or using Advanced Consensus Maker [40]. At each site, the most abundant base was used as consensus, except for potential hypermutable CpG sites, which were curated as 5'-CG-3'. Target site duplication (TSD) patterns and the long terminal repeat (LTR) canonical 5'-TG ... CA-3' ends were used to identify and classify LTR retrotransposons into three groups [41]: endogenous retrovirus superfamily 1 (ERV1, 4 bp TSD), endogenous retrovirus superfamily K/2 (ERV2, 6 bp TSD) and endogenous retrovirus superfamily L/3 (ERV3, 5 bp TSD). The characteristic eight base pair motif [42], 5'-ATTCTRTG-3', was used to identify the 3' ends of CR1 LINEs. CR1 curation proceeded in 5' direction as long as at least three BLASTn hits with high similarity were distinguishable in the alignment.

Manually curated consensus sequences were queried against Repbase using CENSOR [43]. To date, a majority of avian repeats in Repbase are from chicken and zebra finch. SINE and LTR retrotransposons with considerable nucleotide similarity (>80%) across a majority of their lengths (>80%; for at least 80 bp) to a repeat in Repbase or to each other (checked manually), were classified as belonging to the same family. SINE and LTR retrotransposons with hits to Repbase that did not meet these criteria were classified as new families. The criteria used here are based on the TE family 80-80-80 rule cutoff proposed by Wicker et al. [44] in which two TEs belong to the same family if 80% of a novel TE is more than 80% identical for at least 80 bp of an already classified TE, in a BLAST search or similar against a repeat database. By the same classification scheme, a TE subfamily represents a subpopulation of an already identified TE family [44]. We classified novel TEs from the same species as belonging to separate subfamilies if their consensus sequences were less than 95% similar on the nucleotide level. Some blue-capped cordon-bleu consensus sequences were more than 95% similar to zebra finch repeats after manual curation (Table S1). We still consider these as separate subfamilies in our analyses. For all curated LTR retrotransposons that met our criteria for a novel family, we next searched a library of collared flycatcher LTR consensus sequences [22] using BLASTn (E-value = 0.01). We classified a blue-capped cordon-bleu LTR consensus sequence as belonging to a collared flycatcher LTR family if it had considerable nucleotide similarity across the majority of its sequence (see criteria above) to a collared flycatcher LTR consensus. CR1 elements were classified based on a PhyML ver. 3.0 [45,46] maximum likelihood (ML) phylogeny (GTR+G+I substitution model) of all CR1 subfamilies from blue-capped cordon-bleu, chicken, zebra finch and collared flycatcher. The library for the latter three is the same as in Suh et al. [22]. This and another phylogenetic tree of songbird repeats from the TE family TguERVL2_I were depicted using FigTree ver. 1.4.3 (Figures S1 and S2) [47]. TE subfamilies and families were named following previous conventions used in the zebra finch and collared flycatcher repeat annotations [17,22].

2.3. Data Analysis

We created TE landscape plots using the .align files of the RepeatMasker output as described in preceding publications [22,48], except that CpG sites have lower weighting instead of being excluded when counting substitutions (Figure 2). Data presented in Table 2 were obtained from the .tbl file of the RepeatMasker output. We investigated the respective amount of shared and lineage-specific diversity of LTR families and subfamilies using genomes and LTR libraries from in-depth curated birds: chicken—galGal4, zebra finch—taeGut2, collared flycatcher—ficAlb1.5 and blue-capped cordon-bleu, using reciprocal BLASTn searches (E-value cutoff = 10^{-10}) [22] (Figure 3). The zebra finch genome (taeGut2) was masked with two libraries: a library consisting of repeats from the *Aves* category in Repbase and a "full" library where blue-capped cordon-bleu and collared flycatcher repeats were added (Table S4). Statistical analyses of chromosomal content and LTR subfamily number in zebra finch were performed using R ver. 3.5 [49] on a taeGut2 genome assembly acquired from UCSC [50] (Figure 4). Scaffolds with Un* prefix and *random suffix were excluded in the analyses (Figure 4). All repeat libraries were obtained from Repbase [43] except for the collared flycatcher library which was acquired from dfam_consensus [51]. We hypothesized that LTR subfamilies from blue-capped cordon-bleu and collared flycatcher that are more similar to zebra finch LTRs should compete more in masking with zebra finch LTRs in RepeatMasker. Conversely, we predicted that blue-capped cordon-bleu and collared-flycatcher LTR subfamilies that do not belong to a family curated in zebra finch should contribute more to the discovery of previously unannotated repeats in the taeGut2 genome. We tested this prediction by comparing the overlap of chromosomal positions between LTRs from the RepeatMasker output of the *Aves* Repbase library and two sets of LTRs from the output when masking with the "full" library, using the intersect utility in the BEDTools suite [52] (Figure 4c). To annotate a single internal portion of an ERV-like element, we reran the pipeline described above for collecting BLASTn hits along with flanking regions, to obtain more copies of the internal element. We then used the NCBI ORFfinder tool to identify open reading frames [53], NCBI CD-search for characterization of conserved domains [54], and the consensus2genome R script [55] to depict genomic hits (BLASTn) of a concatenated consensus sequence of the ERV internal region and the flanking LTRs (Figure 5).

2.4. Data Deposition

Linked-read data were deposited in Sequence Read Archive (accession number SRR8873500). Both the "pseudohaploid" genome assembly draft and a phased diploid assembly draft were deposited in Dryad (http://dx.doi.org/10.5061/dryad.322gd5p). The newly curated consensus sequences were deposited in dfam_consensus.

3. Results

3.1. Genome Assembly and Karyotype of Blue-Capped Cordon-Bleu

We sequenced the genome of a male blue-capped cordon-bleu using the 10X Genomics Chromium linked-read platform [24] and obtained an average molecule length of 42.4 kb (Table 1). We assembled the genome using Supernova 2.0 and obtained an ~1.1 Gb assembly size, of which 105.6 Mb are "N" gaps, a scaffold N50 of 10.9 Mb, and a contig N50 of 66.3 kb (Table 1). We assessed the completeness of the genome using the aves_odb9 ortholog data set in BUSCO and recovered 90.1% of the genes completely, while 5.9% were fragmented and 4% were missing (Table 1).

Next, we karyotyped male and female blue-capped cordon-bleu using Giemsa staining and C-banding (Figure 1). Like zebra finch, blue-capped cordon-bleu has 2n = 80 [32]. Unlike zebra finch where the W is smaller than the Z [32,56], blue-capped cordon-bleu has sex chromosomes of roughly equal size (Figure 1a). Sex chromosomes were identified as a homomorphic macrochromosome pair in males (ZZ), while in females they were heteromorphic (ZW). Giemsa staining pattern is shown for the largest macrochromosomes and sex chromosomes of a female (Figure 1a). C-banding revealed a highly heterochromatic W chromosome, further confirming its identity (Figure 1b). Constitutive heterochromatin on autosomes is mainly restricted to putatively centromeric regions (Figure 1b).

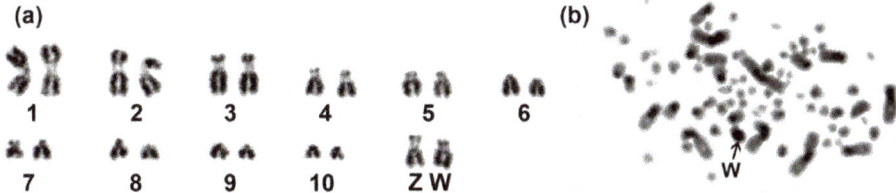

Figure 1. Karyotype of a female blue-capped cordon-bleu. The diploid (2n) chromosome number is 80. Giemsa staining of macrochromosomes showed that the sex chromosomes are approximately equal in size (**a**). C-banding revealed that the W chromosome is enriched in heterochromatin, compared to autosomes in which heterochromatin is restricted to putative centromeric regions (**b**). In panel (**a**), autosomes are numbered from largest to smallest, as proposed by the International System for Standardized Avian Karyotypes [57].

Table 1. Sequencing and assembly statistics for the genome assembly of blue-capped cordon-bleu.

Statistic	Quantity
Assembly size	1099.6 Mb
"N" nucleotides	105.6 Mb
Weighted mean molecule length	42.4 kb
Number of reads	254.2 million
Scaffolds	26,389
Scaffold N50	10.9 Mb
Contigs	51,469
Contig N50	66.3 kb
BUSCO (complete)	90.1%
BUSCO (fragmented)	5.9%
BUSCO (missing)	4%

3.2. The Transposable Element Landscape of Blue-Capped Cordon-Bleu

We identified transposable elements in the genome of blue-capped cordon-bleu using de-novo prediction with RepeatModeler followed by manual curation of all non-redundant and curatable consensus sequences. Masking the genome with RepeatMasker revealed a TE content of 6.44% (Table 2), a number typical for birds [2]. Most transposons were LINEs (132,734 copies) followed by LTR retrotransposons (61,457 copies). However, they have a roughly similar density, indicating that LTR retrotransposons are longer on average (Table 2). In Figure 2, we show three TE landscapes to highlight the difference in results when only relying on previously annotated TEs (Figure 2a), adding a RepeatModeler library (Figure 2b) and when performing in-depth manual curation (Figure 2c). Many repeats were initially classified as unknown by RepeatModeler (compare Figure 2a,b). Our manual curation showed that all curatable "unknown repeats" were in fact solo-LTRs of ERV-like retrotransposons (Table S1, Figure 2c). We used the canonical 5'-TG ... CA-3'-ends and TSDs to identify solo-LTR elements. However, several variations deviating from 5'-TG ... CA-3' were observed (Table S1). Following previous LTR annotations for songbirds [15,20], we classified LTR elements to ERV superfamilies based on the length of their TSDs [41]. A peculiar element—*UcyLTR-Lurtz*—had both 5 and 6 bp target site duplications. In total, 25 new families and 50 new subfamilies of retrovirus-like LTR retrotransposons were curated. Moreover, we identified 16 new CR1 subfamilies and one new CR1-mobilized tRNA-Ile SINE subfamily (Table S1, Figure S1). We found no new curatable DNA transposons, which is perhaps not surprising considering that previous investigations of estrildid finches revealed only a relatively old hAT DNA transposon family, present in low copy numbers in zebra finch [17].

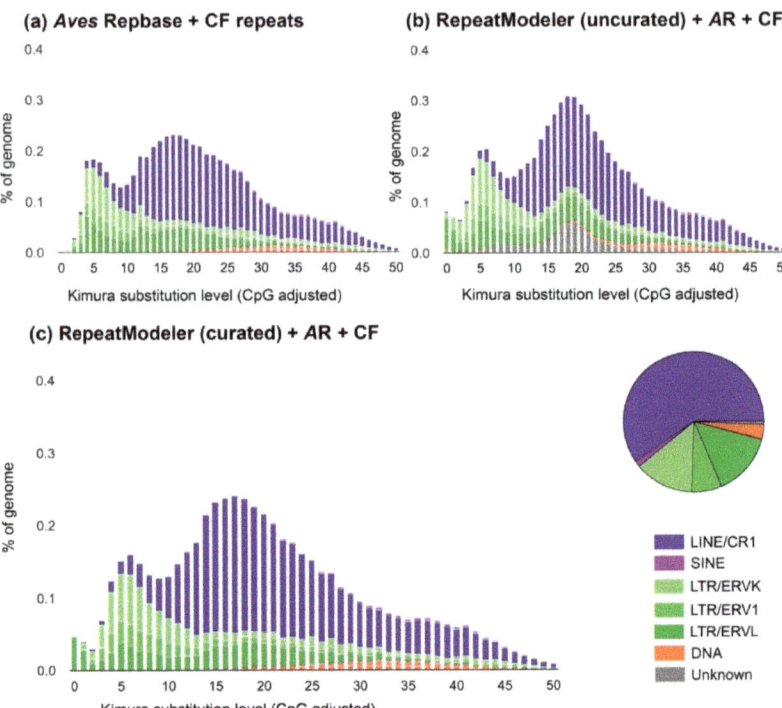

Figure 2. Comparison of transposable element landscapes for the genome of blue-capped cordon-bleu, representing different levels of effort in transposon annotation. Percentage of bp occupied in the genome is plotted against the Kimura 2-parameter (transitions/transversions) distance of each transposable element (TE) copy from its consensus. Panel (**a**) shows the landscape for when avian repeats available in Repbase (*Aves* Repbase, AR) and collared flycatcher (CF) repeats were used for masking the genome. Panel (**b**) is based on de-novo predicted repeats from Repeatmodeler, AR and CF repeats. Note the share of unknown (grey) repeats, a majority of which were identified as solo-long terminal repeats (LTRs) of endogenous retrovirus (ERV)-like retrotransposons when manually curated (**c**). The pie chart specifies the relative abundance of different TEs based on the .tbl file of the RepeatMasker output (Table 2), for the curated, final landscape (**c**).

Table 2. Copy number, total base pair and density of different classes of repetitive elements annotated by RepeatMasker using a library consisting of manually curated blue-capped cordon-bleu and collared flycatcher repeats, and the *Aves* library from Repbase.

Repeat Type	Copies	Total bp	% of Genome
SINE	7163	852,236	0.08
LINE	132,734	37,876,706	3.44
LTR	61,457	29,437,443	2.68
DNA	14,100	2,195,734	0.20
Unclassified	2367	416,198	0.04
Total interspersed repeats	217,821	70,778,317	6.44
Small RNA	1479	199,270	0.02
Satellites	1960	581,825	0.05
Simple repeats	211,440	9,408,016	0.86
Low complexity	43,325	2,238,772	0.20
Total tandem repeats	258,204	12,427,883	1.17
Total repeats	746,059	83,206,200	7.61

3.3. Comparative Genomics Revealed Extensive Shared Diversity of LTRs among Estrildid Finches

From the 50 discovered ERV-like LTR retrotransposons in blue-capped cordon-bleu, we classified 25 as new families based on the lack of extensive nucleotide similarity to LTR elements in Repbase, in collared flycatcher, and to each other. We considered consensus sequences with less than 95% nucleotide identity to each other as separate subfamilies within such a family. To investigate the amount shared LTR diversity between the in-depth curated birds (chicken, collared flycatcher, zebra finch and blue-capped cordon-bleu), we extended the reciprocal BLASTn search of Suh et al. [22] using consensus sequences from blue-capped cordon-bleu. In brief, separate libraries of LTR subfamily consensus sequences from each species were BLASTn searched to each genome, and the presence and absence of LTR families and subfamilies was scored (Tables S2 and S3 and Figure 3). A majority of LTR subfamilies that was curated using the blue-capped cordon-bleu genome is shared between zebra finch and blue-capped cordon-bleu. Thus, 21 of 25 novel ERV-like LTR families are present in the zebra finch genome assembly (taeGut2) but were previously undiscovered. Four families (UcyLTRK7, UcyLTRK15, UcyLTRL6, and UcyLTR-Lurtz) are lineage-specific to blue-capped cordon-bleu (Figure 3). Only TguERV5 is specific to zebra finch (Figure 3).

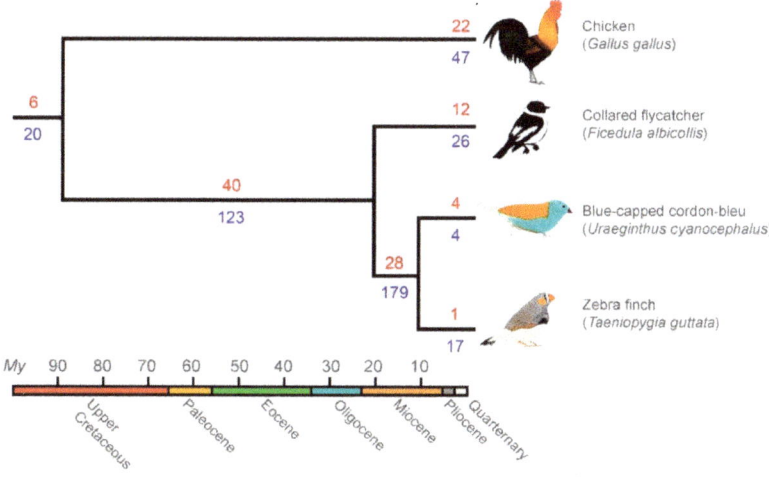

Figure 3. Analysis of LTR diversity along branches in the tree of birds with in-depth curated TE libraries. The number of LTR families and subfamilies on each branch are depicted in red (above branches) and blue (below branches), respectively. Most LTR retrotransposon families are shared between blue-capped cordon-bleu and zebra finch. The previously thoroughly investigated genome of zebra finch contains more lineage-specific TE subfamilies. A large diversity of LTR families and subfamilies are shared among the three songbirds compared with the relative sparse number of LTRs shared with chicken at the root of the tree. Node estimates are based on previously published timetrees [9,15,58].

To understand how heterospecific TE libraries can improve repeat annotation in a model organism and why substantial LTR diversity was previously undetected in the zebra finch, we masked the zebra finch reference genome (taeGut2, based on same isolate as taeGut1) obtained from UCSC [50], using RepeatMasker and two libraries. One library consisted of *Aves* repeats in Repbase only and the other was *Aves* Repbase repeats concatenated with collared flycatcher and blue-capped cordon-bleu repeats. The latter, "full" library masked ~7.5 Mb more repeats than the former, of which ~4.5 Mb are LTR elements and ~2.6 Mb are satellite DNA (Table S4). We visualized the chromosomal content of LTR elements by six different categories and grouped them according to two criteria: (1) songbird species whose genome assembly was used for curation and (2) whether or not the LTR element belongs to a zebra finch LTR family (Figure 4a). One exception is TguLTRL3-L_Ucy, which fulfilled our criteria to be classified as a new family but was highly similar (>75%) at two different parts of its consensus to TguLTRL3. We therefore treated this new family as belonging to the category of zebra finch (ZF) families in these analyses. One LTR subfamily (fAlbLTR1_Ucy) in blue-capped cordon-bleu (BC) belonged to a collared flycatcher (CF) LTR family and was categorized as "Others" along with mostly chicken LTRs. Note that LTR annotation by RepeatMasker includes fragments of elements, which we included in the copy number estimates. Furthermore, five BC and two CF LTR subfamilies curated using respective genome had less than five hits in total and their presence/absence in the ZF genome should thus be considered with caution. The reciprocal BLAST approach should give a more conservative picture of the genomic presence/absence status of specific LTR families and subfamilies (Figure 3).

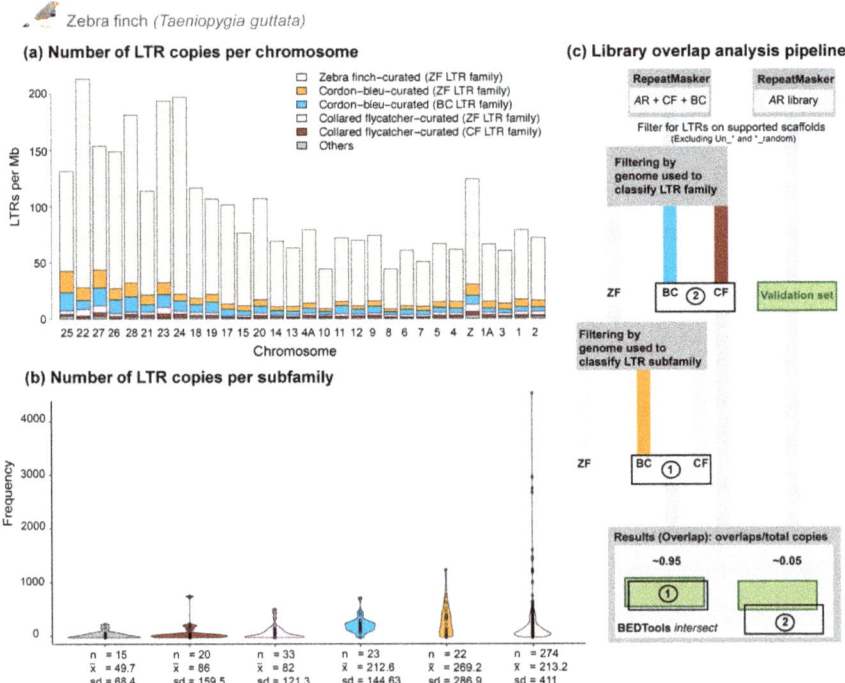

Figure 4. Investigation of LTR subfamily number and diversity in the zebra finch genome. We masked the genome of zebra finch (taeGut2) using RepeatMasker and two repeat libraries. One consisted of *Aves* repeats in Repbase (AR) only, and the other contained Repbase repeats with addition of collared flycatcher [22] and the novel blue-capped cordon-bleu repeats. We found ~4.5 Mb more LTR elements using the latter library (Table S4). Panel (**a**) shows the number of LTRs per Mb per chromosome. LTR copies were grouped according to the genome assembly used for curation and species first used for LTR family definition. Chromosomes are ordered in ascending size and are named according to homology with chicken chromosomes. Panel (**b**) shows the number of LTR copies per subfamily, here depicted as violin distributions. Statistics presented for each group of LTR copies per subfamily are: sample size per category (n), mean ($\bar{x}$) and standard deviation (sd) of copies per subfamily per category. Panel (**c**) shows the library overlap analysis pipeline. Several steps are shared with the other analyses depicted in (**a**) and (**b**). Blue-capped cordon-bleu (BC) and collared flycatcher (CF) LTRs belonging to zebra finch (ZF) LTR families generally map to already annotated repeats (overlaps/total copies ≈ 0.95). LTR copies from families described as novel in respective genome project map to new positions (overlaps/total copies ≈ 0.05).

We observed that BC LTRs were overall more frequent than CF LTRs in the zebra finch genome (Figure 4a). The same pattern was seen for the frequency of LTR copies per subfamily, with a total of 10,811 BC copies and 4427 CF copies (Figure 4b). There were significantly more BC LTR copies per subfamily than CF copies (Welch t-test; p-value = 1.151×10^{-4}). We saw the same trend when we compared LTR subfamilies from BC and CF families, with 4889 copies from BC LTR families and 1719 copies from CF families, and significantly more BC than CF copies per LTR subfamily (Welch t-test; p-value = 9.949×10^{-3}). Furthermore, BC LTRs from BC families constituted significantly more base pairs per chromosome than CF LTRs from CF families (Wilcoxon signed rank test; p-value = 1.863×10^{-9}). BC LTRs from BC families comprised in total 1736 kb compared to 471 kb of CF LTRs from CF families. These results strongly indicate that in-depth curation of LTR families in the more closely related blue-capped cordon-bleu led to annotation of more LTR copies in zebra finch than did the LTR families of the more distantly related collared flycatcher.

Long terminal repeat subfamilies from BC and CF belonging to ZF LTR families have high sequence similarity to zebra finch LTRs and would therefore compete in masking with them. We can call this the "competition-in-masking" hypothesis. A prediction from this hypothesis is that the largest gain in finding previously unannotated LTR elements in zebra finch should be obtained by using consensus sequences from LTR families previously undetected in zebra finch. We tested the "competition-in-masking" hypothesis by counting the number of overlaps between LTRs from the RepeatMasker output using only *Aves* Repbase repeats as library against two sets of LTRs from the "full" library (Figure 4c). The first set consisted of BC and CF LTRs belonging to ZF LTR families (8651 copies), and the second set consisted of BC and CF LTRs belonging to respective BC or CF LTR families (6608 copies). In the first set, 8214 overlaps were found which gave an overlap/copy number ratio of ~0.95. In the second set, only 373 overlaps were counted which results in a ratio of ~0.05. These results strongly confirm the "competition-in-masking" hypothesis and highlight how describing novel LTR families in a non-model relative can uncover hidden LTR diversity in the genome of a model organism.

3.4. Analysis of a Recently Active TE

We were able to curate a full-length LTR retrotransposon subfamily from the ERVL superfamily with complete internal region, in the blue-capped cordon-bleu genome. The copies of this LTR subfamily, TguERVL2_I_Ucy, make up ~1 Mb in total, which is 2.5 times more DNA than the closely related TguERVL2_I in the zebra finch genome (Table S5). The low average divergence (1.7%) to the consensus sequence is a good indication that this TE subfamily was very recently active (Table S5). We did a functional annotation of the consensus sequence of TguERVL2_I_Ucy, which revealed two long ORFs in the same reading frame and intact AP, RT, RH and INT domains, as well as an additional broken RH domain, all of which are canonical for vertebrate ERV-like retrotransposons (Figure 5a) [44,59,60]. However, the AP domain is predicted partially outside of the ORF boundaries (Figure 5a). Curiously, a disrupted envelope (ENV) glycoprotein C domain from the Marek_A superfamily is predicted inside the *gag* ORF (137 amino acid alignment to superfamily member PHA03269, E-value = 7.97×10^{-4}). The Marek_A glycoprotein was originally classified in Marek's disease virus (also known as *Gallid alphaherpesvirus*) [61], a ~174–180 kb dsDNA herpesvirus causing a neoplastic disease in chickens [62,63]. Interestingly, the TguERVL2 family is found in chickens as well (Table S3).

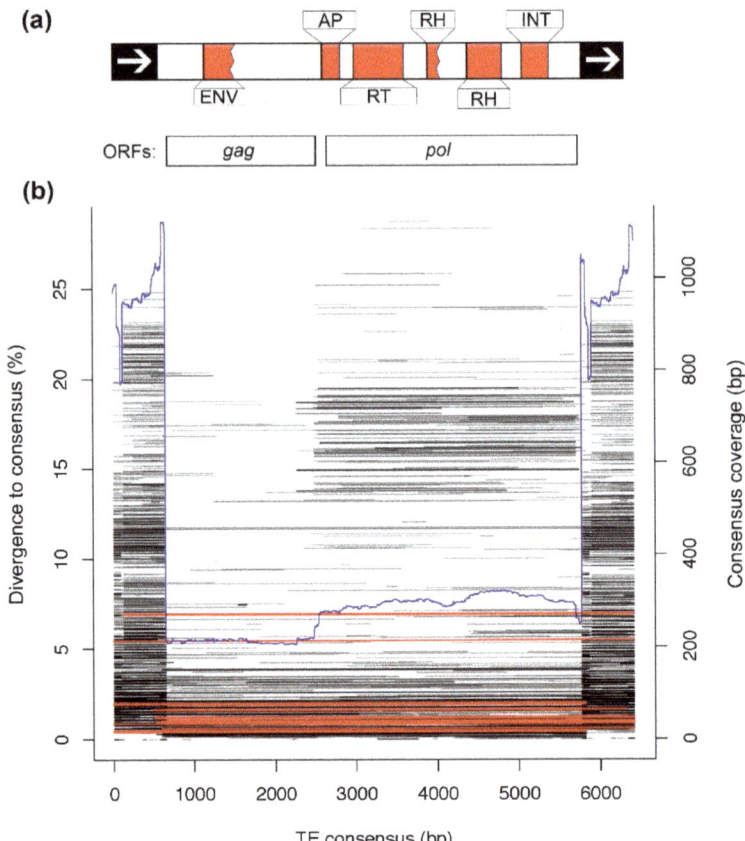

Figure 5. Functional domain annotation and genomic BLASTn hits of TguERVL2_I_Ucy. We predicted conserved domains and open reading frames (ORFs) of the consensus sequence of TguERVL2_I_Ucy (**a**). In addition to the canonical domains (AP, RT, two RH (one partial and one complete), and INT), a disrupted ENV domain was predicted at an upstream position. Panel (**b**) shows the distribution of copies of a pseudo full-length ERV consensus sequence (same LTR flanked by separately classified internal portion) of the TE subfamily TguERVL2_I_Ucy in blue-capped cordon bleu. Most copies in the genome are solo-LTRs and a majority of the full-length copies are less than 5% diverged from the consensus sequence. Hits spanning a majority of the consensus are shown in red and partial hits are black. Blue line represents consensus coverage.

Conserved domain analysis of TguERVL2_I and TguERVL1_I in zebra finch suggests that these have all domains except for this broken ENV (not shown). However, a protein alignment of consensus sequences of TguERVL2_I, TguERVL1_I and members of PHA03269 (Envelope glycoprotein C from *Human alphaherpesvirus 3* and *Cercopithecine alphaherpesvirus 9*) and pfam02124 (various herpesviruses) revealed that all TguERVL2_I subfamilies, but not TguERVL1_I, share similarity in a short region mainly to PHA03269 (Figure S2; Data S2). It is therefore likely that this hit represents an ancestral feature of the TguERVL2 family and not a translocation or recombination with a herpesvirus in the recent history of blue-capped cordon-bleu. Furthermore, this amino acid feature does not mean that TguERVL2_I_Ucy has an intact envelope as has been seen for some invertebrate LTR retrotransposons that likely acquired an entire ENV ORF from dsDNA viruses [64].

We inferred a maximum likelihood phylogeny of internal consensus sequences of TguERVL2_I in blue-capped cordon-bleu, zebra finch and collared flycatcher to analyze the evolutionary history of this LTR family (Figure S3). TguERVL1_I was the most closely related TE in the well-annotated zebra finch genome and was consequently chosen as outgroup. The phylogeny recapitulated the species tree with strong support (99 of 100 bootstrap replicates), indicating that TguERVL2_I has been vertically inherited in the investigated songbirds.

The curated LTR of TguERVL2_I_Ucy was concatenated with both ends of the internal region to create a 6.4 kb "pseudo full-length" ERV consensus, which we subsequently used to characterize consensus coverage of hits in the genome using the consensus2genome R script relying on BLASTn (E-value cutoff = 10^{-7}) [55]. Most copies of this ERV throughout the genome are solo-LTRs, as indicated by the higher coverage of terminal repeats (Figure 5b). We also see a pattern of more copies with intact internal regions being recently diverged from the consensus (Figure 5b). These observations are consistent with the view of deletion of the internal region and one LTR, through within-element non-allelic homologous recombination [65]. Curiously, many hits in the range of 10% to 20% divergence to consensus seem to lack homology for the first ORF containing the broken ENV. These likely represent elements of another LTR subfamily with a similar *pol* ORF but a dissimilar *gag* ORF.

4. Discussion

In this study, we present the third linked-read genome assembly of any bird, to our knowledge. If we compare with one of the 10X genome assemblies published previously for eastern black-eared wheatear (*Oenanthe hispanica melanoleuca*) [66], we obtain a higher weighted average molecule length (42.4 kb vs. 17.5 kb) which most likely contributes to a higher scaffold N50 (10.9 Mb vs. 90 kb), (Table 1). Even higher scaffold N50 may be possible to be obtained by using a subset of reads [23], as shown by Toomey et al. [67] who produced a 10X genome assembly with scaffold N50 of 18.97 Mb for Gouldian finch (*Erythrura gouldiae*) with a read-depth of 60-fold in Supernova. Using the ranking employed by Suh and Kapusta [2], the genome assembly of blue-capped cordon-bleu is of medium quality (scaffold N50 >1 Mb; high quality requiring chromosome-level scaffolds) and has the 11th highest scaffold N50 out of the 77 analyzed bird genomes [2]. We also present the fourth bird genome with a well-curated transposon library and the first that allows comparative TE biology on the within-family level in birds. Previous work has shown that zebra finch has a substantial recent accumulation of ERV-like retrotransposons compared with other bird lineages [2,17,22], but see Mason et al. for a different view of LTR retrotransposon abundance in chickens [68]. The genome of blue-capped cordon-bleu also shows ERV-like LTR retrotransposon accumulation, and notably a recent expansion mostly caused by a single LTR subfamily, TguERVL2_I_Ucy (Table S5). Considering that zebra finch and blue-capped cordon-bleu separated at the deepest node of Estrildidae 10 Mya [15,69], ERV-like LTR retrotransposon accumulation might be ancestral to this clade.

Curiously, a majority (21 out of 25) of ERV-like LTR families described in this study are shared with zebra finch but were not previously described in its repeat annotation [17]. By combining repeats curated from the closely related blue-capped cordon-bleu (BC) and the more distantly related collared flycatcher (CF) with the *Aves* Repbase library, we were able to mask an additional 4.5 Mb (>10% increase) LTR retrotransposons in the zebra finch genome. We found significantly more copies per subfamily and a larger number of base pairs masked per chromosome of BC LTRs from BC families than CF LTRs from CF families (Figure 4a,b). This indicates that phylogenetic relatedness is an important factor when trying to find more repeats in a genome assembly using a TE library from another species. Furthermore, by analyzing the overlap between LTR copies in the RepeatMasker output from of the *Aves* Repbase and "full" libraries, we see that the largest addition of previously unannotated LTR elements in the taeGut2 genome results from novel BC and CF LTR families (Figure 4c). These results indicate that there are more TEs to be found in the reference genomes of model organisms and that they may be discovered by curating the repeatomes of closely related species.

A few novel BC ERV-like LTR retrotransposon families do not occur in zebra finch (Figure 3). Some or all might be unassembled or lost by drift or selection in zebra finch. A more plausible explanation is that they constitute recent germline infiltrations in the blue-capped cordon-bleu lineage. If that was the case for all four novel ERV-like LTR retrotransposon families, then the rate of germline infiltration in the blue-capped cordon-bleu lineage would be one every 2.5 My. This number may be an underestimate considering that research on a recent germline infiltration in koalas (*Phascolarctos cinereus*) show a polymorphic presence/absence pattern and no fixed insertions among individuals [70–73]. In addition, note the 28 LTR families on the short branch shared by zebra finch and blue-capped cordon-bleu (Figure 3). This indicates an even higher rate of germline infiltrations in the common ancestor of estrildid finches. The results presented here give an indication that the repetitive content and diversity of avian genomes may currently be somewhat underestimated. It is likely that the we will see diminishing returns in finding further shared TE diversity as more species are investigated. However, in-depth curation may greatly improve the accuracy of inferring a genome's repeat landscape, especially when in-depth TE libraries from closely related species are missing [74], or when many solo-LTRs are automatically classified as unknown by RepeatModeler as was the case here for blue-capped cordon-bleu (Figure 2b).

In this particular case, variation among species in the effect of sequence modification by TE suppression systems may be increasing LTR sequence diversity in estrildid finches compared with other songbird lineages. It has previously been shown that zebra finch LTR retrotransposons frequently are C→U-modified by APOBEC family proteins leading to a G→A mutation on the antisense strand [75]. Among 111 analyzed vertebrates, APOBEC modification was especially strong in zebra finch [75] and we speculate that it could be one of the most important drivers increasing the genetic diversity of LTR subfamilies in blue-capped cordon-bleu as well as zebra finch. Knisbacher and Levanon [75] observed a much more limited effect of APOBEC in medium ground finch (*Geospiza fortis*) indicating that APOBEC activity varies among songbirds or that the edited sites were more easily detected in zebra finch because of its in-depth curated LTR library. However, it is possible that APOBEC modification mainly affects LTR subfamily diversity, see for example the high number of subfamilies on the zebra finch branch in Figure 3. On the other hand, LTR families with no homology to other repeats in Repbase likely represent previously undiscovered retroviral diversity arising from germline infiltrations. Altogether, genome evolution in Estrildidae may very well be shaped by the expansion of LTR retrotransposons and their strong suppression by APOBEC modification.

The question of shared ERV-like retrotransposon diversity warrants further study, both in Estrildidae and in other songbird clades. Related to the question of shared diversity is the notion that a single LTR subfamily, TguERVL2_I_Ucy, has proliferated very recently in the evolutionary history of the blue-capped cordon-bleu so that it now composes 2.5-fold more DNA than in the genome of its closest relative in zebra finch (Table S5). This number is probably an underestimate considering the difficulty in assembling long repeat sequences with high sequence identity [76]. The fact that a full-length element of 6.4 kb was curatable and the consensus has intact GAG, AP, RT and RH domains suggests that this subfamily is likely still actively retrotransposing. The phylogeny of TguERVL2_I_Ucy and its closest songbird relatives suggests vertical inheritance of this LTR family at least since the common ancestor of Estrildidae (Figure S3). The ultimate cause of this element's high frequency in blue-capped cordon-bleu could be random genetic drift or some molecular feature of its Gag polyprotein—such as the putative Envelope glycoprotein C domain—that has allowed it to escape effective suppression. A horizontal acquisition event may have occurred in either direction between the ancestor of TguERVL2 and an alphaherpesvirus, but we cannot rule out that the similarity to Envelope glycoprotein C is caused by genetic drift or adaptive molecular convergence alone. However, horizontal transfer in both directions between LTR retrotransposons and dsDNA viruses have previously been inferred, which implies that such events do occur successfully [64,77].

Further investigation in Estrildidae is needed to explore the link between ERV-like LTR retrotransposon activity and the high rate of chromosomal inversions observed in this songbird clade [15]. For example, a single insertion of an LTR retrotransposon, *Ty912*, has been shown to increase the rate of gross chromosomal rearrangements (such as inversions) 380-fold in an experimental *Saccharomyces cerevisiae* yeast strain, compared to a wild type strain [12]. The karyotype data we present here indicates that no major interchromosomal rearrangements (i.e., fissions or fusions) have occurred since the divergence of zebra finch and blue-capped cordon-bleu (Figure 1) [32]. Future studies would do service by comparing the number of intrachromosomal rearrangements (especially inversions) in Estrildidae with other bird clades and investigate their likely link with LTR retrotransposon proliferation.

To conclude, we were able to annotate an additional 4.5 Mb of LTR retrotransposons in zebra finch using the in-depth curated LTR libraries of collared flycatcher and, most importantly, blue-capped cordon-bleu. We were also able to uncover a shared estrildid diversity of 21 out of 25 previously undiscovered ERV-like retrotransposon families found in blue-capped cordon-bleu. These results demonstrate the significance of studying close relatives to model organisms.

Supplementary Materials: The following are available online at http://www.mdpi.com/2073-4425/10/4/301/s1, Figure S1: Maximum likelihood phylogeny of CR1 consensuses; Figure S2: Snapshot of protein alignment of TguERVL_I Gag and Envelope glycoprotein C members (PHA03269 and pfam02124) of Marek_A superfamily of alphaherpesviruses; Figure S3: Maximum likelihood phylogeny of songbird TguERVL2_I consensuses; Table S1: Classification sheet of de-novo curated repetitive elements; Table S2: LTR reciprocal BLAST among in-depth curated birds, LTR subfamilies per branch; Table S3: LTR reciprocal BLAST among in-depth curated birds, LTR families per branch; Table S4: Comparison of RepeatMasker output between two different libraries (*Aves* Repbase library vs. "full" merged library) when masking zebra finch; Table S5: Abundance of TguERVL2_I family among in-depth curated songbirds; Data S1: Fasta-formatted consensus sequences of blue-capped cordon-bleu TEs; Data S2: Protein alignment of TguERVL_I Gag and Envelope glycoprotein C members (PHA03269 and pfam02124) of Marek_A superfamily.

Author Contributions: Conceptualization, A.S., J.B.; investigation, J.B., M.d.S.d.S.; resources, C.F.-V., E.H.C.d.O., M.G.; data curation, J.B.; writing—original draft preparation, J.B.; writing—review and editing, J.B., A.S.; visualization, J.B., M.d.S.d.S.; supervision, A.S.; project administration, A.S.; funding acquisition, A.S., M.G., E.H.C.d.O.

Funding: This research was funded by grants to A.S. from the Swedish Research Council Formas (2017-01597), the Swedish Research Council Vetenskapsrådet (2016-05139), and the SciLifeLab Swedish Biodiversity Program (2015-R14). A.S. acknowledges funding from the Knut and Alice Wallenberg Foundation via Hans Ellegren and support from the National Genomics Infrastructure in Stockholm funded by Science for Life Laboratory, the Knut and Alice Wallenberg Foundation and the Swedish Research Council. M.S.S. was supported by a Co-financed Short-Term Research Grant Brazil, 2018 (57417991), according to the joint agreement of the DAAD (German Academic Exchange Service) with CAPES (Coordenaçao de Aperfeiçoamento de Peesoal de Nível Superior, Brazil).

Acknowledgments: We thank Antje Bakker for arranging sample transfer, Martin Irestedt for help with DNA extractions, and Phil Ewels, Remi-André Olsen, and Mattias Ormestad for generating the linked-read data at SciLifeLab Stockholm. We also thank Patric Jern for useful discussions, and Mahwash Jamy and Valentina Peona for providing valuable bioinformatic advice. Boel Olsson and Francisco Ruiz-Ruano provided helpful comments on an earlier version of this manuscript. Computations were performed on resources provided by the Swedish National Infrastructure for Computing (SNIC) through Uppsala Multidisciplinary Center for Advanced Computational Science (UPPMAX). We also would like to thank two anonymous reviewers and editor Maria Xandri Zaragoza for insightful feedback on the manuscript of this paper.

Conflicts of Interest: The authors declare no conflict of interest. The funders had no role in the design of the study; in the collection, analyses, or interpretation of data; in the writing of the manuscript, or in the decision to publish the results.

References

1. Gregory, T.R. Animal Genome Size Database. Available online: http://www.genomesize.com (accessed on 7 December 2018).
2. Kapusta, A.; Suh, A. Evolution of bird genomes—A transposon's-eye view. *Ann. N. Y. Acad. Sci.* **2016**, *1389*, 164–185. [CrossRef]
3. Kapusta, A.; Suh, A.; Feschotte, C. Dynamics of genome size evolution in birds and mammals. *Proc. Natl. Acad. Sci. USA* **2017**, *114*, E1460–E1469. [CrossRef]
4. Hughes, A.L.; Hughes, M.K. Small genomes for better flyers. *Nature* **1995**, *377*, 391. [CrossRef] [PubMed]

5. Gregory, T.R. A Bird's-Eye View of the C-Value Enigma: Genome Size, Cell Size, and Metabolic Rate in the Class Aves. *Evolution* **2002**, *56*, 121–130. [CrossRef] [PubMed]
6. Cavalier-Smith, T. Nuclear volume control by nucleoskeletal DNA, selection for cell volume and cell growth rate, and the solution of the DNA C-value paradox. *J. Cell Sci.* **1978**, *34*, 247–278.
7. Burton, D.W.; Bickham, J.W.; Genoways, H.H. Flow-Cytometric Analyses of Nuclear DNA Content in Four Families of Neotropical Bats. *Evolution* **1989**, *43*, 756–765. [CrossRef]
8. Van den Bussche, R.A.; Longmire, J.L.; Baker, R.J. How bats achieve a small C-value: Frequency of repetitive DNA in Macrotus. *Mamm. Genome* **1995**, *6*, 521–525. [CrossRef] [PubMed]
9. Jarvis, E.D.; Mirarab, S.; Aberer, A.J.; Li, B.; Houde, P.; Li, C.; Ho, S.Y.W.; Faircloth, B.C.; Nabholz, B.; Howard, J.T.; et al. Whole-genome analyses resolve early branches in the tree of life of modern birds. *Science* **2014**, *346*, 1320–1331. [CrossRef] [PubMed]
10. Ricci, M.; Peona, V.; Guichard, E.; Taccioli, C.; Boattini, A. Transposable Elements Activity is Positively Related to Rate of Speciation in Mammals. *J. Mol. Evol.* **2018**, *86*, 303–310. [CrossRef] [PubMed]
11. Lazar, N.H.; Nevonen, K.A.; O'Connell, B.; McCann, C.; O'Neill, R.J.; Green, R.E.; Meyer, T.J.; Okhovat, M.; Carbone, L. Epigenetic maintenance of topological domains in the highly rearranged gibbon genome. *Genome Res.* **2018**, *28*, 983–997. [CrossRef] [PubMed]
12. Chan, J.E.; Kolodner, R.D. A Genetic and Structural Study of Genome Rearrangements Mediated by High Copy Repeat Ty1 Elements. *PLoS Genet.* **2011**, *7*, e1002089. [CrossRef] [PubMed]
13. Farré, M.; Narayan, J.; Slavov, G.T.; Damas, J.; Auvil, L.; Li, C.; Jarvis, E.D.; Burt, D.W.; Griffin, D.K.; Larkin, D.M. Novel Insights into Chromosome Evolution in Birds, Archosaurs, and Reptiles. *Genome Biol. Evol.* **2016**, *8*, 2442–2451. [CrossRef]
14. Wolf, J.B.W.; Ellegren, H. Making sense of genomic islands of differentiation in light of speciation. *Nat. Rev. Genet.* **2017**, *18*, 87–100. [CrossRef]
15. Hooper, D.M.; Price, T.D. Rates of karyotypic evolution in Estrildid finches differ between island and continental clades. *Evolution* **2015**, *69*, 890–903. [CrossRef] [PubMed]
16. Romanov, M.N.; Farré, M.; Lithgow, P.E.; Fowler, K.E.; Skinner, B.M.; O'Connor, R.; Fonseka, G.; Backström, N.; Matsuda, Y.; Nishida, C.; et al. Reconstruction of gross avian genome structure, organization and evolution suggests that the chicken lineage most closely resembles the dinosaur avian ancestor. *BMC Genom.* **2014**, *15*, 1060. [CrossRef]
17. Warren, W.C.; Clayton, D.F.; Ellegren, H.; Arnold, A.P.; Hillier, L.W.; Künstner, A.; Searle, S.; White, S.; Vilella, A.J.; Fairley, S.; et al. The genome of a songbird. *Nature* **2010**, *464*, 757–762. [CrossRef]
18. Garfinkel, D.J.; Boeke, J.D.; Fink, G.R. Ty element transposition: Reverse transcriptase and virus-like particles. *Cell* **1985**, *42*, 507–517. [CrossRef]
19. Ota, N.; Gahr, M.; Soma, M. Tap dancing birds: The multimodal mutual courtship display of males and females in a socially monogamous songbird. *Sci. Rep.* **2015**, *5*, 16614. [CrossRef]
20. Ota, N.; Gahr, M.; Soma, M. Songbird tap dancing produces non-vocal sounds. *Bioacoustics* **2017**, *26*, 161–168. [CrossRef]
21. International Chicken Genome Sequencing Consortium. Sequence and comparative analysis of the chicken genome provide unique perspectives on vertebrate evolution. *Nature* **2004**, *432*, 695–716. [CrossRef]
22. Suh, A.; Smeds, L.; Ellegren, H. Abundant recent activity of retrovirus-like retrotransposons within and among flycatcher species implies a rich source of structural variation in songbird genomes. *Mol. Ecol.* **2017**, *27*, 99–111. [CrossRef]
23. Weisenfeld, N.I.; Kumar, V.; Shah, P.; Church, D.M.; Jaffe, D.B. Direct determination of diploid genome sequences. *Genome Res.* **2017**, *27*, 757–767. [CrossRef]
24. Marks, P.; Garcia, S.; Barrio, A.M.; Belhocine, K.; Bernate, J.; Bharadwaj, R.; Bjornson, K.; Catalanotti, C.; Delaney, J.; Fehr, A.; et al. Resolving the Full Spectrum of Human Genome Variation using Linked-Reads. *Genome Res.* **2019**, *29*, 635–645. [CrossRef]
25. Kinsella, C.M.; Ruiz-Ruano, F.J.; Dion-Côté, A.-M.; Charles, A.J.; Gossmann, T.I.; Cabrero, J.; Kappei, D.; Hemmings, N.; Simons, M.J.P.; Camacho, J.P.M.; et al. Programmed DNA elimination of germline development genes in songbirds. *bioRxiv* **2018**, 444364.
26. Torgasheva, A.A.; Malinovskaya, L.P.; Zadesenets, K.S.; Karamysheva, T.V.; Kizilova, E.A.; Pristyazhnyuk, I.E.; Shnaider, E.P.; Volodkina, V.A.; Saifutdinova, A.F.; Galkina, S.A.; et al. Germline-Restricted Chromosome (GRC) is Widespread among Songbirds. *bioRxiv* **2018**, 414276.

27. Biederman, M.K.; Nelson, M.M.; Asalone, K.C.; Pedersen, A.L.; Saldanha, C.J.; Bracht, J.R. Discovery of the First Germline-Restricted Gene by Subtractive Transcriptomic Analysis in the Zebra Finch, Taeniopygia guttata. *Curr. Biol.* **2018**, *28*, 1620–1627. [CrossRef]
28. Bradnam, K.R.; Fass, J.N.; Alexandrov, A.; Baranay, P.; Bechner, M.; Birol, I.; Boisvert, S.; Chapman, J.A.; Chapuis, G.; Chikhi, R.; et al. Assemblathon 2: Evaluating de novo methods of genome assembly in three vertebrate species. *GigaScience* **2013**, *2*, 10. [CrossRef]
29. Simão, F.A.; Waterhouse, R.M.; Ioannidis, P.; Kriventseva, E.V.; Zdobnov, E.M. BUSCO: Assessing genome assembly and annotation completeness with single-copy orthologs. *Bioinformatics* **2015**, *31*, 3210–3212. [CrossRef]
30. Sumner, A.T. A simple technique for demonstrating centromeric heterochromatin. *Exp. Cell Res.* **1972**, *75*, 304–306. [CrossRef]
31. Sasaki, M.; Ikeuchi, T.; Makino, S. A feather pulp culture technique for avian chromosomes, with notes on the chromosomes of the peafowl and the ostrich. *Experientia* **1968**, *24*, 1292–1293. [CrossRef]
32. Dos Santos, M.D.S.; Kretschmer, R.; Frankl-Vilches, C.; Bakker, A.; Gahr, M.; Ferguson-Smith, M.A.; De Oliveira, E.H. Comparative Cytogenetics between Two Important Songbird, Models: The Zebra Finch and the Canary. *PLoS ONE* **2017**, *12*, e0170997.
33. de Oliveira Furo, I.; Kretschmer, R.; dos Santos, M.S.; de Lima Carvalho, C.A.; Gunski, R.J.; O'Brien, P.C.; Ferguson-Smith, M.A.; Cioffi, M.B.; de Oliveira, E.H. Chromosomal Mapping of Repetitive DNAs in *Myiopsitta monachus* and *Amazona aestiva* (Psittaciformes, Psittacidae) with Emphasis on the Sex Chromosomes. *Cytogenet. Genome Res.* **2017**, *151*, 151–160. [CrossRef]
34. Smit, A.F.A.; Hubley, R. RepeatModeler. Available online: http://www.repeatmasker.org/RepeatModeler/ (accessed on 21 August 2018).
35. Smit, A.F.A.; Hubley, R.; Green, P. RepeatMasker Open-4.0 2013–2015. Available online: http://www.repeatmasker.org (accessed on 6 December 2018).
36. Altschul, S.F.; Gish, W.; Miller, W.; Myers, E.W.; Lipman, D.J. Basic local alignment search tool. *J. Mol. Biol.* **1990**, *215*, 403–410. [CrossRef]
37. Katoh, K.; Standley, D.M. MAFFT Multiple Sequence Alignment Software Version 7: Improvements in Performance and Usability. *Mol. Biol. Evol.* **2013**, *30*, 772–780. [CrossRef]
38. Larsson, A. AliView: A fast and lightweight alignment viewer and editor for large datasets. *Bioinformatics* **2014**, *30*, 3276–3278. [CrossRef] [PubMed]
39. Hall, T.A. BioEdit: A user-friendly biological sequence alignment editor and analysis program for Windows 95/98/NT. *Nucleic Acids Symp. Ser.* **1999**, *41*, 95–98.
40. Advanced Consensus Maker. Available online: https://www.hiv.lanl.gov/content/sequence/CONSENSUS/AdvCon.html (accessed on 30 October 2018).
41. Kapitonov, V.V.; Jurka, J. A universal classification of eukaryotic transposable elements implemented in Repbase. *Nat. Rev. Genet.* **2008**, *9*, 411–412. [CrossRef] [PubMed]
42. Suh, A. The Specific Requirements for CR1 Retrotransposition Explain the Scarcity of Retrogenes in Birds. *J. Mol. Evol.* **2015**, *81*, 18–20. [CrossRef]
43. Bao, W.; Kojima, K.K.; Kohany, O. Repbase Update, a database of repetitive elements in eukaryotic genomes. *Mob. DNA* **2015**, *6*, 11. [CrossRef] [PubMed]
44. Wicker, T.; Sabot, F.; Hua-Van, A.; Bennetzen, J.L.; Capy, P.; Chalhoub, B.; Flavell, A.; Leroy, P.; Morgante, M.; Panaud, O.; et al. A unified classification system for eukaryotic transposable elements. *Nat. Rev. Genet.* **2007**, *8*, 973–982. [CrossRef]
45. Guindon, S.; Dufayard, J.-F.; Lefort, V.; Anisimova, M.; Hordijk, W.; Gascuel, O. New algorithms and methods to estimate maximum-likelihood phylogenies: Assessing the performance of PhyML 3.0. *Syst. Biol.* **2010**, *59*, 307–321. [CrossRef]
46. Lefort, V.; Longueville, J.-E.; Gascuel, O. SMS: Smart Model Selection in PhyML. *Mol. Biol. Evol.* **2017**, *34*, 2422–2424. [CrossRef] [PubMed]
47. Rambaut, A. FigTree. Available online: http://tree.bio.ed.ac.uk/software/figtree/ (accessed on 30 December 2017).
48. Suh, A.; Churakov, G.; Ramakodi, M.P.; Platt, R.N.; Jurka, J.; Kojima, K.K.; Caballero, J.; Smit, A.F.; Vliet, K.A.; Hoffmann, F.G.; et al. Multiple Lineages of Ancient CR1 Retroposons Shaped the Early Genome Evolution of Amniotes. *Genome Biol. Evol.* **2015**, *7*, 205–217. [CrossRef]

49. R Core Team. *R: A Language and Environment for Statistical Computing*; R Foundation for Statistical Computing: Vienna, Austria, 2018.
50. Casper, J.; Zweig, A.S.; Villarreal, C.; Tyner, C.; Speir, M.L.; Rosenbloom, K.R.; Raney, B.J.; Lee, C.M.; Lee, B.T.; Karolchik, D.; et al. The UCSC Genome Browser database: 2018 update. *Nucleic Acids Res.* **2018**, *46*, D762–D769. [PubMed]
51. Smit, A.; Hubley, R.; Wheeler, T. dfam_consensus. Available online: https://dfam.org/home (accessed on 10 April 2019).
52. Quinlan, A.R.; Hall, I.M. BEDTools: A flexible suite of utilities for comparing genomic features. *Bioinformatics* **2010**, *26*, 841–842. [CrossRef]
53. Wheeler, D.L.; Church, D.M.; Federhen, S.; Lash, A.E.; Madden, T.L.; Pontius, J.U.; Schuler, G.D.; Schriml, L.M.; Sequeira, E.; Tatusova, T.A.; et al. Database resources of the National Center for Biotechnology. *Nucleic Acids Res.* **2003**, *31*, 28–33. [CrossRef]
54. Marchler-Bauer, A.; Bo, Y.; Han, L.; He, J.; Lanczycki, C.J.; Lu, S.; Chitsaz, F.; Derbyshire, M.K.; Geer, R.C.; Gonzales, N.R.; et al. CDD/SPARCLE: Functional classification of proteins via subfamily domain architectures. *Nucleic Acids Res.* **2017**, *45*, D200–D203. [CrossRef] [PubMed]
55. Goubert, C. consensus2genome. Available online: https://github.com/clemgoub/consensus2genome (accessed on 8 September 2018).
56. Christidis, L. Chromosomal evolution within the family Estrildidae (Aves) I. The Poephilae. *Genetica* **1986**, *71*, 81–97. [CrossRef]
57. Ladjali-Mohammedi, K.; Bitgood, J.J.; Tixier-Boichard, M.; de Leon, F.A.P. International System for Standardized Avian Karyotypes (ISSAK): Standardized banded karyotypes of the domestic fowl (*Gallus domesticus*). *Cytogenet. Genome Res.* **1999**, *86*, 271–276. [CrossRef] [PubMed]
58. Moyle, R.G.; Oliveros, C.H.; Andersen, M.J.; Hosner, P.A.; Benz, B.W.; Manthey, J.D.; Travers, S.L.; Brown, R.M.; Faircloth, B.C. Tectonic collision and uplift of Wallacea triggered the global songbird radiation. *Nat. Commun.* **2016**, *7*, 12709. [CrossRef]
59. Arkhipova, I.R. Using bioinformatic and phylogenetic approaches to classify transposable elements and understand their complex evolutionary histories. *Mob. DNA* **2017**, *8*, 19. [CrossRef] [PubMed]
60. Malik, H.S. Ribonuclease H evolution in retrotransposable elements. *Cytogenet. Genome Res.* **2005**, *110*, 392–401. [CrossRef]
61. Schmidt, J.; Klupp, B.G.; Karger, A.; Mettenleiter, T.C. Adaptability in herpesviruses: Glycoprotein D-independent infectivity of pseudorabies virus. *J. Virol.* **1997**, *71*, 17–24.
62. Churchill, A.E.; Biggs, P.M. Agent of Marek's Disease in Tissue Culture. *Nature* **1967**, *215*, 528–530. [CrossRef] [PubMed]
63. Lee, L.F.; Wu, P.; Sui, D.; Ren, D.; Kamil, J.; Kung, H.J.; Witter, R.L. The complete unique long sequence and the overall genomic organization of the GA strain of Marek's disease virus. *Proc. Natl. Acad. Sci. USA* **2000**, *97*, 6091–6096. [CrossRef] [PubMed]
64. Malik, H.S.; Henikoff, S.; Eickbush, T.H. Poised for Contagion: Evolutionary Origins of the Infectious Abilities of Invertebrate Retroviruses. *Genome Res.* **2000**, *10*, 1307–1318. [CrossRef] [PubMed]
65. Devos, K.M.; Brown, J.K.M.; Bennetzen, J.L. Genome Size Reduction through Illegitimate Recombination Counteracts Genome Expansion in Arabidopsis. *Genome Res.* **2002**, *12*, 1075–1079. [CrossRef]
66. Schweizer, M.; Warmuth, V.; Kakhki, N.A.; Aliabadian, M.; Förschler, M.; Shirihai, H.; Suh, A.; Burri, R. Parallel plumage colour evolution and introgressive hybridization in wheatears. *J. Evol. Biol.* **2019**, *32*, 100–110. [CrossRef]
67. Toomey, M.B.; Marques, C.I.; Andrade, P.; Araújo, P.M.; Sabatino, S.; Gazda, M.A.; Afonso, S.; Lopes, R.J.; Corbo, J.C.; Carneiro, M. A non-coding region near Follistatin controls head colour polymorphism in the Gouldian finch. *Proc. R. Soc. B Biol. Sci.* **2018**, *285*, 20181788. [CrossRef]
68. Mason, A.S.; Fulton, J.E.; Hocking, P.M.; Burt, D.W. A new look at the LTR retrotransposon content of the chicken genome. *BMC Genom.* **2016**, *17*, 688. [CrossRef] [PubMed]
69. Sorenson, M.D.; Balakrishnan, C.N.; Payne, R.B. Clade-Limited Colonization in Brood Parasitic Finches (Vidua spp.). *Syst. Biol.* **2004**, *53*, 140–153. [CrossRef]

70. Löber, U.; Hobbs, M.; Dayaram, A.; Tsangaras, K.; Jones, K.; Alquezar-Planas, D.E.; Ishida, Y.; Meers, J.; Mayer, J.; Quedenau, C.; et al. Degradation and remobilization of endogenous retroviruses by recombination during the earliest stages of a germ-line invasion. *Proc. Natl. Acad. Sci. USA* **2018**, *115*, 8609–8614. [CrossRef] [PubMed]
71. Stoye, J.P. Koala retrovirus: A genome invasion in real time. *Genome Biol.* **2006**, *7*, 241. [CrossRef] [PubMed]
72. Simmons, G.S.; Young, P.R.; Hanger, J.J.; Jones, K.; Clarke, D.; McKee, J.J.; Meers, J. Prevalence of koala retrovirus in geographically diverse populations in Australia. *Aust. Vet. J.* **2012**, *90*, 404–409. [CrossRef] [PubMed]
73. Tarlinton, R.; Meers, J.; Hanger, J.; Young, P. Real-time reverse transcriptase PCR for the endogenous koala retrovirus reveals an association between plasma viral load and neoplastic disease in koalas. *J. Gen. Virol.* **2005**, *86*, 783–787. [CrossRef]
74. Platt, R.N.; Blanco-Berdugo, L.; Ray, D.A. Accurate Transposable Element Annotation Is Vital When Analyzing New Genome Assemblies. *Genome Biol. Evol.* **2016**, *8*, 403–410. [CrossRef]
75. Knisbacher, B.A.; Levanon, E.Y. DNA Editing of LTR Retrotransposons Reveals the Impact of APOBECs on Vertebrate Genomes. *Mol. Biol. Evol.* **2016**, *33*, 554–567. [CrossRef]
76. Peona, V.; Weissensteiner, M.H.; Suh, A. How complete are "complete" genome assemblies?—An avian perspective. *Mol. Ecol. Resour.* **2018**, *18*, 1188–1195. [CrossRef]
77. Aswad, A.; Katzourakis, A. Convergent capture of retroviral superantigens by mammalian herpesviruses. *Nat. Commun.* **2015**, *6*, 8299. [CrossRef] [PubMed]

© 2019 by the authors. Licensee MDPI, Basel, Switzerland. This article is an open access article distributed under the terms and conditions of the Creative Commons Attribution (CC BY) license (http://creativecommons.org/licenses/by/4.0/).

Review

Sequence Expression of Supernumerary B Chromosomes: Function or Fluff?

Elena Dalla Benetta [1,2], Omar S. Akbari [2] and Patrick M. Ferree [1,*]

1. W. M. Keck Science Department of Claremont McKenna, Pitzer, and Scripps Colleges, Claremont, CA 91711, USA; edallabenetta@ucsd.edu
2. Division of Biological Sciences, Section of Cell and Developmental Biology, University of California, San Diego, La Jolla, CA 92093, USA; oakbari@ucsd.edu
* Correspondence: pferree@kecksci.claremont.edu; Tel.: +1-909-607-8304

Received: 16 January 2019; Accepted: 5 February 2019; Published: 8 February 2019

Abstract: B chromosomes are enigmatic heritable elements found in the genomes of numerous plant and animal species. Contrary to their broad distribution, most B chromosomes are non-essential. For this reason, they are regarded as genome parasites. In order to be stably transmitted through generations, many B chromosomes exhibit the ability to "drive", i.e., they transmit themselves at super-Mendelian frequencies to progeny through directed interactions with the cell division apparatus. To date, very little is understood mechanistically about how B chromosomes drive, although a likely scenario is that expression of B chromosome sequences plays a role. Here, we highlight a handful of previously identified B chromosome sequences, many of which are repetitive and non-coding in nature, that have been shown to be expressed at the transcriptional level. We speculate on how each type of expressed sequence could participate in B chromosome drive based on known functions of RNA in general chromatin- and chromosome-related processes. We also raise some challenges to functionally testing these possible roles, a goal that will be required to more fully understand whether and how B chromosomes interact with components of the cell for drive and transmission.

Keywords: B chromosomes; PSR (Paternal sex ratio); genome elimination; ncRNAs (non coding RNAs); selfish elements; super-Mendelian; repeated elements

1. Introduction

Since the time when microscopy first allowed visualization of hereditary material, researchers have observed peculiar chromosome variants in the genomes of higher eukaryotes. For any given species, the core of the genome consists of a certain number of A chromosomes, which carry all genes needed collectively for the organism's development, metabolism, and reproduction. Thus, without the complete set of A chromosomes, the organism cannot survive. However, extra or supernumerary non-essential chromosomes, termed B chromosomes, have been detected in numerous plant and animal species [1–4]. The frequency of B chromosomes within a given population can range from very low to complete fixation among all individuals, and a carrying individual can contain one to as many as ten or more B chromosome copies in each nucleus [1–3]. Interestingly, with little exception, B chromosomes are not essential for the organism. Indeed, the fact that B chromosomes can persist without providing any measurable benefit has contributed to the view that they are genome parasites [5]. But if they do not help the organism, how then do B chromosomes persist, thereby defying the expectation that non-essential genetic elements are eventually lost?

Previous work in various B-carrying organisms has provided some insights regarding this question. For example, in several grasshopper species, certain B chromosomes are transmitted to progeny through both parents, and they segregate with very high efficiency to daughter cells during mitotic and meiotic divisions [6,7]. In these organisms, the B chromosomes appear to behave similarly to the A chromosomes. In contrast, B chromosomes in other organisms exhibit "drive"—that is, they behave in specific ways that defy normal Mendelian transmission patterns in order to ensure their inheritance in subsequent generations [8]. For example, one B chromosome in a rye species counters its tendency to loss through a form of mitotic drive. Specifically, the sister B chromatids fail to separate (i.e., non-disjoin) in the division that produces the vegetative (non-gametic) cell and pollen (gametic) cell [9,10]. Moreover, the unseparated B chromatids tend to segregate toward the side of the spindle that will give rise to the pollen cell. This tropism of the B chromatid pair for the future pollen cell, which can lead to an accumulation of multiple B chromosome copies in offspring, is thought to occur by the B chromatids utilizing an inherent asymmetry in the makeup of the spindle apparatus [9]. Indeed, B chromosomes may tend to drive in this way in plants because of the universally asymmetric spindle at the pollen production stage [11].

As a remarkably different example of drive, the jewel wasp *Nasonia vitripennis* harbors a B chromosome known as PSR (for Paternal Sex Ratio) that is transmitted via the sperm (i.e., paternally) to progeny [12]. Interestingly, this B chromosome causes complete loss of the sperm's hereditary material but not itself during the first mitotic division of the embryo [13]. Interestingly, the PSR chromosome does not eliminate itself, but instead associates with the functional egg-derived chromosomes, and successfully segregates with them. Due to the fact that in wasps and other hymenopteran insects, males normally develop as haploids from unfertilized eggs while females develop as diploids from fertilized eggs, this genome elimination event converts fertilized embryos, which should become female, into haploid B chromosome-carrying males, thereby ensuring B chromosome transmission.

Even though we know what happens at the descriptive level in these cases of drive, what remains to be determined is how mechanistically B chromosomes drive in their resident genomes. Two general possibilities exist: a B chromosome may either (i) act passively, being transmitted and/or driving due to the intrinsic properties of its own DNA sequences [14], or (ii) it may operate actively through expression of its DNA sequences [9,15]. Here, we focus on the second possibility by reviewing a number of different types of B chromosome-linked DNA sequences, many of which are repetitive and non-coding in nature, that have been identified through genetic and genomic analyses. We highlight a subset of these DNA sequences that are known to be expressed, and we propose possible roles for each. Several recent studies have demonstrated that B chromosomes can influence the expression of A-linked genes [16–19] and epigenetic marks of A chromatin, reviewed in [20,21]. However, given the non-essential nature of most B chromosomes, these effects may not substantially affect the biology of the organism. Thus, we limit our attention here to B-expressed sequences and their potential roles in B chromosome drive and transmission. Finally, we raise some challenges to functionally testing these possible roles, a goal that will be required to more fully understand whether and how B chromosomes interact with components of the cell.

2. B Chromosomes Are Mosaics of Protein-Coding and Repetitive, Non-Coding Sequences

Genetic and genomic studies have been performed in a number of different organisms, including but not exclusive to maize [22], rye [19,23], grasshoppers [24], wasps [25,26], cichlids [27,28], raccoon dogs [29] and fungi [30,31], in order to identify specific DNA sequences carried by B chromosomes (see Table 1). The repertoire of B-linked DNA sequences includes both protein-coding and non-coding sequences. The origin of any given B-linked sequence may date back to the beginning of the B chromosome itself, or the sequence may have arisen subsequently as a copy of another B-linked gene or of an ancestral gene located on an A chromosome that was moved via transposable element (TE) activity, inter-chromosomal meiotic recombination, or imperfect DNA repair [19].

The majority of known B-linked protein-coding genes match genes located on the A chromosomes and belong to nearly all protein function categories [19,29,32–35]. Most B-linked protein-coding genes are degenerate; they can be present as partial gene copies, such as truncated forms or missing exons, or they can show low sequence similarity across their entire lengths to the ancestral sequences [35].

For this reason, many B-linked protein-coding genes are considered to be pseudogenes [19]. The few B-linked protein-coding genes that do show high sequence similarity to their ancestral copies are likely not needed for the organism because the B chromosomes themselves are non-essential. Thus, given the lack of functional constraint on such B-linked protein-coding genes, high sequence similarity may indicate that the origin of the B-linked copy from its ancestral gene was a relatively recent event.

Despite the presence of protein-coding sequences, it appears that most B chromosomes consist primarily of non-coding sequences including TEs, simple satellites, and complex satellite-like repeats [3,36–40]. Such highly repetitive DNA sequences are known to be enriched in the heterochromatin that surrounds the centromeres of the A chromosomes [40]. For this reason, it has been proposed that B chromosome formation begins with duplication of an A chromosome followed by the loss of its euchromatic chromosome arms, thus producing a nascent B chromosome consisting mainly of a centromere and its pericentromeric regions. Over time, TE activity can move genes from the A chromosomes onto the B chromosome; these sequences may then undergo mutational decay, further duplication through replication slippage, or rearrangement events such as intra-chromosomal inversion, deletion, and translocation [40]. We should mention here that the PSR chromosome present in the jewel wasp contains transposon-like sequences that appear to be absent from the wasp genome but to match sequences present in another wasp species [41,42]. Such patterns open up the possibility that this B chromosome derives from a chromosome of another species that moved into the wasp genome through interspecific hybridization or by parasites or food sources [43–46].

3. Expression of B-Linked DNA Sequences

Despite our current knowledge of some DNA sequences carried by B chromosomes, only a handful of studies have addressed which of them are expressed. Certainly, transcription of any given DNA sequence alone does not guarantee that it is functional; work in different model organisms suggests that much of the non-coding part of the A genome may be transcribed without function [47]. Nevertheless, a reasonable (and perhaps obvious) assumption is that a locus that is functional will at least be expressed at the RNA level. To date, several studies have identified RNAs produced by B chromosomes, either by examination of individual B-linked sequences or through whole genome approaches such as RNA-Seq. To our knowledge, no individual B-linked sequence has yet been tested for functionality through genetic manipulation. However, all B-linked sequences producing RNA should be considered as potential candidates for involvement in B chromosome dynamics and drive. Here, we highlight several examples of B-linked sequences that are known to express RNAs, and we speculate on the possible functions of these sequences in light of the types of RNAs that they produce and the underlying B chromosome biology in each case.

3.1. Copies of Protein-Coding Genes

Of the previously identified B-linked DNA sequences that are copies of A-linked protein-coding genes, few are known to be expressed (Table 1) [19,23,24,27,28,48,49]. For example, a B chromosome in the cichlid fish *Astatotilapia latifasciata* was shown to express multiple different protein-coding genes. Three of these expressed genes derive from the ancestral A-linked genes encoding Separin, Tubulin B1 (*TUBB1*), and *KIF11* [27,28]. Interestingly, these A-linked genes play important roles in chromosomal segregation during cell division: *TUBB1* is involved in microtubule organization [50], *KIF11* functions in centrosome behavior and spindle assembly [50], and Separin mediates the release of sister chromatids at the onset of anaphase [50]. It has been proposed that because these genes are implicated in different aspects of chromosome segregation, expression of the B-linked variants may somehow promote B chromosome transmission such that they are inherited to over 50% of the gametes [27,28]. Indeed, a number of B-linked genes that derive from protein-coding genes involved in various aspects of the cell cycle and cell division have been detected in other organisms [19,24,29,34]. However, in most of these other cases, it is not yet known which of the protein-coding gene variants are expressed.

We point out here an important consideration: that any expressed protein-coding gene that affects aspects of cell division would likely impact the A chromosomes in addition to the B chromosome, potentially having large costs to the organism. Thus, proteins that play roles in B chromosome drive may be expected to specifically affect the B chromosome. Certain chromatin-associated proteins, which could have affinity for specific DNA sequences found uniquely on B chromosomes, would fulfill such an expectation. This idea is consistent with previously proposed models invoking co-evolving centromere repeats and their chromatin proteins as agents of centromere/meiotic drive [51]. Copies of histones H3 and H4 are known to be expressed from B chromosomes in different grasshopper species [52,53]. However, it remains to be determined whether variants of these conventional histones or other non-histone chromatin proteins are carried and expressed by B chromosomes.

Table 1. Protein-coding genes located on B chromosomes in different species. (NK = not known).

Functional Group	Gene Name	Organism	Transcribed	Gene Integrity	References
Cell Division and Microtubules	Tubulin beta-1 (*TUBB1*)	Cichlid fishes	yes	High	[27]
	Tubulin beta-5 (*TUBB5*)	Cichlid fishes	yes	High	[27]
	Spindle and kinetochore-associated protein-1 (*SKA-1*)	Cichlid fishes	yes	High	[27]
	Kinesin-like protein-11 (*KIFF11*)	Cichlid fishes	yes	High	[27]
	Centromere-associated protein-E (*CENP-E*)	Cichlid fishes	yes	High	[27]
	Centromere-associated protein-N (*CENP-N*)	Red fox	NK	NK	[34]
	Cytoskeleton-associated protein 2 (*CKAP2*)	Grasshoppers	yes	truncated	[24]
	Condensin I complex subunit G (*CAP-G*)	Grasshoppers	yes	truncated	[24]
	E3 ubiquitin-protein ligase *MYCBP2*	Grasshoppers	yes	truncated	[24]
	Kinesin-like protein *KIF20A*	Grasshoppers	yes	High	[24]
	DNA topoisomerase 2-alpha (*TOP2A*)	Grasshoppers	yes	truncated	[24]
	Kinesin-3-like	Rye	yes	High	[19,23]
	Shortage in chiasmata gene (*SHOC 1*)	Rye	yes	High	[19,23]
	Chromosome-associated kinesin *KIF4A*-like	Rye	yes	high + truncated	[19,23]
	Aurora kinase-B (*AURK*)	Cichlid fishes	yes	High	[27]
	Separin-like protein	Cichlid fishes	yes	High	[27]
	Coiled-coil and C2 domain Containing 2A (*CC2D2A*)	Deer	NK	NK	[34]
	Ecotropic viral integration site 5-like (*EVI5*)	Deer	NK	NK	[34]
	E3 ubiquitin-protein ligase *CHFR*	Deer	NK	NK	[34]
	G1/S-specific cyclin-D2 (*CCND2*)	Deer	NK	NK	[34]
	Tripartite motif-containing 67 (*TRIM67*)	Deer	NK	NK	[34]
	Palladin (*PALLD*)	Deer	NK	NK	[34]
	Cdc42 effector protein 4 (*CDC42EP4*)	Deer	NK	NK	[34]

Table 1. Cont.

Functional Group	Gene Name	Organism	Transcribed	Gene Integrity	References
Differentiation, Proliferation	v-kit Hardy-Zuckerman 4 feline sarcoma viral oncogene homolog (C-KIT)	Red fox	NK	NK	[34]
		Raccoon dogs	NK	NK	[34]
		Deer	NK	NK	[34]
	Kinase insert domain receptor (KDR)	Raccoon dogs	NK	NK	[34]
	Low density lipoprotein receptor-related protein 1B (LRP1B)	Raccoon dogs	NK	NK	[29,34]
	AICDA	Raccoon dogs	NK	NK	[34]
	RET	Raccoon dogs	NK	NK	[34]
	APOBEC1	Raccoon dogs	NK	NK	[34]
	ARNTL	Raccoon dogs	NK	NK	[34]
	BARX2	Raccoon dogs	NK	NK	[34]
	BTBD10	Raccoon dogs	NK	NK	[34]
	COL4A3BP	Raccoon dogs	NK	NK	[34]
	CXCR4	Raccoon dogs	NK	NK	[34]
	ENPP1	Raccoon dogs	NK	NK	[34]
	GDF3	Raccoon dogs	NK	NK	[34]
	GNAS	Raccoon dogs	NK	NK	[34]
	HMGCR	Raccoon dogs	NK	NK	[34]
	JAG1	Raccoon dogs	NK	NK	[34]
	MDM4	Raccoon dogs	NK	NK	[34]
	TNNI3K	Deer	NK	NK	[34]
	ZNF268	Deer	NK	NK	[34]
	ACVR2B	Deer	NK	NK	[34]
	BCL6	Deer	NK	NK	[34]
	BST1	Deer	NK	NK	[34]
	CD38	Deer	NK	NK	[34]
	DHCR7	Deer	NK	NK	[34]
	DLEC1	Deer	NK	NK	[34]
	EOMES	Deer	NK	NK	[34]
	FBXL5	Deer	NK	NK	[34]
	FGFBP1	Deer	NK	NK	[34]
	FNIP1	Deer	NK	NK	[34]
	GABRB1	Deer	NK	NK	[34]
	GFI1	Deer	NK	NK	[34]
	HPSE	Deer	NK	NK	[34]
	MYD88	Deer	NK	NK	[34]
	PLCD1	Deer	NK	NK	[34]
	SDK2	Deer	NK	NK	[34]
	SERPINB9	Deer	NK	NK	[34]
	SSBP3	Deer	NK	NK	[34]
	SST	Deer	NK	NK	[34]
	SSTR2	Deer	NK	NK	[34]
	TXK	Deer	NK	NK	[34]
	CIP2A (CIP2A protein)	Grasshopper	yes	High	[34]
Neuron Synapse, Cell Junction	Cadherin-associated protein-2 (CTNND2)	Red Fox	NK	NK	[34]
	LRRC7	Raccoon dogs	NK	NK	[34]
	CXCR4	Raccoon dogs	NK	NK	[34]
	ARHGAP32	Raccoon dogs	NK	NK	[34]
	SDK1 and 2	Deer	NK	NK	[34]
	GABRA4 and GABRB1	Deer	NK	NK	[34]
	LPP	Deer	NK	NK	[34]
	SHANK2	Deer	NK	NK	[34]
Recombination and Repair	DNA repair protein XRCC2	Cichlid fishes	yes	High	[27]
	SC protein-2 (SYCP-2)	Cichlid fishes	yes	High	[27]
	Regulator of telomere elongation helicase (RTEL)	Cichlid fishes	ye	High	[27]
Regulation of Transcription	Peroxisome proliferator-activated receptor gamma coactivator-1 (PPRC1)	Cichlid fishes	yes	Low	[27]
	Mesogenin-1 (MSGN1)	Cichlid fishes	yes	Low	[28]
	C-Myc-binding protein (MYCBP)	Cichlid fishes	yes	Low	[28]
	Nuclear receptor-subfamily 2-group F-member 6 (NR2F6)	Cichlid fishes	yes	Low	[28]
	Zinc finger protein-596 (ZNF596)	Cichlid fishes	yes	High	[28]
	DEAD-box ATP-dependent RNA helicase 7	Maize	yes	High	[48]
	Myb-like DNA-binding domain	Maize	yes	High	[48]
	Conserved mid region of cactin	Maize	yes	High	[48]
	Argonaute-likeprotein (AGO4)	Rye	yes	High	[23]
	DNA (cytosine-5-)-methyltransferase	Rye	yes	High	[19]
	Ubiquitin ligase sinat5	Rye	yes	High	[19]
	histone-lysine n-methyltransferase	Rye	yes	High	[19]
	protein kinase subfamily lrk10l-2	Rye	yes	High	[19]
Sex determination and Differentiation	Wilms tumor gene	cichlid fishes	yes	Low	[27]
	pre-B-cell leukemia transcription factor 1	cichlid fishes	yes	Low	[27]
	FKBP4	cichlid fishes	yes	Low	[27]
	FNDC3A	cichlid fishes	yes	Low	[27]

Table 1. Cont.

Functional Group	Gene Name	Organism	Transcribed	Gene Integrity	References
Metabolism Regulation	Fucose-1-phosphate guanylyltransferase (FPGT)	Siberian Roe deer	yes	High	[49]
		Raccoon dogs	NK	NK	[29,34]
	Lysosomal alpha-mannosidase	Raccoon dogs	NK	NK	[29,34]
	Hydroxypyruvate isomerase (HYI)	Grasshoppers	yes	truncated	[24]
	Putative aldose reductase-related protein	Maize	yes	High	[48]
	Leucine-rich repeat- containing protein 23 (LRC23)	Cichlid fishes	yes	Low	[27,28]
Leucin-Rich Protein	Acidic leucine-rich nuclear phosphoprotein 32 family member E (Cpd1)	Cichlid fishes	yes	High	[27]
	Leucine-rich repeats and immunoglobulin-like domains 1(LRIG1)	Raccoon dogs	NK	NK	[29,34]
	Leucine-rich repeat and IQ domain-containing protein 3 (LRRIQ3)	Siberian Roe deer	yes	High	[49]
Olfactory Receptors	Olfactory receptors 5F1 (or OR11-10)	Cichlid fishes	yes	High	[27]
	Olfactory receptor 6C4 (or OR12-10)	Cichlid fishes	yes	High	[27]
	Olfactory receptor 6N1 (or OR6N1)	Cichlid fishes	yes	High	[27]
	Olfactory receptor 51E1 (or OR51E1)	Cichlid fishes	yes	High	[27]
Ribonucleotide Binding	GTP-binding protein 6 (GTPB6)	Grasshoppers	yes	High	[24]
	Mitochondrial GTPase 1 (MTG1)	Grasshoppers	yes	High	[24]
Development	Indian hedgehog homolog b (IHHB)	Raccoon dogs	NK	NK	[29]
Immune Responses	Rnasel 2 (Ribonuclease-like 2)	Raccoon dogs	NK	NK	[29]
Cell-cell Signalling and Cellular Response to Stimuli	VPS10 domain receptor protein SORCS 3–like	Raccoon dogs	NK	NK	[29]
	SLIT	Grasshoppers	yes	truncated	[34]
Histones	H3 and H4	Migratory locust	NK	it varies among copies	[52]
		Grasshoppers	NK	NK	[53]

3.2. Transposable Elements

DNA TEs, retro-TEs, and TE-like elements are abundant in higher eukaryotes, making up anywhere from ~5 to as much as 50% of a given eukaryotic genome [54,55]. It is, therefore, no surprise that these elements have been found to be carried and expressed by B chromosomes in a number of organisms including rye [15,33], maize [56], fishes [39,57], and the jewel wasp [41,42]. It is difficult to imagine a scenario in which TE expression per se could enhance B chromosome drive. Moreover, a substantial amount of cellular energy is devoted to the silencing of TE expression, and unsilenced TEs can lead to severe genome instability [58]. However, TEs may play secondary but important roles in B chromosome drive. Given the potential of TEs to mobilize and amplify within single generations, it is expected that these genetic elements move frequently between the A and B chromosomes over short periods of evolutionary time; as mentioned above, such movement likely serves as a mechanism for the transfer of gene copies between the A and B chromosomes [39,40]. Additionally, TEs that have moved onto B chromosomes may themselves degenerate and become pseudogenes, they may fuse with non-TE sequence to form new genes [39,40], or they may decay over time and become tandemly copied to form arrays of complex satellite-like repeats [59]. Any such TE-derived sequence may itself be expressed through the transcriptional regulatory sequences of transposase or other TE-associated genes, or it may induce the expression of adjacent sequences that would otherwise be transcriptionally silent.

3.3. Long Non-Coding RNAs

An interesting class of candidates for involvement in B chromosome transmission and drive consists of long non-coding RNAs (lncRNAs). Bioinformatically, lncRNAs are challenging to identify from RNA expression datasets for a number of reasons. For one, it is difficult to identify secondary structural domains that suggest potential function of a putative lncRNA. Additionally, long RNAs that function as structural molecules may contain cryptic, unused open reading frames, leading to ambiguity in bioinformatically assigning such RNAs as coding or non-coding. Despite these challenges, previous work has led to the identification of putative lncRNAs expressed from B chromosomes in the jewel wasp and in cichlids [25,60].

In the jewel wasp, comparison of testis transcriptomes between wild type and B chromosome-carrying (PSR+) males led to the identification of ten transcripts, ranging between ~500–1500 nucleotides in length, that are present only in the PSR + genotype [25,26]. These transcripts represent the highest-expressed sequences from the PSR chromosome. Fluorescence in situ hybridization (FISH) and PCR of genomic DNA were used to demonstrate that the cognate DNA sequences of these transcripts are located exclusively on the PSR chromosome [25,26]. A couple of these transcripts contain potential, short open reading frames, but the majority of them were bioinformatically predicted to be non-coding [25]. A different study in the cichlid *A. latifasciata* identified a transcript corresponding to a non-coding DNA repeat represented in multiple copies on a B chromosome in this organism [60]. This transcript was shown to be expressed in multiple different fish tissues [60].

A central question is whether such non-coding RNAs are functional, especially with regard to B chromosome drive. While no studies have yet demonstrated functionality of these RNAs, some interesting speculations stem from examples of lncRNA function in non-B systems. It has become apparent that lncRNAs span a wide range of cellular and developmental processes [61,62], but those of particular interest here pertain to chromatin and chromosome dynamics. In particular, two well-studied groups of lncRNA pertain to the X chromosome. In the fruit fly *Drosophila melanogaster* the roX1 and roX2 lncRNAs associate with the male-specific lethal (MSL) proteins to form the dosage compensation complex (DCC) in young male embryos [63]. This complex localizes to "entry" sites located along the male's single X chromosome. There, the DCC spreads to other regions on the X, where it ultimately induces transcriptional upregulation of most X-linked genes [64]. This effect involves remodeling of X chromatin through acetylation of Lysine residue 16 of histone H4 (H4K16ac) and phosphorylation of Serine residue 10 of histone H3 (H3S10p), each by a different enzymatic activity of DCC-associated components (reviewed in [65]). In this case, the roX lncRNAs play an indispensable role as a structural "glue" that scaffolds together the DCC proteins [66]. In mammals, a different lncRNA known as Xist is expressed initially from both X chromosomes during early embryogenesis but its expression is eventually turned off on one of the two X chromosomes (reviewed in [67]). Xist coats the X chromosome that continues to express it, an effect that leads to the association of the Polycomb Repressive Complex (PRC) and its trimethylation of Lysine residue 27 of histone H3 (H3K27me3). This histone mark leads to the facultative heterochromatinization of this Xist-expressing X chromosome, leaving the other X in a transcriptionally active state [68]. lcnRNA function is not limited to the X chromosome; these molecules also facilitate chromatin remodeling elsewhere in the genome, and they function in other aspects of chromatin dynamics (reviewed in [69,70]).

Taken together, these examples demonstrate the potential for lncRNAs to associate with specific chromosome regions and not others, as well as their ability to facilitate specific alterations of chromatin. Thus, it is intriguing to speculate that B chromosome-expressed lncRNAs may play unique roles in B chromosome drive in certain cases through such chromatin interactions. For example, in both rye and maize, B chromosomes that are deleted for a small region of repetitive DNA lose their ability to drive by nondisjunction at the pollen mitosis stage [9,10]. Currently it is not known if these repeats are transcribed from the undeleted B chromosomes. However, if they are expressed, then their encoded RNAs could associate with the centromeric regions where they may recruit enzymes that interact with the cohesin machinery. Such an effect could, in turn, retard the separation of the sister B chromatids during anaphase so that both B chromatids end up in the gamete (Figure 1).

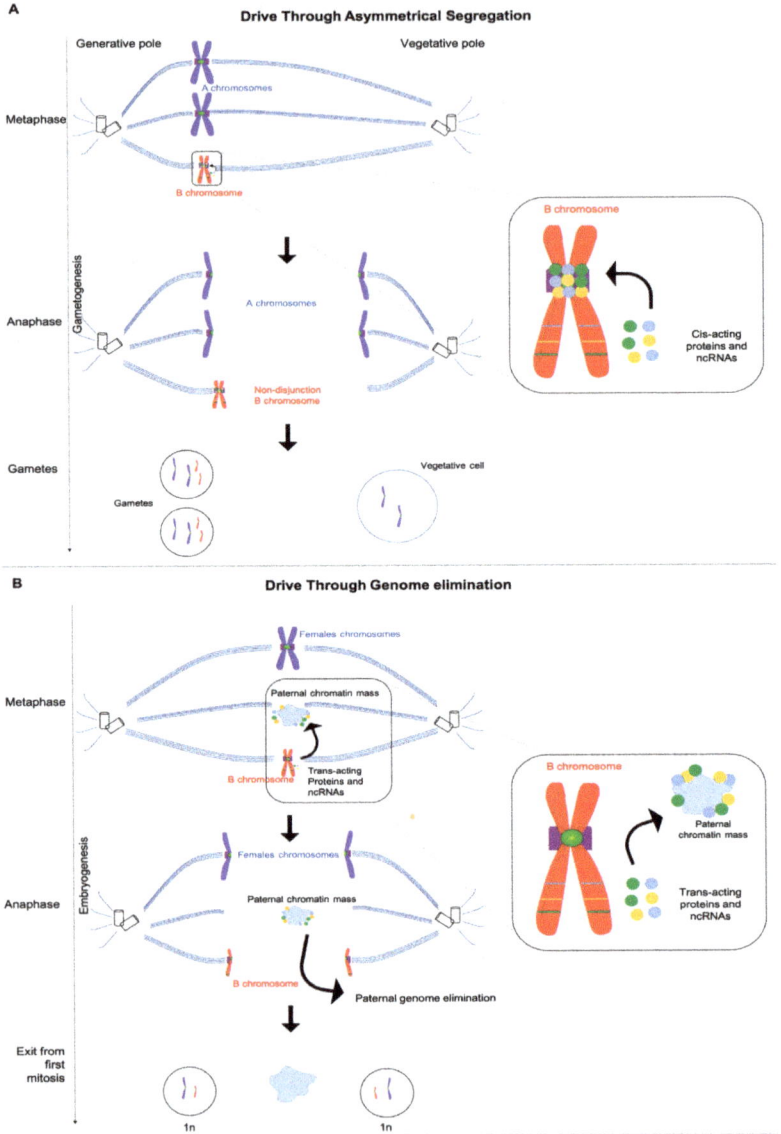

Figure 1. Possible roles for B-expressed sequences in B drive. (**a**) B chromosome drive through asymmetrical segregation and non-disjunction could involve cis-acting B specific products (proteins or ncRNAs) that retard release of the two sister chromatids at the kinetochore. The sister B chromatid pair then migrates preferentially toward the generative pole due to an intrinsic asymmetry of the spindle apparatus. As a result, multiple copies of the B chromosome accumulate in progeny over multiple generations. (**b**) B chromosome drive through genome elimination, such as occurs by the PSR chromosome in the jewel wasp *Nasonia vitripennis*, occurs during the first mitotic division of the newly fertilized embryo. In this model, B-chromosome-expressed protein or RNA could localize preferentially with the paternal chromatin, recruiting chromatin-remodeling enzymes that disrupt normal chromatin remodeling dynamics through abnormal histone modification. As a consequence, the paternal chromatin forms a condensed mass that is unable to resolve into chromosomes and segregate properly.

It has been proposed that the putative lncRNAs expressed by the PSR chromosome in the jewel wasp may underlie the elimination of the paternally-inherited half of the wasp genome [26]. Previous work demonstrated that certain histone marks (H3K9me3, H3K27me1, and H4K20me1) appeared in abnormal patterns on the paternal chromatin immediately before its elimination [71]. The abnormal placement of these histone marks may block subsequent chromatin remodeling events, such as histone phosphorylation, that are essential for normal condensation of chromatin into chromosomes during mitosis [71]. An intriguing possibility is that PSR induces the abnormal histone marks through one or more of the identified lncRNAs. For example, one or more of these molecules may associate with the paternal chromatin and recruit chromatin-remodeling enzymes that disrupt normal chromatin dynamics [71]. Regardless of the mechanism, PSR must possess some way of sparing itself from this abnormal chromatin remodeling [71] (Figure 1).

3.4. Small Non-Coding RNAs

So far, little is known about the potential for B chromosomes to express small RNAs, non-coding molecules that typically range between ~21–33 nucleotides in length (reviewed in [61]). A multitude of studies have characterized the functional roles of the three major classes of small RNAs and their corresponding pathways: micro-RNAs (miRNAs), which block translation of their cognate mRNA targets, endogenous small interfering RNAs (endo-siRNAs), which inhibit translation by inducing degradation of target mRNAs, and PIWI-associated RNAs (piRNAs), which facilitate the transcriptional silencing of chromatin through the association of certain chromatin-remodeling enzymes (reviewed in [72]). The functions of these different small RNA classes are not completely distinct from one another since there is evidence of some crossover between small RNA pathways [73]. To our knowledge, only one study, conducted in the jewel wasp, has detected small RNAs expressed from a B chromosome [26]. In this insect, several different small RNAs were found to be produced by PSR at expression levels matching those of more abundant small RNAs expressed from the A chromosomes [26]. Interestingly, the most abundant PSR-specific small RNA exhibits peculiar properties, having a length (32–33 nt) and starting in a uracil similar to piRNAs while appearing to be processed from a hairpin precursor like endo-siRNAs [26]. More work will be required to better understand to which class this and other PSR-expressed small RNAs belong. However, this work demonstrates that B chromosomes can, indeed, express this type of non-coding RNA. Additionally, given the link of certain small RNAs in chromatin remodeling, it should be strongly considered that B chromosomes like PSR, whose drive involves chromatin remodeling, may drive at least in part through the actions of small RNAs.

4. Functional Testing of Expressed B Loci and Some Challenges

Previous studies aimed at identifying functional B-specific sequences have been restricted to deletion analysis in rye and maize [9,10], jewel wasp, [14] and grasshoppers [74]. Although certain deletions of B chromosomes elicited a loss of drive, it is still unclear in each of these cases which individual sequence(s) within the deleted regions underlie drive and transmission [9,10,14,74]. Until only recently have studies begun to uncover individual RNAs that are expressed by B-linked sequences. Given that most known B chromosomes are not essential for the organism, it may be that much of B chromosome expression may be nothing more than noise. A fundamental question is whether any B expressed loci are functional, and if so, which ones. The ideas presented here may serve as some basis for deciding which candidate loci to prioritize within each B chromosome system. But one thing is certain: fully understanding if and how a given locus is involved in B chromosome transmission or drive will ultimately require some form of genetic manipulation. Such a goal has been challenging due to the fact that most studied B chromosomes reside in non-model organisms that lack traditional genetic tools. However, the development of CRISPR/Cas9 genome editing has made genetic manipulation of individual loci possible in almost any organism, model or not. In principle, this method promises to allow "knock out" of target loci on B chromosomes or, alternatively, the transgenic expression of B chromosome-derived sequences in a non-B genotype, in order to test for functionality.

Just as there is strong promise for CRISPR/Cas9 in achieving these goals, there are some substantial obstacles that will need to be tackled. For example, unlike essential genes located on the A chromosomes, which provide lethal or semi-lethal phenotypes when altered, mutant alleles created by the editing of B-linked loci would not provide any overt phenotype to follow. Contrarily, any such induced mutant allele that affects B chromosome drive would likely lead to quick loss of the B chromosome under study. Another difficulty would be mutagenesis of candidate sequences that are present in multiple copy number, such as the complex repeats that express putative lncRNAs in the jewel wasp [26]. A less problematic goal may be the expression of candidate B linked sequences from transgenes inserted through CRISPR/Cas9 and homology-dependent recombination (HDR). A consideration of this approach will be whether transgenic expression of multiple different B-linked sequences simultaneously is required to cause a phenotype of interest. Despite these obstacles, genome editing provides a very promising means for finally understanding how B chromosomes mediate their own transmission and drive at the mechanistic level.

Funding: This work was funded by the U. S. National Science Foundation, grant number NSF-1451839.

Conflicts of Interest: The authors declare no conflict of interest.

References

1. Jones, R.N. B-Chromosome drive. *Am. Nat.* **1991**, *137*, 430–442. [CrossRef]
2. Jones, R.N. B chromosomes in plants. *New Phytol.* **1995**, *131*, 411–434. [CrossRef]
3. Camacho, J.P.; Sharbel, T.F.; Beukeboom, L.W. B-chromosome evolution. *Philos. Trans. R. Soc. Lond. B. Biol. Sci.* **2000**, *355*, 163–178. [CrossRef] [PubMed]
4. Werren, J.H.; Stouthamer, R. PSR (paternal sex ratio) chromosomes: the ultimate selfish genetic elements. *Genetica* **2003**, *117*, 85–101. [CrossRef]
5. Hurst, G.D.D.; Werren, J.H. The role of selfish genetic elements in eukaryotic evolution. *Nat. Rev. Genet.* **2001**, *2*, 597–606. [CrossRef] [PubMed]
6. Hewitt, G.M.; John, B. The B-chromosome system of *Myrmeleotettix macculatus* (Thunb.). *Chromosoma* **1967**, *21*, 140–162. [CrossRef]
7. Fontana, P.G.; Vickery, V.R. Segregation-distortion in the B-chromosome system of *Tettigidea lateralis* (Say) (Orthoptera: Tetrigidae). *Chromosoma* **1973**, *43*, 75–100. [CrossRef] [PubMed]
8. Kimura, M.; Kayano, H. The maintenance of super-numerary chromosomes in wild populations of *Lilium callosum* by preferential segregation. *Genetics* **1961**, *46*, 1699–1712.
9. Banaei-Moghaddam, A.M.; Schubert, V.; Kumke, K.; Weiß, O.; Klemme, S.; Nagaki, K.; Macas, J.; González-Sánchez, M.; Heredia, V.; Gómez-Revilla, D.; et al. Nondisjunction in favor of a chromosome: The mechanism of rye B chromosome drive during pollen mitosis. *Plant Cell* **2012**, *24*, 4124–4134. [CrossRef]
10. Han, F.; Lamb, J.C.; Yu, W.; Gao, Z.; Birchler, J.A. Centromere function and nondisjunction are independent components of the maize B chromosome accumulation mechanism. *Plant Cell* **2007**, *19*, 524–533. [CrossRef]
11. Müntzing, A. Chromosome number, nuclear volume and pollen grain size in *Galeopsis*. *Hereditas* **2010**, *10*, 241–260. [CrossRef]
12. Reed, K.M. Cytogenetic analysis of the paternal sex ratio chromosome of *Nasonia vitripennis*. *Genome* **1993**, *36*, 157–161. [CrossRef] [PubMed]
13. Reed, K.M.; Werren, J.H. Induction of paternal genome loss by the paternal-sex-ratio chromosome and cytoplasmic incompatibility bacteria (Wolbachia): A comparative study of early embryonic events. *Mol. Reprod. Dev.* **1995**, *40*, 408–418. [CrossRef] [PubMed]
14. Beukeboom, L.W.; Werren, J.H. Deletion analysis of the selfish B chromosome, paternal sex ratio (PSR), in the parasitic wasp *Nasonia vitripennis*. *Genetics* **1993**, *133*, 637–648. [PubMed]
15. Klemme, S.; Banaei-Moghaddam, A.M.; Macas, J.; Wicker, T.; Novák, P.; Houben, A. High-copy sequences reveal distinct evolution of the rye B chromosome. *New Phytol.* **2013**, *199*, 550–558. [CrossRef] [PubMed]

16. Navarro-Domínguez, B.; Martín-Peciña, M.; Ruiz-Ruano, F.J.; Cabrero, J.; Corral, J.M.; López-León, M.D.; Sharbel, T.F.; Camacho, J.P.M. Gene expression changes elicited by a parasitic B chromosome in the grasshopper *Eyprepocnemis plorans* are consistent with its phenotypic effects. *Chromosoma* **2019**. [CrossRef] [PubMed]
17. Navarro-Domínguez, B.; Ruiz-Ruano, F.J.; Camacho, J.P.M.; Cabrero, J.; López-León, M.D. Transcription of a B chromosome CAP-G pseudogene does not influence normal Condensin Complex genes in a grasshopper. *Sci. Rep.* **2017**, *7*, 17650. [CrossRef] [PubMed]
18. Carchilan, M.; Kumke, K.; Mikolajewski, S.; Houben, A. Rye B chromosomes are weakly transcribed and might alter the transcriptional activity of A chromosome sequences. *Chromosoma* **2009**, *118*, 607–616. [CrossRef]
19. Banaei-Moghaddam, A.M.; Meier, K.; Karimi-Ashtiyani, R.; Houben, A. Formation and expression of pseudogenes on the B chromosome of rye. *Plant Cell* **2013**, *25*, 2536–2544. [CrossRef]
20. Delgado, M.; Caperta, A.; Ribeiro, T.; Viegas, W.; Jones, R.N.; Morais-Cecílio, L. Different numbers of rye B chromosomes induce identical compaction changes in distinct A chromosome domains. *Cytogenet. Genome Res.* **2004**, *106*, 320–324. [CrossRef]
21. Rubtsov, N.B.; Borisov, Y.M. Sequence composition and evolution of mammalian B chromosomes. *Genes* **2018**, *9*, 490. [CrossRef] [PubMed]
22. Ward, E.J. Nondisjunction: localization of the controlling site in the maize B chromosome. *Genetics* **1973**, *73*, 387–391. [PubMed]
23. Ma, W.; Gabriel, T.S.; Martis, M.M.; Gursinsky, T.; Schubert, V.; Vrána, J.; Doležel, J.; Grundlach, H.; Altschmied, L.; Scholz, U.; et al. Rye B chromosomes encode a functional Argonaute-like protein within vitroslicer activities similar to its A chromosome paralog. *New Phytol.* **2016**, *213*, 916–928. [CrossRef] [PubMed]
24. Navarro-Domínguez, B.; Ruiz-Ruano, F.J.; Cabrero, J.; Corral, J.M.; López-León, M.D.; Sharbel, T.F.; Camacho, J.P.M. Protein-coding genes in B chromosomes of the grasshopper *Eyprepocnemis plorans*. *Sci. Rep.* **2017**, *7*, 45200. [CrossRef] [PubMed]
25. Akbari, O.S.; Antoshechkin, I.; Hay, B.A.; Ferree, P.M. Transcriptome profiling of *Nasonia vitripennis* testis reveals novel transcripts expressed from the selfish B chromosome, paternal sex ratio. *G3* **2013**, *3*, 1597–1605. [CrossRef] [PubMed]
26. Li, Y.; Jing, X.A.; Aldrich, J.C.; Clifford, C.; Chen, J.; Akbari, O.S.; Ferree, P.M. Unique sequence organization and small RNA expression of a "selfish" B chromosome. *Chromosoma* **2017**, *126*, 753–768. [CrossRef] [PubMed]
27. Valente, G.T.; Conte, M.A.; Fantinatti, B.E.A.; Cabral-de-Mello, D.C.; Carvalho, R.F.; Vicari, M.R.; Kocher, T.D.; Martins, C. Origin and evolution of B chromosomes in the cichlid fish *Astatotilapia latifasciata* based on integrated genomic analyses. *Mol. Biol. Evol.* **2014**, *31*, 2061–2072. [CrossRef]
28. Valente, G.T.; Nakajima, R.T.; Fantinatti, B.E.A.; Marques, D.F.; Almeida, R.O.; Simões, R.P.; Martins, C. B chromosomes: from cytogenetics to systems biology. *Chromosoma* **2017**, *126*, 73–81. [CrossRef]
29. Becker, S.E.D.; Thomas, R.; Trifonov, V.A.; Wayne, R.K.; Graphodatsky, A.S.; Breen, M. Anchoring the dog to its relatives reveals new evolutionary breakpoints across 11 species of the Canidae and provides new clues for the role of B chromosomes. *Chromosome Res.* **2011**, *19*, 685–708. [CrossRef]
30. Coleman, J.J.; Rounsley, S.D.; Rodriguez-Carres, M.; Kuo, A.; Wasmann, C.C.; Grimwood, J.; Schmutz, J.; Taga, M.; White, G.J.; Zhou, S.; et al. The genome of *Nectria haematococca*: contribution of supernumerary chromosomes to gene expansion. *PLoS Genet.* **2009**, *5*, e1000618. [CrossRef]
31. Goodwin, S.B.; M'barek, S.B.; Dhillon, B.; Wittenberg, A.H.J.; Crane, C.F.; Hane, J.K.; Foster, A.J.; Van der Lee, T.A.J.; Grimwood, J.; Aerts, A.; et al. Finished genome of the fungal wheat pathogen *Mycosphaerella graminicola* reveals dispensome structure, chromosome plasticity, and stealth pathogenesis. *PLoS Genet.* **2011**, *7*, e1002070. [CrossRef] [PubMed]
32. Houben, A.; Banaei-Moghaddam, A.M.; Klemme, S.; Timmis, J.N. Evolution and biology of supernumerary B chromosomes. *Cell. Mol. Life Sci.* **2014**, *71*, 467–478. [CrossRef] [PubMed]
33. Martis, M.M.; Klemme, S.; Banaei-Moghaddam, A.M.; Blattner, F.R.; Macas, J.; Schmutzer, T.; Scholz, U.; Gundlach, H.; Wicker, T.; Simkova, H.; et al. Selfish supernumerary chromosome reveals its origin as a mosaic of host genome and organellar sequences. *Proc. Natl. Acad. Sci.* **2012**, *109*, 13343–13346. [CrossRef] [PubMed]

34. Makunin, A.; Romanenko, S.; Beklemisheva, V.; Perelman, P.; Druzhkova, A.; Petrova, K.; Prokopov, D.; Chernyaeva, E.; Johnson, J.; Kukekova, A.; et al. Sequencing of supernumerary chromosomes of red fox and raccoon dog confirms a non-random gene acquisition by B Chromosomes. *Genes* **2018**, *9*, 405. [CrossRef] [PubMed]
35. Ruban, A.; Schmutzer, T.; Scholz, U.; Houben, A. How next-generation sequencing has aided our understanding of the sequence composition and origin of B chromosomes. *Genes* **2017**, *8*, 294. [CrossRef] [PubMed]
36. Cheng, Y.-M.; Lin, B.-Y. Cloning and characterization of maize B chromosome sequences derived from microdissection. *Genetics* **2003**, *164*, 299–310. [PubMed]
37. Bugrov, A.G.; Karamysheva, T.V.; Perepelov, E.A.; Elisaphenko, E.A.; Rubtsov, D.N.; Warchałowska-Śliwa, E.; Tatsuta, H.; Rubtsov, N.B. DNA content of the B chromosomes in grasshopper Podisma kanoi Storozh. (Orthoptera, Acrididae). *Chromosome Res.* **2007**, *15*, 315–325. [CrossRef]
38. Ruiz-Ruano, F.J.; Cabrero, J.; López-León, M.D.; Sánchez, A.; Camacho, J.P.M. Quantitative sequence characterization for repetitive DNA content in the supernumerary chromosome of the migratory locust. *Chromosoma* **2018**, *127*, 45–57. [CrossRef]
39. Coan, R.; Martins, C. Landscape of transposable elements focusing on the B chromosome of the cichlid fish *Astatotilapia latifasciata*. *Genes* **2018**, *9*, 269. [CrossRef]
40. Marques, A.; Klemme, S.; Houben, A. Evolution of plant B chromosome enriched sequences. *Genes* **2018**, *9*, 515. [CrossRef]
41. McAllister, B.F. Isolation and characterization of a retroelement from B chromosome (PSR) in the parasitic wasp *Nasonia vitripennis*. *Insect Mol. Biol.* **1995**, *4*, 253–262. [CrossRef] [PubMed]
42. McAllister, B.F.; Werren, J.H. Hybrid origin of a B chromosome (PSR) in the parasitic wasp *Nasonia vitripennis*. *Chromosoma* **1997**, *106*, 243–253. [CrossRef] [PubMed]
43. Perfectti, F.; Werren, J.H. The interspecific origin of B chromosomes: experimental evidence. *Evolution* **2001**, *55*, 1069–1073. [CrossRef]
44. McVean, G.T. Fractious chromosomes: hybrid disruption and the origin of selfish genetic elements. *Bioessays* **1995**, *17*, 579–582. [CrossRef] [PubMed]
45. Schartl, M.; Nanda, I.; Schlupp, I.; Wilde, B.; Epplen, J.T.; Schmid, M.; Parzefall, J. Incorporation of subgenomic amounts of DNA as compensation for mutational load in a gynogenetic fish. *Nature* **1995**, *373*, 68–71. [CrossRef]
46. Banaei-Moghaddam, A.M.; Martis, M.M.; Macas, J.; Gundlach, H.; Himmelbach, A.; Altschmied, L.; Mayer, K.F.X.; Houben, A. Genes on B chromosomes: Old questions revisited with new tools. *Biochim. Biophys. Acta* **2015**, *1849*, 64–70. [CrossRef] [PubMed]
47. Palazzo, A.F.; Lee, E.S. Non-coding RNA: what is functional and what is junk? *Front. Genet.* **2015**, *6*, 2. [CrossRef] [PubMed]
48. Huang, W.; Du, Y.; Zhao, X.; Jin, W. B chromosome contains active genes and impacts the transcription of A chromosomes in maize (*Zea mays* L.). *BMC Plant Biol.* **2016**, *16*, 88. [CrossRef]
49. Trifonov, V.A.; Dementyeva, P.V.; Larkin, D.M.; O'Brien, P.C.M.; Perelman, P.L.; Yang, F.; Ferguson-Smith, M.A.; Graphodatsky, A.S. Transcription of a protein-coding gene on B chromosomes of the Siberian roe deer (*Capreolus pygargus*). *BMC Biol.* **2013**, *11*, 90. [CrossRef]
50. Gaudet, P.; Livstone, M.S.; Lewis, S.E.; Thomas, P.D. Phylogenetic-based propagation of functional annotations within the Gene Ontology consortium. *Brief. Bioinform.* **2011**, *12*, 449–462. [CrossRef]
51. Malik, H.S.; Bayes, J.J. Genetic conflicts during meiosis and the evolutionary origins of centromere complexity. *Biochem. Soc. Trans.* **2006**, *34*, 569–573. [CrossRef] [PubMed]
52. Teruel, M.; Cabrero, J.; Perfectti, F.; Camacho, J.P.M. B chromosome ancestry revealed by histone genes in the migratory locust. *Chromosoma* **2010**, *119*, 217–225. [CrossRef] [PubMed]
53. Oliveira, N.L.; Cabral-de-Mello, D.C.; Rocha, M.F.; Loreto, V.; Martins, C.; Moura, R.C. Chromosomal mapping of rDNAs and H3 histone sequences in the grasshopper *Rhammatocerus brasiliensis* (Acrididae, gomphocerinae): extensive chromosomal dispersion and co-localization of 5S rDNA/H3 histone clusters in the A complement and B chromosome. *Mol. Cytogenet.* **2011**, *4*, 24. [CrossRef] [PubMed]
54. SanMiguel, P.; Tikhonov, A.; Jin, Y.K.; Motchoulskaia, N.; Zakharov, D.; Melake-Berhan, A.; Springer, P.S.; Edwards, K.J.; Lee, M.; Avramova, Z.; et al. Nested retrotransposons in the intergenic regions of the maize genome. *Science* **1996**, *274*, 765–768. [CrossRef] [PubMed]

55. Tang, W.; Mun, S.; Joshi, A.; Han, K.; Liang, P. Mobile elements contribute to the uniqueness of human genome with 15,000 human-specific insertions and 14 Mbp sequence increase. *DNA Res.* **2018**, *25*, 521–533. [CrossRef]
56. Cheng, Y.-M.; Lin, B.-Y. Molecular organization of large fragments in the maize B chromosome: indication of a novel repeat. *Genetics* **2004**, *166*, 1947–1961. [CrossRef] [PubMed]
57. Ziegler, C.G.; Lamatsch, D.K.; Steinlein, C.; Engel, W.; Schartl, M.; Schmid, M. The giant B chromosome of the cyprinid fish *Alburnus alburnus harbours* a retrotransposon-derived repetitive DNA sequence. *Chromosome Res.* **2003**, *11*, 23–35. [CrossRef]
58. Gross, L. Transposon silencing keeps jumping genes in their place. *PLoS Biol.* **2006**, *4*, e353. [CrossRef]
59. McGurk, M.P.; Barbash, D.A. Double insertion of transposable elements provides a substrate for the evolution of satellite DNA. *Genome Res.* **2018**, *28*, 714–725. [CrossRef]
60. Ramos, É.; Cardoso, A.L.; Brown, J.; Marques, D.F.; Fantinatti, B.E.A.; Cabral-de-Mello, D.C.; Oliveira, R.A.; O'Neill, R.J.; Martins, C. The repetitive DNA element BncDNA, enriched in the B chromosome of the cichlid fish *Astatotilapia latifasciata*, transcribes a potentially noncoding RNA. *Chromosoma* **2016**, *126*, 313–323. [CrossRef]
61. Perry, R.B.-T.; Ulitsky, I. The functions of long noncoding RNAs in development and stem cells. *Development* **2016**, *143*, 3882–3894. [CrossRef]
62. Ulitsky, I.; Bartel, D.P. lincRNAs: genomics, evolution, and mechanisms. *Cell* **2013**, *154*, 26–46. [CrossRef] [PubMed]
63. Park, Y.; Oh, H.; Meller, V.H.; Kuroda, M.I. Variable splicing of non-coding roX2 RNAs influences targeting of MSL dosage compensation complexes in *Drosophila*. *RNA Biol.* **2005**, *2*, 157–164. [CrossRef] [PubMed]
64. Park, Y. Extent of chromatin spreading determined by roX RNA Recruitment of MSL proteins. *Science* **2002**, *298*, 1620–1623. [CrossRef]
65. Lucchesi, J.C.; Kuroda, M.I. dosage compensation in *Drosophila*. *Cold Spring Harb. Perspect. Biol.* **2015**, *7*, a019398. [CrossRef] [PubMed]
66. Ilik, I.A.; Quinn, J.J.; Georgiev, P.; Tavares-Cadete, F.; Maticzka, D.; Toscano, S.; Wan, Y.; Spitale, R.C.; Luscombe, N.; Backofen, R.; et al. Tandem stem-loops in roX RNAs act together to mediate X chromosome dosage compensation in *Drosophila*. *Mol. Cell* **2013**, *51*, 156–173. [CrossRef] [PubMed]
67. Sahakyan, A.; Yang, Y.; Plath, K. The Role of Xist in X-chromosome dosage compensation. *Trends Cell Biol.* **2018**, *28*, 999–1013. [CrossRef] [PubMed]
68. Rougeulle, C.; Chaumeil, J.; Sarma, K.; Allis, C.D.; Reinberg, D.; Avner, P.; Heard, E. Differential histone H3 Lys-9 and Lys-27 methylation profiles on the X chromosome. *Mol. Cell. Biol.* **2004**, *24*, 5475–5484. [CrossRef]
69. Han, P.; Chang, C.-P. Long non-coding RNA and chromatin remodeling. *RNA Biol.* **2015**, *12*, 1094–1098. [CrossRef] [PubMed]
70. Böhmdorfer, G.; Wierzbicki, A.T. Control of chromatin structure by long noncoding RNA. *Trends Cell Biol.* **2015**, *25*, 623–632. [CrossRef]
71. Aldrich, J.C.; Leibholz, A.; Cheema, M.S.; Ausió, J.; Ferree, P.M. A "selfish" B chromosome induces genome elimination by disrupting the histone code in the jewel wasp *Nasonia vitripennis*. *Sci. Rep.* **2017**, *7*, 42551. [CrossRef] [PubMed]
72. Zhang, C. Novel functions for small RNA molecules. *Curr. Opin. Mol. Ther.* **2009**, *11*, 641–651. [PubMed]
73. Chapman, E.J.; Carrington, J.C. Specialization and evolution of endogenous small RNA pathways. *Nat. Rev. Genet.* **2007**, *8*, 884–896. [CrossRef]
74. López-León, M.D.; Cabrero, J.; Pardo, M.C.; Viseras, E.; Camacho, J.P.M.; Santos, J.L. Generating high variability of B chromosomes in *Eyprepocnemis plorans* (grasshopper). *Heredity* **1993**, *71*, 352–362. [CrossRef]

© 2019 by the authors. Licensee MDPI, Basel, Switzerland. This article is an open access article distributed under the terms and conditions of the Creative Commons Attribution (CC BY) license (http://creativecommons.org/licenses/by/4.0/).

Review

Centromere Repeats: Hidden Gems of the Genome

Gabrielle Hartley [1] and Rachel J. O'Neill [2,*]

1. Department of Molecular and Cell Biology, University of Connecticut, Storrs, CT 06269, USA; gabrielle.hartley@uconn.edu
2. Department of Molecular and Cell Biology and Institute for Systems Genomics, University of Connecticut, Storrs, CT 06269, USA
* Correspondence: rachel.oneill@uconn.edu; Tel.: +1-860-486-6031

Received: 8 February 2019; Accepted: 11 March 2019; Published: 16 March 2019

Abstract: Satellite DNAs are now regarded as powerful and active contributors to genomic and chromosomal evolution. Paired with mobile transposable elements, these repetitive sequences provide a dynamic mechanism through which novel karyotypic modifications and chromosomal rearrangements may occur. In this review, we discuss the regulatory activity of satellite DNA and their neighboring transposable elements in a chromosomal context with a particular emphasis on the integral role of both in centromere function. In addition, we discuss the varied mechanisms by which centromeric repeats have endured evolutionary processes, producing a novel, species-specific centromeric landscape despite sharing a ubiquitously conserved function. Finally, we highlight the role these repetitive elements play in the establishment and functionality of de novo centromeres and chromosomal breakpoints that underpin karyotypic variation. By emphasizing these unique activities of satellite DNAs and transposable elements, we hope to disparage the conventional exemplification of repetitive DNA in the historically-associated context of 'junk'.

Keywords: satellite; transposable element; repetitive DNA; chromosome evolution; centromere drive; genetic conflict; CENP-A; centromeric transcription

1. Introduction

Specific types of repetitive segments within eukaryotic genomes are now recognized as critical to maintaining subspecialized genomic functions. Common elements within repetitive segments include both transposable elements (TEs) and satellite DNA [1], collectively representing a large portion of eukaryotic genomes [2,3]. Unlike TEs that are capable of moving within a genome and thus are often found dispersed (albeit not randomly; reviewed in [4]), satellite DNA consists of short stationary DNA sequences that tandemly repeat to form a larger array, often restricted to specific sub-regions of chromosomes [1,5]. Ranging from just a few base pairs to several megabases in length, satellite repetitive units comprise up to 10% of the human genome [6]; across eukaryotes, variation in copy number and satellite family diversity contributes to differences in total satellite DNA content among taxa, often with dramatic total satellite content differentials [5]. Despite the high degree of variation among species in both sequence diversity and overall content, satellite DNAs are collectively found most highly concentrated in the centromeric and pericentromeric regions of chromosomes [7]. While the exact functions of satellite DNA have not been fully realized, this incommensurate distribution of satellite DNA within the genome highlights the importance of satellite DNA in chromosome inheritance through participation in centromere function.

First described in the context of DNA content in eukaryotes by Kit et al. [8] and Seuoka et al. [9] in 1961, satellite DNA was discovered via ultracentrifugation of genomic DNA—Note: the first use of the term satellite as a genetic descriptor is attributed to Sergius Navashin in his 1912 study of secondary constrictions on the chromosomes of a hyacinth [10]. Following the centrifugation of DNA from animal tissue extracts across a cesium chloride density gradient, Kit et al. [8] described a satellite band that was clearly differentiated from the major band of genomic DNA. Due to the repetitive nature of the DNA within this band, this fraction displayed an observable shift in density and led to the first description of satellite DNA. Despite this traditional descriptor, the phrase satellite DNA has been used more broadly to describe all tandemly arranged repetitive DNA sequences [11] irrespective of their resolution on density gradients. Since their discovery, a number of different methods have been used to characterize tandem repeats, including C_0T analysis, in which the rate of re-association of complementary DNA strands is used to identify the frequency of repetitive elements [12], and separation following restriction endonuclease treatment, in which digested genomic DNA is separated via electrophoresis on an agarose gel [13]. Modern molecular techniques including next-generation sequencing (NGS) and fluorescence in situ hybridization (FISH) have provided additional clarity in the identification and physical characterization of satellite DNA sequences. This review includes emerging discoveries about satellite array characteristics and the other types of repeats found within, model systems proven useful for studying the role of satellite DNA in genome evolution, and the intimate relationship between satellite DNA and TEs. In addition, this review examines the paradoxical link between divergent satellite DNA and conserved centromere function as well as the connection between repeats and the emergence of new centromeres during chromosome evolution.

2. A Brief Primer on Satellite DNA in a Chromosomal Context

While the term satellite DNA encompasses all tandem nucleotide repeats, this large category can be further divided into a number of different subcategories and families. In addition to larger tandem repeats, one such grouping of smaller repeats can be created based on the number of nucleotides existing in the core repetitive segment. Microsatellites, for example [14], consist of repeating units less than 10 nucleotides in length and constitute up to 3% of the human genome [15]. Minisatellites, often referred to as variable number tandem repeats (VNTRs) [16], consist of a 10 to 100 nucleotide unit repeating up to several hundred times. With several thousand minisatellite loci distributed throughout the human genome [17], minisatellites are found at a high frequency in telomeric regions [18]. Telomeres are also enriched for a specific microsatellite, $(TTAGGG)_n$, which constitutes the bulk of telomeric sequences, extending for 9–15 kb on human chromosomes [19,20]. Nucleoproteins (TRF1, TRF2, and POT1) bind to these telomeric satellites to form the shelterin complex [21], which interacts with the ribonucleoprotein telomerase that contains the enzyme component telomerase reverse transcriptase (TERT) [22], and an RNA (TERRA) [23]. The resulting 'cap' distinguishes chromosome ends from DNA breaks requiring repair and thus protects the chromosome from end-degradation and interchromosomal fusions [24].

Perhaps the most notable satellite families in the human genome are those located at both pericentromeric and centromeric regions: α satellites. α satellites, found ubiquitously at all human centromeres, are a ~171 base pair unit, known as a monomer, with sequences that are 50–80% identical among all monomers within an array (repeated monomers in tandem) [25]. The core of the centromere, where the kinetochore will form and mediate microtubule attachment and faithful chromosome segregation, is functionally defined by the assembly of centromeric nucleosomes containing the centromere-specific histone 3, CENP-A [26]. In humans, this core is enriched for α satellite DNA [25]. While found as solo repetitive units scattered among other satellites in the pericentromeric regions of human chromosomes without higher organization, α satellites within human centromeres are tandemly repeated to form a block of satellites, called a higher order repeat (HOR). HORs are comprised of a set number of monomers that varies from 2 to 34 monomers ([25,27–30] and reviewed in [31]) in a largely chromosome-specific arrangement (Figure 1a). For example, the α satellite HOR blocks on chromosome 1 consist of 2 monomers [32], referred to as 2-mers, the HOR blocks on chromosome 7 consist of 6-mers [33], and the HOR blocks on the Y chromosome are 34-mers [34]. These HOR blocks are further repeated to form HOR arrays than can span megabases. Because of the highly repetitive nature of these centromeric HOR arrays with identities among HOR blocks nearing 99% in some cases [35] and high HOR copy numbers [36], centromeric regions have historically been refractive to characterization, at least in the context of genome assemblies [6].

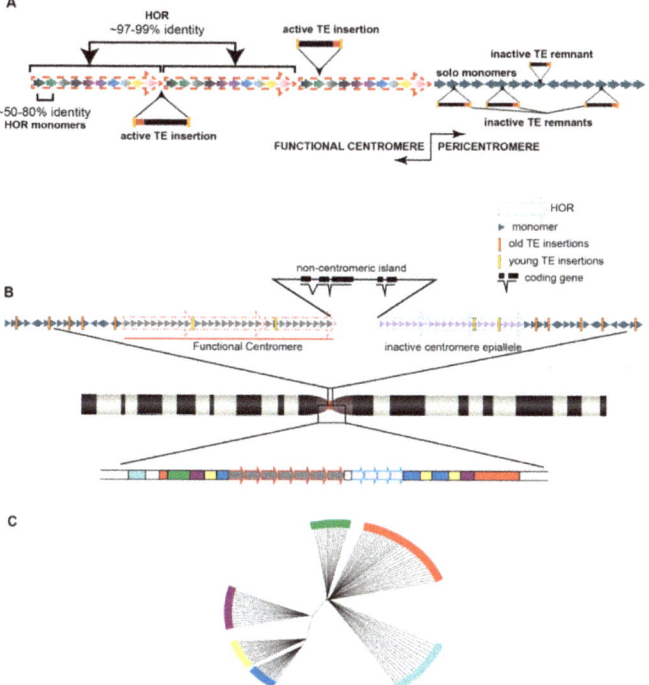

Figure 1. Overview of satellite DNA structure in a human centromere/pericentromere. (**a**) α satellite monomers (colored solid arrows) are organized into a repeating unit, called a higher order repeat (HOR) (red dashed arrows). In this example, 10 monomers are in each HOR (10-mers). HOR units are repeated in a chromosome-specific manner 100–1000 s of times within a functional centromere core. Within a single HOR, monomers share anywhere from 50–80% sequence identity with one another. The same monomer within different HORs in the same array may share up to 99% identity. Solo monomers (solid arrows) are found in the pericentromeric region and are highly variable in terms of sequence and orientation. Within the centromere, transposable elements (TE) insertions typically include recently active or active (hot) elements, while the TE insertions found in the pericentromere are older, inactive elements. (**b**) The core centromere structure (red dot, chromosome schematic) of human chromosomes (a generic chromosome ideogram is indicated, middle) consists of different α satellite arrays arranged in HORs (dashed arrows). Each HOR array may contain a different monomer number; in this example, the functional centromere (i.e., assembles CENP-A nucleosomes) at a 10-mer HOR (red dashed arrows). A 7-mer HOR is found nearby but is an inactive epiallele. Both HORs are separated by non-centromeric DNA, which may contain genes. α satellites are also found throughout the pericentromere (bottom schematic, different colored blocks). (**c**) Representative cladogram of the phylogenetic relationship of the non-HOR α monomers shown in (b). In this example, strata of newer satellites are closer to the HOR arrays, while older satellites are found more distally. Relative age of satellites is indicated by tree branch length; shorter branches are younger elements and deeper branches are older.

Despite challenges associated with characterizing highly repetitive stretches of DNA, groups are uncovering variation in satellite DNA, both within the human reference genome and among different individuals, and identifying the functional consequences of these variants. α satellite monomers are classified into 12 consensus monomers (J1, J2, D1, D2, W1, W2, W3, W4, W5, M1, R1, and

R2) [29,30,37–39], which are further grouped into five suprachromosomal families (SF1-5) [29,30,40]. A specific strata of satellites within each human chromosome was revealed by fine-scale mapping and sequence annotation of monomers and HORs [30,40], wherein highly homogenized and recently derived monomers are organized into HORs within the functional core of the centromere and the older, divergent monomers are organized further from the core and into the pericentromere (Figure 1b). In other words, the closer a satellite stratum is to the functionally defined core of the centromere, the younger and more homogenized the monomers within those HORs will be. It has thus been proposed that the α satellite strata are a phylogenetic record of the evolution of human centromeres, with the younger and more homogenized monomers closer to the functionally defined core of the centromere and older centromere remnants orbiting the central core, indicating the location and/or abandoned sequence of long-dead centromeres shared with our primate ancestors [40] (Figure 1c).

Once a satellite variant becomes dominant in a species, there is subsequent intrachromosomal homogenization that further distinguishes chromosome-specific arrays. Recent work in humans has also revealed that there is variation of chromosome-specific α satellite arrays among different individuals in the human population [41,42]. Aldrup-MacDonald et al. [41] describe variation within the α satellite DNA arrays of human chromosome 17 first characterized by several groups over the past few decades [40,43–45]. At this chromosome, the centromeric region contains three unique α satellite arrays arranged adjacently: D17Z1, D17Z1-B, and D17Z1-C (Figure 2). Among these three arrays, only one acts as the functional centromere and recruits CENP-A histones; thus, multiple, potentially functional arrays on one chromosome are known as epialleles [46]. In roughly 70% of individuals, the centromere is assembled at the 16-mer D17Z1 locus, while the remaining 30% of individuals display differential centromere assembly at the D17Z1 locus of one homolog and the 14-mer D17Z1-B locus of the other [46] (Figure 2a). While the D17Z1-B epiallele can support centromere assembly in human artificial chromosomes, no individual homozygous for this allele has yet been identified. Because of this, it is purported that those homozygous for the D17Z1-B epiallele represent a rare, yet functionally viable, variant in the human population [41]. Similarly, Miga et al. [42] have identified size and sequence satellite array variants on human chromosomes X and Y via their utilization of whole-genome shotgun sequencing in efforts to create centromeric reference models [47]. This ongoing work continues to build upon the foundational understanding of satellite array variation that has been characterized by others [48,49] and suggests that centromeric HOR variants are not a phenomenon exclusive to human chromosome 17.

While the underlying molecular foundation for the formation of centromeric epialleles remains unknown, Aldrup-MacDonald et al. [41] propose, based on their work with somatic cell hybrid lines, that genomic variation of satellite DNA is an influential factor dictating which epiallele will assemble centromeric nucleosomes [46]. Using restriction enzyme digestion and Southern blotting to identify variation in D17Z1, D17Z1-B, and D17Z1-C epialleles, Aldrup-MacDonald et al. [41] determined that larger D17Z1 satellite arrays were more likely to be both homogenous (wild type for the canonical 16-mer HOR) and the active site of centromere assembly. By using cytogenetic techniques like fluorescence in situ hybridization (FISH) and monitoring chromosome stability, it was determined that centromeres assembling at a highly variant D17Z1 array locus (containing a number of different HOR variants) were unstable while those assembling at the D17Z1-B locus remained stable despite D17Z1 variability [41] (Figure 2b). Furthermore, it was determined that these unstable centromeric locations had about half of the amount of centromeric proteins CENP-A and CENP-C present in comparison to stable centromeres [41]. These studies suggest that variant chromosome 17 epialleles do not perform equally and thus highlight the important role variation of satellite DNA might play in the maintenance of proper chromosome segregation.

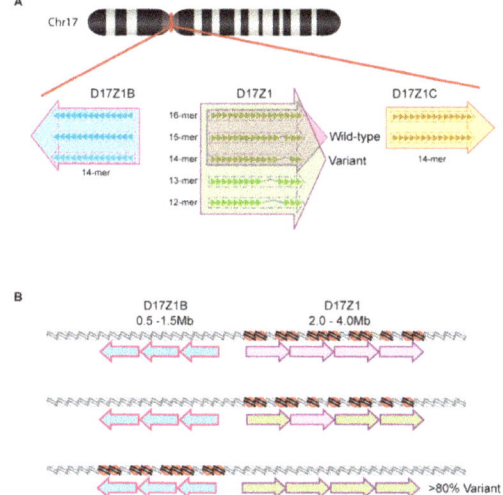

Figure 2. Chromosome 17 epialleles. (**a**) Ideogram of chromosome 17 (top). Zoom inset of epialleles showing monomer number for HORs and orientation. D17Z1B HORs carry 14 monomers, as do D17Z1C HORs. D17Z1 HORs are variable in the human population, with wild type epialleles containing 16-mer, 15-mer, and 14-mer HORs (pink) and variant epialleles containing wild type HORs in addition to 13-mer and 12-mer HORs (green). (**b**) Variation of the D17Z1 epiallele is linked to centromere activity. When the variation in D17Z1 increases, CENP-A nucleosomes (red) decrease; when variation exceeds 80%, the centromere assembles on the D17Z1B epiallele.

3. Centromere Repeats Endure Unique Evolutionary Processes

Although the presence of satellite DNA in centromeres is a shared characteristic found among many eukaryotic taxonomic groups, as is the protein cascade required for faithful chromosome segregation mediated by CENP-A, the underlying sequence of this satellite DNA is highly variable and largely species-specific [50–55]. Tandemly arrayed satellites within a single chromosome experience high rates of sequence turnover via concerted evolution, a non-independent process of molecular drive [56] (Figure 3a). Several mechanisms have been invoked to explain this observation, including nonhomologous and/or unequal crossing over [57], replication slippage [58], gene conversion [59], and rolling circle amplification and subsequent reinsertion ([60], reviewed in [61]). Such mechanisms impact sequence homogenization across an array as well as variation in overall array length.

While tandemly arrayed sequences are not capable of transposition, a family of arrays appears to spread from one chromosome to another, rendering the centromere repeats of non-homologous chromosomes within a karyotype highly similar and phylogenetically closely related. For example, several pairs of human chromosomes share the same satellite arrays: chromosomes 1, 5, and 19 [62,63], 13 and 21 [64], and 14 and 22 [65]. Interestingly, chromosomes 13 and 21 in the chimpanzee share the same satellite array as is observed on the homologous chromosomes 13 and 21 in humans, but the 13/21 arrays of these two species are not orthologous [65,66], indicating some chromosomes efficiently evoke inter-chromosomal recombination in independent lineages [28]. How this occurs or why this appears restricted to a subset of chromosomes is not known.

Homogenization of arrays is not linked specifically to the presence of tandem repeats. In fact, the only stratum of satellites across the centromere/pericentromere that experience forces of homogenization across an array, and thus carry HORs and high identity repeat units, is that of the recently derived and functional core [40] (Figure 3a). In other words, only the satellites that serve as the foundation for the kinetochore undergo continual homogenization, linking the assembly of the kinetochore to the homogenization process [37], and consequently, rapid evolution. It has been proposed that proteins facilitating homogenization, known as a kinetochore-associated recombination machine (KARM), have become integrated into the kinetochore complex, fostering this core-satellite specific homogenization process [28,40]. One candidate for this machine is topoisomerase II [40], a DNA decatenating enzyme that resides in the kinetochore during mitosis and initiates homologous recombination following the induction of DNA breaks [67].

What is the source material for new satellite arrays that seed within older arrays, eventually pushing them to outer, non-homogenized and highly variable strata? The library hypothesis [68] provides one explanation for how satellite DNA content at the centromere may diverge rapidly among closely related species (Figure 3a). In this scenario, extant but distinct centromeric repeats, representing a satellite library for a species, may independently expand or contract in copy number in different evolutionary lineages (be they chromosomes or species within a complex). If a repeat from this library finds itself in the core of the centromere, associated homogenization and expansion could result in the establishment of what appears to be a new satellite array [69–74]. In some cases, the seeding of a centromere from such a library is facilitated by chromosome rearrangement [75–77].

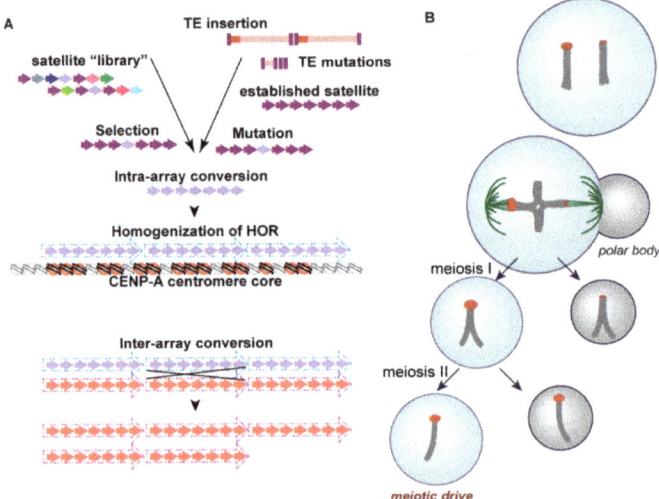

Figure 3. Schematic of the evolutionary mechanisms that impact centromere repeats. (**a**) Two models for the derivation of species-specific satellites are shown: (left) A satellite array evolves from a library of satellites, culminating in a dominant satellite; (right) TE insertion(s) followed by mutations, such as deletions, lead to the evolution of new satellites. In both cases, a homogenized array evolves through molecular drive mechanisms, such as intra-array concerted evolution. Stabilization of the arrays into HOR arrays defines the active centromere core, where CENP-A nucleosomes (red) are assembled. Other events, such as inter-array conversion, can lead to the spread of new HORs or changes in HOR copy number (bottom). (**b**) Two homologous chromosomes share the same satellite repeat (red), but one homolog experiences an expansion of that repeat through de novo mutations. During female meiosis, the larger centromere attracts more microtubules, resulting in the loss of the homolog with the weaker centromere into the polar body during meiosis I. The larger centromere is preferentially driven to the viable egg following unequal distribution of chromatids during meiosis II.

Another mechanism proposed to give rise to the variability of satellite sequences in different species is a meiotic drive model, known as centromere drive [78] (Figure 3b). As predicted by this model, satellite arrays attract more microtubules during female meiosis if the arrays experience accretion [79,80]. Preferentially sorted into the egg, these expanded satellite sequences are predicted to promote increased rates of evolution of centromere proteins, particularly CENP-A, which directly interacts with satellite DNA, through genetic conflict. Eventually, these divergent centromere proteins become highly prevalent in the population as they evolve to restore parity in meiosis [52,81]. In fact, this model is supported not only by the rapid evolution and variability of satellites in a variety of species, but by the positive selection of nucleic-acid interacting centromere proteins like CENP-A and CENP-C in plants, primates, and others [82–86]. This model is further supported by evidence that Robertsonian fusions with a single centromere are preferentially segregated due to a higher recruitment of CENP-A, Ncd80, and microtubules than their unfused mates [87]. Heterozygosity for these fusions has been observed to reduce male fertility, creating a selective pressure for the fixation of a new karyotype via a fitness cost ([88], reviewed in [89]).

Despite the ability of the centromere drive model to explain the high variation observed in satellite sequence from one species to the next, this model does not offer a complete mechanism by which chromosomal evolution and karyotypic changes may occur, particularly when considering the circumstances of de novo centromere formation. Described in human patients presenting with an abnormal karyotype (reviewed in [90]), a neocentromere forms on an ectopic site on a chromosome when the original centromere is lost or inactivated, or the entire karyotype is unstable, as in cancer (e.g., [91]) —Note: It has been argued that the term neocentromere is incorrectly used to describe de novo centromeres that are kinetochore-competent [92]. The original use of the term neocentromere is attributed to describe subtelomeric heterochromatin blocks that behave similarly to centromeres but do not build a traditional kinetochore [93]. While stable neocentromeres can be fully functional in kinetochore assembly and thus maintain proper chromosome segregation, most lack the typical satellite DNA characteristic of centromeric regions [94–99]. Not only are functional neocentromeres devoid of satellite DNA, but in some cases, the original inactive centromere retains satellite arrays yet they no longer recruit centromere proteins (and thus are rendered non-functional) (reviewed in [90,100]). The identification of functional neocentromeres lacking satellite DNA spawned the prediction that satellite DNA is neither sufficient nor required for centromere function [101], despite its apparent ubiquity across taxonomic groups.

Neocentromeres are not restricted to clinical cases of chromosome instability; shifts in centromere location with no discernable change in intervening gene order distinguish species-specific karyotypes in many eukaryotic taxa. Formerly referred to as centric shifts [102,103], these evolutionary new centromeres (ENCs) [104] (Figure 4) have been characterized in primates, horses, cattle, marsupials, plants, insects, and many other species complexes (see [102–108] for examples). Moreover, several groups have noted a lack of higher order satellite arrays in newly emerged, functional centromeres, indicating that the formation of homogenized arrays succeeds centromere fixation in a population [109]. It has been proposed that following the fixation of a novel centromere in a species, satellite arrays accumulate to further stabilize the centromere [110,111]. Successive interchromosomal homogenization further support the establishment of large, stable regional centromeres that are rendered species-specific [109,112–115]. Thus, ENCs accumulate satellite DNA arrays across successive generations as they phylogenetically age, while their immature counterparts lack these types of repetitive sequences (Figure 4).

Not only has it been established that some recently emerged centromeres lack the higher order satellite arrays characteristic of functional centromeres in a wide variety of organisms, but species in the *Equus* genus carry several centromeres that lack satellite DNA altogether [110,111,116]. Included in those devoid of satellite DNA are ENCs, repositioned to a non-centromeric location following the loss of function at the original centromere [108]. Based on the emerging ENC hypothesis, the recently diverged *Equus* genus, estimated to share a last common ancestor with other genera just 2–3 million years ago despite considerable karyotypic variation, would be predicted to contain de novo centromeres helping drive karyotypic variation that lack satellite DNA. Immuno-FISH experiments using satellite DNA and antibodies against CENP-A completed by Piras et al. [111] identified both functional centromeres lacking satellite DNA as well as satellite repeats present at non-centromeric locations, suggesting the presence of both immature centromeres and ancestral yet inactive centromeric locations, respectively. The identification of a fixed, satellite-free centromere on chromosome 11 in *Equus caballus* presented a distinctive opportunity to test whether there was detectable variability in kinetochore assembly localization on an ENC. ChIP-on-chip analyses in five *Equus* individuals using an antibody against CENP-A revealed at least seven functional centromere epialleles on chromosome 11 dispersed across a region of 500 kb and extending between 80 to 160 kb [117]. The results of these experiments, and recent work in *Equus asinus* [110], demonstrate significant plasticity in CENP-A binding domains among individuals and suggest the potential for centromeres across mammalian species to positionally 'slide', resulting in the formation of variable functional epialleles [110,111].

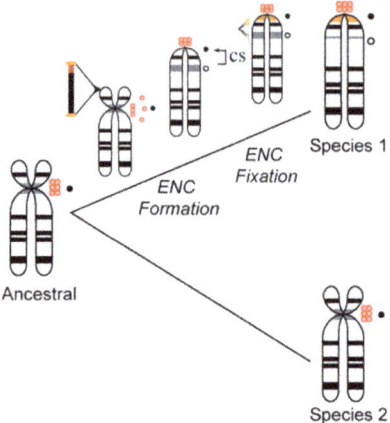

Figure 4. The hypothetical evolution of new centromeres. The ancestral chromosome in this example is submetacentric (the centromere is indicated with red 'nucleosomes'). The active locus (black dot) carries satellite arrays. Some individual(s) in a population experience the destabilization of the active centromere and formation of a neocentromere, perhaps through the activation of a new TE, resulting in a centric shift (CS). The new centromere is indicated with a black dot, while the latent centromere is indicated with an open circle. The new centromere becomes fixed in a population and eventually gains new satellite arrays (orange), either by interchromosomal seeding from the old centromere (grey) or from the TE itself. Over time, the latent centromere loses its HORs while the new centromere becomes stabilized. In some cases, the ENC can lead to a new species karyotype.

Genome sequencing efforts have further revealed that many eukaryotic species lack centromeres enriched for satellite arrays. For example, sequencing following chromatin immunoprecipitation with antibodies to centromeric proteins CENP-A and CREST, Johnson et al. [118] report a lack of satellite arrays in the centromeres of the recently characterized koala (*Phascolarctos cinereus*) genome, an observation also described in gibbon centromeres and suggestive of the recent evolution of new elements associated with centromere function [119]. Furthermore, this observation has also been documented in a number of other species with small centromeres, ranging from plant species like rice [120] and potato [74] to marsupials like the tammar wallaby [121,122], and fungal species such as *Candida albicans* [123,124]. Taken collectively, new centromere formation is likely not initiated by satellite DNAs; however, satellite DNA is a shared feature of regional centromeres and thus likely promotes their stability. While the introduction of α satellite arrays in human cells can result in the formation of a functional neocentromere, supporting the proposal that satellite DNA is foundational to centromere activity [125,126], the seeding of new ectopic neocentromeres appears to occur in the absence of satellite DNA.

4. Satellites and Their Party Friends—Transposable Elements

While satellite DNA is pervasive in the stable, regional centromeres of many species, another class of repetitive element is found within satellite-rich centromeres, ENCs, and neocentromeres: TEs. TEs are repetitive sequences that are able to alter their location in the genome and thus are often considered selfish elements [1,127,128]. Originally characterized by cytogeneticist Barbara McClintock [129], transposable elements can be divided into two categories based on mobility; transposons alter their position directly via a cut and paste mechanism, while retrotransposons move via a copy and paste mechanism through which an RNA intermediate is first created before being reverse transcribed into an identical DNA sequence inserted at a particular genomic locus [130,131].

Transposons moving via a cut and paste mechanism, also called type II transposable elements, require a self-encoded enzyme, transposase, in order to move from one locus to another [130,131]. The transposon, flanked by terminal inverted repeats, is recognized by transposase which removes the transposon before reintegrating it at a target location. The gap left behind by transposon excision can be repaired either with, or without, the addition of a replacement transposon. Dissimilarly, retrotransposons, also called type I transposable elements, rely on the transcription of an RNA intermediate as part of their transposition [130,131]. Following transcription, retrotransposon RNA intermediates are reverse transcribed into identical DNA sequences and integrated into a target locus [130,131]. Unlike transposase-mediated mobility, the number of retrotransposons present in a genome increases in number each time they undergo transposition.

Like satellite DNA, transposable elements form a significant portion of eukaryotic genomes. In fact, due to the ability for many subfamilies to multiply during retrotransposition, TEs can occupy a significant majority of eukaryotic genomes [132–134], constituting up to 85% of the maize genome [134] and nearly 50% of the human genome [135]. Historically believed to simply self-propagate, it is now understood that these elements not only comprise a bulk of eukaryotic DNA but also contribute significantly to a wide range of regulatory functions within a genome. Unsilenced TEs have been observed to contain *cis*-regulatory sequences that, due to their motility, have been dispersed broadly throughout the human genome [136,137]. These *cis*-regulatory elements have been shown by several groups to act as promoters, enhancers, and repressors of transcription [138–142]. Using human and mouse cell lines, Sundaram et al. [136] found that 20% of transcription factor binding sites were embedded within transposable elements. Similarly, Cao et al. [142] identified widespread enhancer-like repeats throughout the human genome, many of which were enriched in the mammalian-wide interspersed repeat (MIR) family of short interspersed nuclear elements (SINEs) and the L2 family of long interspersed nuclear elements (LINEs). Moreover, Makarevitch et al. [143] suggest the potential for TEs to provide a mechanism for the upregulation of particular genetic transcripts following abiotic stress in maize via their enhancer-like activity. These studies represent just a fraction of the mounting evidence suggesting that TEs provide necessary regulatory functions within a genome (e.g., see [4,144,145] for reviews).

Despite the high frequency of transposable elements in both human and other eukaryotic genomes, the majority of transposable elements are not actively moving from one genomic locus to another. While mutations have rendered many transposable elements inactive, some have been epigenetically silenced through various mechanisms, such as post transcriptional modifications via RNA interference, DNA and chromatin modifications, and germline silencing. Epigenetic silencing prevents TEs from producing the proteins required for mobility despite a lack of change to the underlying DNA sequence (reviewed in [109,146,147]).

While satellite DNA is characteristic of centromeres across eukaryotic organisms, the surrounding regions of pericentric heterochromatin are often enriched in TE content. For example, while human centromere cores are enriched for tandem repeat stretches of α satellite DNA, the surrounding heterochromatin regions consist of shorter satellites (e.g., satellites I and II) and primarily two different types of retrotransposons: LINEs and SINEs. Emerging models of centromeric contigs have shown that TE insertions are also found within HOR arrays of the centromere core of all human chromosomes [148–150]. This characteristic, coupled with the observations that TEs are often found at both neocentromeres [99,151,152] and ENCs that are devoid of any satellite content [117,118,153,154], suggests a potential role for TEs in centromere function independent of resident satellite DNA.

While the exact role TEs play in centromere function is not currently known, several features of centromeric TEs have been revealed. For example, epigenetic silencing of transposable elements appears to be critical in maintaining proper centromere function and chromosome segregation [147,155,156]. In mice, activating regularly silenced long terminal repeat (LTR) and non-LTR retrotransposons at centromeric regions has led to defects in both meiosis and chromosomal segregation, suggesting the necessity of epigenetically silent transposable elements for appropriate centromere function [156]. Undermethylation of centromeric retroelements in interspecific hybrids led to centromere destabilization and chromosome instability [157,158], indicating that tight regulation of TE activity underlies centromere stability. Moreover, studies have suggested a link between centromeric retrotransposons and the silencing of satellite DNA in the centromere, as well as a link between satellite DNA and the silencing of retrotransposons. May et al. [159] describe this relationship in *Arabidopsis thaliana*, in which satellite-derived transcripts are epigenetically silenced in part due to the insertion of transposable elements. Phylogenetic analyses and TE annotations have led to the observation that species-specific [118] and recently active [160] or hot TEs [152,161] are often the type of element found within centromere cores, while divergent and ancestral TEs are relegated to the hypermutated satellites [30,40,149] found in the outer strata of the centromere and pericentromere (Figure 1a,b).

It has been suggested that a close evolutionary relationship exists between centromeric TEs and the birth of new satellite families (Figure 3a). In the plant species *Aegilops speltoides*, a 250 base pair repeat satellite array family is present at centromeres [162]. While not identical to that of a transposable element, this satellite DNA sequence shares high similarity to portions of a transposable element: Ty3/gypsy-like retrotransposons. Furthermore, this phenomenon has been observed in other model species as well, including members of the *Arabidopsis* [163], *Drosophila* [164], and *Cetacean* [165] genera. A recent study observed that tandem dimers of TEs form during bursts of TE activity and may serve as fodder for the evolution of satellite arrays, as was found for the *hobo* element in Drosophila [166]. It has been proposed that large-scale mutations, insertions and deletions within centromeric TEs followed by unequal crossing over or even seeding across chromosomes, may give rise to novel tandem repetitive elements found highly enriched at centromeres [112,167–170]. These processes are thought to act as part of the host-defense mechanisms to inactivate mobile elements ([171,172] but see [173]) or prevent non-allelic homologous recombination ([174,175] and reviewed in [109,176,177]).

Within plants, allopolyploidy presents a unique opportunity for the evolution of centromeric sequences from resident TEs. Following allopolyploidization, and during the genomic instability that ensues, centromeric TEs from the different progenitor genomes may become activated [178]. Evidence has been found in *Gossypium* (cottons) that such activation likely occurred, resulting in the integration of TEs from one genome into another, and subsequent proliferation among centromeres [179]. This activity, coupled with the exposure of new genomic material in the polyploid state, provides an opportunity for competition among multiple, newly emerged centromere repeats and the possible selection for repeats that are more conducive to supporting centromere nucleosome structure [180].

5. Transcription in the Centromere—Let's Get the Party Started!

While a function for satellite DNAs in kinetochore assembly and/or stability has been inferred since their discovery (e.g., [181–183]), a common misconception has been that these sequences were not actively transcribed into RNAs (but see [184] and references therein from the 1960s). Undoubtedly, the discovery that satellite DNAs are transcriptionally viable has led to a shift in how we view centromeric chromatin [185,186]. Soon after the discovery of satellite DNA in cesium gradients, electron microscopy revealed RNA at plant and animal centromeres [187,188], although satellites themselves were not directly linked to active transcription. Furthermore, examination of the linear organization of histones within centromeres using chromatin fiber FISH revealed that CENP-A nucleosome domains were interrupted by nucleosomes containing H3K4me2 [189,190] and H3K36me2 [191], epigenetic marks of active transcription. Using chromatin immunoprecipitation, Choi et al. reported the detection of RNA polymerase II at centromeres in the fission yeast *Schizosaccharomyces pombe* [192]. Further analyses have identified the presence of RNA polymerase II at centromeres in humans [193], flies [194,195], and budding yeast [196], among others. The presence of RNA polymerase II at these sites suggested active transcription occurring from the DNA present at centromeres: satellites and TEs.

Transcripts originating from centromeric satellite DNA and TEs have now been observed in a variety of species across eukaryotes [122,159,186,194,196–204], while some species, such as *S. pombe*, exhibit transcription of boundary elements (e.g., tRNAs [205]). Thus, centromeric RNAs are a conserved component of the centromere, despite a lack of sequence conservation across these regions. Recent work implicates centromere transcription as integral to centromere function, impacting the pivotal event in centromere assembly: the loading of newly synthesized CENP-A histones. For example, in budding yeast [196] and human artificial chromosomes [191,206,207], transcriptional silencing of centromeric DNA has been shown to lead to a failure to maintain proper centromere function. In human artificial chromosomes, this malfunction was attributed to the inability to load new CENP-A during mitosis to G1. Conversely, upregulation of satellite transcripts is also detrimental to centromere function, causing the removal of the CENP-A histone variant [196,200,207,208] as well as cellular instability [196,200,209–213]. Intriguingly, several proteins involved in the kinetochore assembly cascade are either RNA binding proteins or have been demonstrated to associate with RNAs in a complex, including CENP-A [199], CENP-C [202,214], and KNL2/M18BP1 [215]. While the transcriptional framework underlying centromere assembly is not fully understood (but see [109,186]), several mechanisms have been proposed that can promote transcription within regional centromeres. Early work in plants [199] and marsupials [122] supported the hypothesis that centromeric TEs promote transcription, and their ability to transcribe neighboring satellites is implied by the presence of bi-directional promoters within these TEs [200,216–218]. More recently, it has been hypothesized that non-B form DNA facilitated by dyad symmetries and CENP-B binding within centromeres may facilitate transcription [219]. While there is a clear connection between transcription and centromere nucleosome assembly [194,195,199,200,220,221], how and when this occurs during the cell cycle remains elusive.

Transcription has also been linked to the emergence of new centromeres. In a human neocentromere case, a L1 was found transcribed and actively demarcating the CENP-A domain of the new centromere [152,222]. Given earlier work demonstrating that demethylation of centromeric TEs led to increased activity [158], release of ectopic TEs from a silenced state may facilitate their transcription and subsequent recruitment of CENP-A nucleosomes, leading to the rescue of acentric chromosome fragments following the inactivation of the native centromere. How an ectopic site becomes activated, enabling the recruitment of CENP-A nucleosomes in the absence of chromosome damage, as is implied by centromere repositioning events, is unknown. If multiple inversion events, insertion events by active TEs, or simply deletions of part of an HOR array lead to the interruption of the native satellite array, destabilization of the kinetochore assembly cascade may follow, necessitating a rescue centromere elsewhere on the chromosome. Perhaps the most recent TE insertions in a genome allow ectopic centromere formation as such elements have yet to experience silencing by host defense mechanisms. Under this model, some mechanism must prevent the activation of ectopic centromeres at these hot elements when native centromeres are still functional to prevent the formation of dicentric chromosomes and subsequent breakage-fusion-bridge cycles [223–225].

6. Conclusions

The influence of repeated DNAs on eukaryotic genomes is often presented in the framework of the logical fallacy that repeated DNA should no longer be considered inconsequential 'junk DNA'. Contextualizing repeated DNAs under such as false descriptor, even when presented as an oft challenged and subsequently defeated cliché, undercuts not only the long-standing validity of studying repeated DNAs, but the growing impact the field of repeat DNA biology has had on our understanding of eukaryotic genome biology and evolution. The repeats found at centromeres are an excellent case in point. There is little doubt that centromeric repeats, including both satellites and TEs, are integral to centromere function and stability as well as the evolution of novel karyotypes. The models discussed herein are not all-inclusive yet demonstrate the unique processes that have allowed for significant species-specific variation among repetitive DNAs despite a simultaneously foundational role in genome stability and regulation. As we gain an understanding of the evolutionary forces that influence the constitution of centromeric DNA, we can start to unravel the impact centromeric sequences have on both maintaining chromosome stability within a species and karyotypic change during species evolution.

Author Contributions: Contributions to this review are as follows: conceptualization, writing and editing, G.H. and R.O.; funding acquisition, R.O.

Funding: This research was funded by the National Science Foundation, grant number 1613806.

Acknowledgments: The authors thank Kate Castellano, Judy Brown, and Mike O'Neill for critical comments on the manuscript. Chromosome ideogram schematic in Figures 1 and 2 made by Mysid, based on http://ghr.nlm.nih.gov/.

Conflicts of Interest: The authors declare no conflict of interest. The funders had no role in the design of the study; in the collection, analyses, or interpretation of data; in the writing of the manuscript, or in the decision to publish the results.

References

1. Biscotti, M.A.; Olmo, E.; Heslop-Harrison, J.S. Repetitive DNA in eukaryotic genomes. *Chromosome Res.* **2015**, *23*, 415–420. [CrossRef] [PubMed]
2. Charlesworth, B.; Sniegowski, P.; Stephan, W. The evolutionary dynamics of repetitive DNA in eukaryotes. *Nature* **1994**, *371*, 215–220. [CrossRef] [PubMed]
3. Lopez-Flores, I.; Garrido-Ramos, M.A. The repetitive DNA content of eukaryotic genomes. *Genome Dyn.* **2012**, *7*, 1–28. [CrossRef] [PubMed]
4. Bourque, G.; Burns, K.H.; Gehring, M.; Gorbunova, V.; Seluanov, A.; Hammell, M.; Imbeault, M.; Izsvak, Z.; Levin, H.L.; Macfarlan, T.S.; et al. Ten things you should know about transposable elements. *Genome Biol.* **2018**, *19*, 199. [CrossRef] [PubMed]
5. Garrido-Ramos, M.A. Satellite DNA: An Evolving Topic. *Genes* **2017**, *8*, 230. [CrossRef] [PubMed]
6. Sullivan, L.L.; Chew, K.; Sullivan, B.A. α satellite DNA variation and function of the human centromere. *Nucleus* **2017**, *8*, 331–339. [CrossRef]
7. Jagannathan, M.; Cummings, R.; Yamashita, Y.M. A conserved function for pericentromeric satellite DNA. *eLife* **2018**, *7*, e34122. [CrossRef]
8. Kit, S. Equilibrium sedimentation in density gradients of DNA preparations from animal tissues. *J. Mol. Biol.* **1961**, *3*, 711–716. [CrossRef]
9. Sueoka, N. Variation and heterogeneity of base composition of deoxyribonucleic acids: A compilation of old and new data. *J. Mol. Biol.* **1961**, *3*, 31–40. [CrossRef]
10. Navashin, S. On the nuclear dimorphism in somatic cells of *Galtonia candicans*. *Bull. Acad. Imp. Sci* **1912**, *6*, 375–385.
11. Singer, M.F. Highly repeated sequences in mammalian genomes. *Int. Rev. Cytol.* **1982**, *76*, 67–112. [PubMed]
12. Waring, M.; Britten, R.J. Nucleotide sequence repetition: A rapidly reassociating fraction of mouse DNA. *Science* **1966**, *154*, 791–794. [CrossRef]
13. Horz, W.; Zachau, H.G. Characterization of distinct segments in mouse satellite DNA by restriction nucleases. *Eur. J. Biochem.* **1977**, *73*, 383–392. [CrossRef] [PubMed]
14. Vieira, M.L.C.; Santini, L.; Diniz, A.L.; Munhoz Cde, F. Microsatellite markers: What they mean and why they are so useful. *Genet. Mol. Biol.* **2016**, *39*, 312–328. [CrossRef] [PubMed]
15. Subramanian, S.; Mishra, R.K.; Singh, L. Genome-wide analysis of microsatellite repeats in humans: Their abundance and density in specific genomic regions. *Genome Biol.* **2003**, *4*, R13. [CrossRef] [PubMed]
16. Ramel, C. Mini- and microsatellites. *Environ Health Perspect* **1997**, *105*, 781–789. [PubMed]

17. Naslund, K.; Saetre, P.; von Salome, J.; Bergstrom, T.F.; Jareborg, N.; Jazin, E. Genome-wide prediction of human VNTRs. *Genomics* **2005**, *85*, 24–35. [CrossRef] [PubMed]
18. Vergnaud, G.; Gauguier, D.; Schott, J.J.; Lepetit, D.; Lauthier, V.; Mariat, D.; Buard, J. Detection, cloning, and distribution of minisatellites in some mammalian genomes. *EXS* **1993**, *67*, 47–57.
19. O'Sullivan, R.J.; Karlseder, J. Telomeres: Protecting chromosomes against genome instability. *Nat. Rev. Mol. Cell Biol.* **2010**, *11*, 171. [CrossRef]
20. Moyzis, R.K.; Buckingham, J.M.; Cram, L.S.; Dani, M.; Deaven, L.L.; Jones, M.D.; Meyne, J.; Ratliff, R.L.; Wu, J.R. A highly conserved repetitive DNA sequence, (TTAGGG)n, present at the telomeres of human chromosomes. *Proc. Natl. Acad. Sci. USA* **1988**, *85*, 6622–6626. [CrossRef]
21. Bandaria, J.N.; Qin, P.; Berk, V.; Chu, S.; Yildiz, A. Shelterin protects chromosome ends by compacting telomeric chromatin. *Cell* **2016**, *164*, 735–746. [CrossRef]
22. Wyatt, H.D.; West, S.C.; Beattie, T.L. InTERTpreting telomerase structure and function. *Nucleic Acids Res.* **2010**, *38*, 5609–5622. [CrossRef] [PubMed]
23. Cusanelli, E.; Chartrand, P. Telomeric repeat-containing RNA TERRA: A noncoding RNA connecting telomere biology to genome integrity. *Front. Genet.* **2015**, *6*, 143. [CrossRef]
24. Maddar, H.; Ratzkovsky, N.; Krauskopf, A. Role for telomere cap structure in meiosis. *Mol. Biol. Cell.* **2001**, *12*, 3191–3203. [CrossRef] [PubMed]
25. Willard, H.F. Chromosome-specific organization of human α satellite DNA. *Am. J. Hum. Genet.* **1985**, *37*, 524–532. [PubMed]
26. Van Hooser, A.A.; Ouspenski, I.I.; Gregson, H.C.; Starr, D.A.; Yen, T.J.; Goldberg, M.L.; Yokomori, K.; Earnshaw, W.C.; Sullivan, K.F.; Brinkley, B.R. Specification of kinetochore-forming chromatin by the histone H3 variant CENP-A. *J. Cell Sci.* **2001**, *114*, 3529–3542. [PubMed]
27. Willard, H.F.; Waye, J.S.; Skolnick, M.H.; Schwartz, C.E.; Powers, V.E.; England, S.B. Detection of restriction fragment length polymorphisms at the centromeres of human chromosomes by using chromosome-specific α satellite DNA probes: Implications for development of centromere-based genetic linkage maps. *Proc. Natl. Acad. Sci. USA* **1986**, *83*, 5611–5615. [CrossRef]
28. Alexandrov, I.; Kazakov, A.; Tumeneva, I.; Shepelev, V.; Yurov, Y. α-satellite DNA of primates: Old and new families. *Chromosoma* **2001**, *110*, 253–266. [CrossRef] [PubMed]
29. Alexandrov, I.A.; Medvedev, L.I.; Mashkova, T.D.; Kisselev, L.L.; Romanova, L.Y.; Yurov, Y.B. Definition of a new α satellite suprachromosomal family characterized by monomeric organization. *Nucleic Acids Res.* **1993**, *21*, 2209–2215. [CrossRef] [PubMed]
30. Shepelev, V.A.; Uralsky, L.I.; Alexandrov, A.A.; Yurov, Y.B.; Rogaev, E.I.; Alexandrov, I.A. Annotation of suprachromosomal families reveals uncommon types of α satellite organization in pericentromeric regions of hg38 human genome assembly. *Genom. Data* **2015**, *5*, 139–146. [CrossRef]
31. McNulty, S.M.; Sullivan, B.A. α satellite DNA biology: Finding function in the recesses of the genome. *Chromosome Res* **2018**, *26*, 115–138. [CrossRef] [PubMed]
32. Carine, K.; Jacquemin-Sablon, A.; Waltzer, E.; Mascarello, J.; Scheffler, I.E. Molecular characterization of human minichromosomes with centromere from chromosome 1 in human-hamster hybrid cells. *Somat. Cell Mol. Genet.* **1989**, *15*, 445–460. [CrossRef] [PubMed]
33. Waye, J.S.; England, S.B.; Willard, H.F. Genomic organization of α satellite DNA on human chromosome 7: Evidence for two distinct alphoid domains on a single chromosome. *Mol. Cell. Biol.* **1987**, *7*, 349–356. [CrossRef]
34. Tyler-Smith, C.; Brown, W.R. Structure of the major block of alphoid satellite DNA on the human Y chromosome. *J. Mol. Biol.* **1987**, *195*, 457–470. [CrossRef]
35. Roizès, G. Human centromeric alphoid domains are periodically homogenized so that they vary substantially between homologues. Mechanism and implications for centromere functioning. *Nucleic Acids Res.* **2006**, *34*, 1912–1924. [CrossRef] [PubMed]
36. Aldrup-Macdonald, M.E.; Sullivan, B.A. The past, present, and future of human centromere genomics. *Genes* **2014**, *5*, 33–50. [CrossRef]
37. Alexandrov, I.A.; Mashkova, T.D.; Akopian, T.A.; Medvedev, L.I.; Kisselev, L.L.; Mitkevich, S.P.; Yurov, Y.B. Chromosome-specific α satellites: Two distinct families on human chromosome 18. *Genomics* **1991**, *11*, 15–23. [CrossRef]

38. Alexandrov, I.A.; Mitkevich, S.P.; Yurov, Y.B. The phylogeny of human chromosome specific α satellites. *Chromosoma* **1988**, *96*, 443–453. [CrossRef]
39. Rosandic, M.; Paar, V.; Basar, I.; Gluncic, M.; Pavin, N.; Pilas, I. CENP-B box and pJalpha sequence distribution in human α satellite higher-order repeats (HOR). *Chromosome Res* **2006**, *14*, 735–753. [CrossRef]
40. Shepelev, V.A.; Alexandrov, A.A.; Yurov, Y.B.; Alexandrov, I.A. The evolutionary origin of man can be traced in the layers of defunct ancestral α satellites flanking the active centromeres of human chromosomes. *PLoS Genet.* **2009**, *5*, e1000641. [CrossRef]
41. Aldrup-MacDonald, M.E.; Kuo, M.E.; Sullivan, L.L.; Chew, K.; Sullivan, B.A. Genomic variation within α satellite DNA influences centromere location on human chromosomes with metastable epialleles. *Genome Res.* **2016**, *26*, 1301–1311. [CrossRef] [PubMed]
42. Miga, K.H.; Newton, Y.; Jain, M.; Altemose, N.; Willard, H.F.; Kent, W.J. Centromere reference models for human chromosomes X and Y satellite arrays. *Genome Res.* **2014**, *24*, 697–707. [CrossRef] [PubMed]
43. Waye, J.S.; Willard, H.F. Structure, organization, and sequence of α satellite DNA from human chromosome 17: Evidence for evolution by unequal crossing-over and an ancestral pentamer repeat shared with the human X chromosome. *Mol. Cell. Biol.* **1986**, *6*, 3156–3165. [CrossRef]
44. Rudd, M.K.; Willard, H.F. Analysis of the centromeric regions of the human genome assembly. *Trends Genet.* **2004**, *20*, 529–533. [CrossRef]
45. Warburton, P.E.; Willard, H.F. Interhomologue sequence variation of α satellite DNA from human chromosome 17: Evidence for concerted evolution along haplotypic lineages. *J. Mol. Evol.* **1995**, *41*, 1006–1015. [CrossRef] [PubMed]
46. Maloney, K.A.; Sullivan, L.L.; Matheny, J.E.; Strome, E.D.; Merrett, S.L.; Ferris, A.; Sullivan, B.A. Functional epialleles at an endogenous human centromere. *Proc. Natl. Acad. Sci. USA* **2012**, *109*, 13704–13709. [CrossRef]
47. Jain, M.; Olsen, H.E.; Turner, D.J.; Stoddart, D.; Bulazel, K.V.; Paten, B.; Haussler, D.; Willard, H.F.; Akeson, M.; Miga, K.H. Linear assembly of a human centromere on the Y chromosome. *Nat. Biotechnol.* **2018**, *36*, 321–323. [CrossRef] [PubMed]
48. Durfy, S.J.; Willard, H.F. Molecular analysis of a polymorphic domain of α satellite from the human X chromosome. *Am. J. Hum. Genet.* **1987**, *41*, 391–401. [PubMed]
49. Schindelhauer, D.; Schwarz, T. Evidence for a fast, intrachromosomal conversion mechanism from mapping of nucleotide variants within a homogeneous α-satellite DNA array. *Genome Res.* **2002**, *12*, 1815–1826. [CrossRef]
50. Alkan, C.; Cardone, M.F.; Catacchio, C.R.; Antonacci, F.; O'Brien, S.J.; Ryder, O.A.; Purgato, S.; Zoli, M.; Della Valle, G.; Eichler, E.E.; et al. Genome-wide characterization of centromeric satellites from multiple mammalian genomes. *Genome Res.* **2011**, *21*, 137–145. [CrossRef]
51. Henikoff, S.; Ahmad, K.; Malik, H. The centromere paradox: Stable inheritance with rapidly evolving DNA. *Science* **2001**, *293*, 1098–1102. [CrossRef] [PubMed]
52. Malik, H.S.; Henikoff, S. Conflict begets complexity: The evolution of centromeres. *Curr. Opin. Genet. Dev.* **2002**, *12*, 711–718. [CrossRef]
53. Melters, D.P.; Bradnam, K.R.; Young, H.A.; Telis, N.; May, M.R.; Ruby, J.G.; Sebra, R.; Peluso, P.; Eid, J.; Rank, D.; et al. Comparative analysis of tandem repeats from hundreds of species reveals unique insights into centromere evolution. *Genome Biol.* **2013**, *14*, R10. [CrossRef]
54. Plohl, M.; Mestrovic, N.; Mravinac, B. Centromere identity from the DNA point of view. *Chromosoma* **2014**, *123*, 313–325. [CrossRef]
55. Plohl, M.; Mestrovic, N.; Mravinac, B. Satellite DNA evolution. *Genome Dyn.* **2012**, *7*, 126–152. [CrossRef] [PubMed]
56. Dover, G. Molecular drive: A cohesive mode of species evolution. *Nature* **1982**, *299*, 111–117. [CrossRef] [PubMed]
57. Smith, G.P. Evolution of repeated DNA sequences by unequal crossover. *Science* **1976**, *191*, 528–535. [CrossRef]
58. Walsh, J.B. Persistence of tandem arrays: Implications for satellite and simple-sequence DNAs. *Genetics* **1987**, *115*, 553–567.
59. Shi, J.; Wolf, S.E.; Burke, J.M.; Presting, G.G.; Ross-Ibarra, J.; Dawe, R.K. Widespread gene conversion in centromere cores. *PLoS Biol.* **2010**, *8*, e1000327. [CrossRef]

60. Bertelsen, A.H.; Humayun, M.Z.; Karfopoulos, S.G.; Rush, M.G. Molecular characterization of small polydisperse circular deoxyribonucleic acid from an African green monkey cell line. *Biochemistry* **1982**, *21*, 2076–2085. [CrossRef]
61. Gaubatz, J.W. Extrachromosomal circular DNAs and genomic sequence plasticity in eukaryotic cells. *Mutat. Res.* **1990**, *237*, 271–292. [CrossRef]
62. Baldini, A.; Smith, D.I.; Rocchi, M.; Miller, O.J.; Miller, D.A. A human alphoid DNA clone from the EcoRI dimeric family: Genomic and internal organization and chromosomal assignment. *Genomics* **1989**, *5*, 822–828. [CrossRef]
63. Pironon, N.; Puechberty, J.; Roizès, G. Molecular and evolutionary characteristics of the fraction of human α satellite DNA associated with CENP-A at the centromeres of chromosomes 1, 5, 19, and 21. *BMC Genom.* **2010**, *11*, 195. [CrossRef]
64. Greig, G.M.; Warburton, P.E.; Willard, H.F. Organization and evolution of an α satellite DNA subset shared by human chromosomes 13 and 21. *J. Mol. Evol.* **1993**, *37*, 464–475. [CrossRef] [PubMed]
65. Jorgensen, A.L.; Kolvraa, S.; Jones, C.; Bak, A.L. A subfamily of alphoid repetitive DNA shared by the NOR-bearing human chromosomes 14 and 22. *Genomics* **1988**, *3*, 100–109. [CrossRef]
66. Jorgensen, A.L.; Laursen, H.B.; Jones, C.; Bak, A.L. Evolutionarily different alphoid repeat DNA on homologous chromosomes in human and chimpanzee. *Proc. Natl. Acad. Sci. USA* **1992**, *89*, 3310–3314. [CrossRef]
67. Sabourin, M.; Nitiss, J.L.; Nitiss, K.C.; Tatebayashi, K.; Ikeda, H.; Osheroff, N. Yeast recombination pathways triggered by topoisomerase II-mediated DNA breaks. *Nucleic Acids Res.* **2003**, *31*, 4373–4384. [CrossRef] [PubMed]
68. Salser, W.; Bowen, S.; Browne, D.; el-Adli, F.; Fedoroff, N.; Fry, K.; Heindell, H.; Paddock, G.; Poon, R.; Wallace, B.; et al. Investigation of the organization of mammalian chromosomes at the DNA sequence level. *Fed. Proc.* **1976**, *35*, 23–35.
69. Cacheux, L.; Ponger, L.; Gerbault-Seureau, M.; Loll, F.; Gey, D.; Richard, F.A.; Escude, C. The targeted sequencing of α satellite DNA in *Cercopithecus pogonias* provides new insight into the diversity and dynamics of centromeric repeats in old world monkeys. *Genome Biol. Evol.* **2018**, *10*, 1837–1851. [CrossRef] [PubMed]
70. da Silva, E.L.; Busso, A.F.; Parise-Maltempi, P.P. Characterization and genome organization of a repetitive element associated with the nucleolus organizer region in Leporinus elongatus (*Anostomidae: Characiformes*). *Cytogenet. Genome Res.* **2013**, *139*, 22–28. [CrossRef] [PubMed]
71. Mestrovic, N.; Plohl, M.; Mravinac, B.; Ugarkovic, D. Evolution of satellite DNAs from the genus Palorus–experimental evidence for the "library" hypothesis. *Mol. Biol. Evol.* **1998**, *15*, 1062–1068. [CrossRef] [PubMed]
72. Lee, H.R.; Zhang, W.; Langdon, T.; Jin, W.; Yan, H.; Cheng, Z.; Jiang, J. Chromatin immunoprecipitation cloning reveals rapid evolutionary patterns of centromeric DNA in Oryza species. *Proc. Natl. Acad. Sci. USA* **2005**, *102*, 11793–11798. [CrossRef] [PubMed]
73. Faravelli, M.; Moralli, D.; Bertoni, L.; Attolini, C.; Chernova, O.; Raimondi, E.; Giulotto, E. Two extended arrays of a satellite DNA sequence at the centromere and at the short-arm telomere of Chinese hamster chromosome 5. *Cytogenet. Cell Genet.* **1998**, *83*, 281–286. [CrossRef]
74. Gong, Z.; Wu, Y.; Koblizkova, A.; Torres, G.A.; Wang, K.; Iovene, M.; Neumann, P.; Zhang, W.; Novak, P.; Buell, C.R.; et al. Repeatless and repeat-based centromeres in potato: Implications for centromere evolution. *Plant Cell* **2012**, *24*, 3559–3574. [CrossRef] [PubMed]
75. Bulazel, K.; Ferreri, G.C.; Eldridge, M.D.; O'Neill, R.J. Species-specific shifts in centromere sequence composition are coincident with breakpoint reuse in karyotypically divergent lineages. *Genome Biol.* **2007**, *8*, R170. [CrossRef] [PubMed]
76. Chaves, R.; Adega, F.; Heslop-Harrison, J.S.; Guedes-Pinto, H.; Wienberg, J. Complex satellite DNA reshuffling in the polymorphic t(1;29) Robertsonian translocation and evolutionarily derived chromosomes in cattle. *Chromosome Res.* **2003**, *11*, 641–648. [CrossRef] [PubMed]
77. Chaves, R.; Guedes-Pinto, H.; Heslop-Harrison, J.; Schwarzacher, T. The species and chromosomal distribution of the centromeric α-satellite I sequence from sheep in the tribe Caprini and other Bovidae. *Cytogenet. Cell Genet.* **2000**, *91*, 62–66. [CrossRef] [PubMed]
78. Malik, H.S. The centromere-drive hypothesis: A simple basis for centromere complexity. *Prog. Mol. Subcell. Biol.* **2009**, *48*, 33–52. [CrossRef]

79. Iwata-Otsubo, A.; Dawicki-McKenna, J.M.; Akera, T.; Falk, S.J.; Chmatal, L.; Yang, K.; Sullivan, B.A.; Schultz, R.M.; Lampson, M.A.; Black, B.E. Expanded satellite repeats amplify a discrete CENP-A nucleosome assembly site on chromosomes that drive in female meiosis. *Curr. Biol.* **2017**, *27*, 2365–2373.e8. [CrossRef]
80. Drpic, D.; Almeida, A.C.; Aguiar, P.; Renda, F.; Damas, J.; Lewin, H.A.; Larkin, D.M.; Khodjakov, A.; Maiato, H. Chromosome segregation is biased by kinetochore size. *Curr. Biol.* **2018**, *28*, 1344–1356. [CrossRef]
81. Zwick, M.E.; Salstrom, J.L.; Langley, C.H. Genetic variation in rates of nondisjunction: Association of two naturally occurring polymorphisms in the chromokinesin nod with increased rates of nondisjunction in *Drosophila melanogaster*. *Genetics* **1999**, *152*, 1605–1614. [PubMed]
82. Hirsch, C.D.; Wu, Y.; Yan, H.; Jiang, J. Lineage-specific adaptive evolution of the centromeric protein CENH3 in diploid and allotetraploid Oryza species. *Mol. Biol. Evol.* **2009**, *26*, 2877–2885. [CrossRef] [PubMed]
83. Schueler, M.G.; Swanson, W.; Thomas, P.J.; Green, E.D. Adaptive evolution of foundation kinetochore proteins in primates. *Mol. Biol. Evol.* **2010**, *27*, 1585–1597. [CrossRef] [PubMed]
84. Talbert, P.B.; Masuelli, R.; Tyagi, A.P.; Comai, L.; Henikoff, S. Centromeric localization and adaptive evolution of an *Arabidopsis* histone H3 variant. *Plant Cell* **2002**, *14*, 1053–1066. [CrossRef] [PubMed]
85. Talbert, P.B.; Bryson, T.D.; Henikoff, S. Adaptive evolution of centromere proteins in plants and animals. *J. Biol.* **2004**, *3*, 18. [CrossRef] [PubMed]
86. Zedek, F.; Bures, P. Evidence for centromere drive in the holocentric chromosomes of Caenorhabditis. *PLoS ONE* **2012**, *7*, e30496. [CrossRef] [PubMed]
87. Chmatal, L.; Gabriel, S.I.; Mitsainas, G.P.; Martinez-Vargas, J.; Ventura, J.; Searle, J.B.; Schultz, R.M.; Lampson, M.A. Centromere strength provides the cell biological basis for meiotic drive and karyotype evolution in mice. *Curr. Biol.* **2014**, *24*, 2295–2300. [CrossRef] [PubMed]
88. Pardo-Manuel de Villena, F.; Sapienza, C. Female meiosis drives karyotypic evolution in mammals. *Genetics* **2001**, *159*, 1179–1189. [PubMed]
89. Rosin, L.F.; Mellone, B.G. Centromeres drive a hard bargain. *Trends Genet.* **2017**, *33*, 101–117. [CrossRef] [PubMed]
90. Scott, K.C.; Sullivan, B.A. Neocentromeres: A place for everything and everything in its place. *Trends Genet.* **2014**, *30*, 66–74. [CrossRef]
91. Garsed, D.W.; Marshall, O.J.; Corbin, V.D.; Hsu, A.; Di Stefano, L.; Schroder, J.; Li, J.; Feng, Z.P.; Kim, B.W.; Kowarsky, M.; et al. The architecture and evolution of cancer neochromosomes. *Cancer Cell* **2014**, *26*, 653–667. [CrossRef] [PubMed]
92. Schubert, I. What is behind "centromere repositioning"? *Chromosoma* **2018**, *127*, 229–234. [CrossRef] [PubMed]
93. Rhoades, M.M.; Vilkomerson, H. On the anaphase movement of chromosomes. *Proc. Natl. Acad. Sci. USA* **1942**, *28*, 433–436. [CrossRef] [PubMed]
94. Alonso, A.; Fritz, B.; Hasson, D.; Abrusan, G.; Cheung, F.; Yoda, K.; Radlwimmer, B.; Ladurner, A.G.; Warburton, P.E. Co-localization of CENP-C and CENP-H to discontinuous domains of CENP-A chromatin at human neocentromeres. *Genome Biol.* **2007**, *8*, R148. [CrossRef] [PubMed]
95. Alonso, A.; Mahmood, R.; Li, S.; Cheung, F.; Yoda, K.; Warburton, P.E. Genomic microarray analysis reveals distinct locations for the CENP-A binding domains in three human chromosome 13q32 neocentromeres. *Hum. Mol. Genet.* **2003**, *12*, 2711–2721. [CrossRef] [PubMed]
96. Voullaire, L.E.; Slater, H.R.; Petrovic, V.; Choo, K.H. A functional marker centromere with no detectable α-satellite, satellite III, or CENP-B protein: Activation of a latent centromere? *Am. J. Hum. Genet.* **1993**, *52*, 1153–1163.
97. du Sart, D.; Cancilla, M.R.; Earle, E.; Mao, J.I.; Saffery, R.; Tainton, K.M.; Kalitsis, P.; Martyn, J.; Barry, A.E.; Choo, K.H. A functional neo-centromere formed through activation of a latent human centromere and consisting of non-α-satellite DNA. *Nat. Genet.* **1997**, *16*, 144–153. [CrossRef] [PubMed]
98. Amor, D.J.; Bentley, K.; Ryan, J.; Perry, J.; Wong, L.; Slater, H.; Choo, K.H. Human centromere repositioning "in progress". *Proc. Natl. Acad. Sci. USA* **2004**, *101*, 6542–6547. [CrossRef] [PubMed]
99. Barry, A.E.; Howman, E.V.; Cancilla, M.R.; Saffery, R.; Choo, K.H. Sequence analysis of an 80 kb human neocentromere. *Hum. Mol. Genet.* **1999**, *8*, 217–227. [CrossRef]
100. Amor, D.J.; Choo, K.H.A. Neocentromeres: Role in human disease, evolution, and centromere study. *Am. J. Hum. Genet.* **2002**, *71*, 695–714. [CrossRef]

101. Warburton, P.E. Chromosomal dynamics of human neocentromere formation. *Chromosome Res.* **2004**, *12*, 617–626. [CrossRef] [PubMed]
102. Eldridge, M.D.; Close, R.L. Radiation of chromosome shuffles. *Curr. Opin. Genet. Dev.* **1993**, *3*, 915–922. [CrossRef]
103. Suja, J.A.; Camacho, J.P.M.; Cabrero, J.; Rufas, J.S. Analysis of a centric shift in the S11 chromosome of *Aiolopus strepens* (Orthoptera: Acrididae). *Genetica* **1986**, *70*, 211–216. [CrossRef]
104. Ventura, M.; Archidiacono, N.; Rocchi, M. Centromere emergence in evolution. *Genome Res.* **2001**, *11*, 595–599. [CrossRef]
105. Iannuzzi, L.; Di Meo, G.P.; Perucatti, A.; Incarnato, D.; Schibler, L.; Cribiu, E.P. Comparative FISH mapping of bovid X chromosomes reveals homologies and divergences between the subfamilies bovinae and caprinae. *Cytogenet. Cell Genet.* **2000**, *89*, 171–176. [CrossRef] [PubMed]
106. Wang, K.; Wu, Y.; Zhang, W.; Dawe, R.K.; Jiang, J. Maize centromeres expand and adopt a uniform size in the genetic background of oat. *Genome Res.* **2014**, *24*, 107–116. [CrossRef] [PubMed]
107. Rothfels, K.H.; Mason, G.F. Achiasmate meiosis and centromere shift in Eusimulium aureum (Diptera-Simuliidae). *Chromosoma* **1975**, *51*, 111–124. [CrossRef] [PubMed]
108. Carbone, L.; Nergadze, S.G.; Magnani, E.; Misceo, D.; Francesca Cardone, M.; Roberto, R.; Bertoni, L.; Attolini, C.; Francesca Piras, M.; de Jong, P.; et al. Evolutionary movement of centromeres in horse, donkey, and zebra. *Genomics* **2006**, *87*, 777–782. [CrossRef]
109. Klein, S.J.; O'Neill, R.J. Transposable elements: Genome innovation, chromosome diversity, and centromere conflict. *Chromosome Res.* **2018**, *26*, 5–23. [CrossRef]
110. Nergadze, S.G.; Piras, F.M.; Gamba, R.; Corbo, M.; Cerutti, F.; McCarter, J.G.W.; Cappelletti, E.; Gozzo, F.; Harman, R.M.; Antczak, D.F.; et al. Birth, evolution, and transmission of satellite-free mammalian centromeric domains. *Genome Res.* **2018**, *28*, 789–799. [CrossRef]
111. Piras, F.M.; Nergadze, S.G.; Magnani, E.; Bertoni, L.; Attolini, C.; Khoriauli, L.; Raimondi, E.; Giulotto, E. Uncoupling of satellite DNA and centromeric function in the genus Equus. *PLoS Genet.* **2010**, *6*, e1000845. [CrossRef] [PubMed]
112. Birchler, J.A.; Presting, G.G. Retrotransposon insertion targeting: A mechanism for homogenization of centromere sequences on nonhomologous chromosomes. *Genes Dev.* **2012**, *26*, 638–640. [CrossRef] [PubMed]
113. Rocchi, M.; Archidiacono, N.; Schempp, W.; Capozzi, O.; Stanyon, R. Centromere repositioning in mammals. *Heredity* **2012**, *108*, 59–67. [CrossRef] [PubMed]
114. Rocchi, M.; Stanyon, R.; Archidiacono, N. Evolutionary new centromeres in primates. *Prog. Mol. Subcell. Biol.* **2009**, *48*, 103–152. [CrossRef] [PubMed]
115. O'Neill, R.J.; Eldridge, M.D.; Metcalfe, C.J. Centromere dynamics and chromosome evolution in marsupials. *J. Hered.* **2004**, *95*, 375–381. [CrossRef] [PubMed]
116. Wade, C.; Giulotto, E.; Sigurdsson, S.; Zoli, M.; Gnerre, S.; Imsland, F.; Lear, T.; Adelson, D.; Bailey, E.; Bellone, R.; et al. Genome sequence, comparative analysis and population genetics of the domestic horse (*Equus caballus*). *Science* **2009**, *326*, 865–867. [CrossRef] [PubMed]
117. Purgato, S.; Belloni, E.; Piras, F.M.; Zoli, M.; Badiale, C.; Cerutti, F.; Mazzagatti, A.; Perini, G.; Della Valle, G.; Nergadze, S.G.; et al. Centromere sliding on a mammalian chromosome. *Chromosoma* **2015**, *124*, 277–287. [CrossRef] [PubMed]
118. Johnson, R.N.; O'Meally, D.; Chen, Z.; Etherington, G.J.; Ho, S.Y.W.; Nash, W.J.; Grueber, C.E.; Cheng, Y.; Whittington, C.M.; Dennison, S.; et al. Adaptation and conservation insights from the koala genome. *Nat. Genet.* **2018**, *50*, 1102–1111. [CrossRef]
119. Carbone, L.; Harris, R.A.; Gnerre, S.; Veeramah, K.R.; Lorente-Galdos, B.; Huddleston, J.; Meyer, T.J.; Herrero, J.; Roos, C.; Aken, B.; et al. Gibbon genome and the fast karyotype evolution of small apes. *Nature* **2014**, *513*, 195–201. [CrossRef] [PubMed]
120. Nagaki, K.; Cheng, Z.; Ouyang, S.; Talbert, P.B.; Kim, M.; Jones, K.M.; Henikoff, S.; Buell, C.R.; Jiang, J. Sequencing of a rice centromere uncovers active genes. *Nat. Genet.* **2004**, *36*, 138–145. [CrossRef]
121. Renfree, M.B.; Papenfuss, A.T.; Deakin, J.E.; Lindsay, J.; Heider, T.; Belov, K.; Rens, W.; Waters, P.D.; Pharo, E.A.; Shaw, G.; et al. Genome sequence of an Australian kangaroo, Macropus eugenii, provides insight into the evolution of mammalian reproduction and development. *Genome Biol.* **2011**, *12*, R81. [CrossRef] [PubMed]

122. Carone, D.; Longo, M.; Ferreri, G.; Hall, L.; Harris, M.; Shook, N.; Bulazel, K.; Carone, B.; Obergfell, C.; O'Neill, M.; et al. A new class of retroviral and satellite encoded small RNAs emanates from mammalian centromeres. *Chromosoma* **2009**, *118*, 113–125. [CrossRef] [PubMed]
123. Sanyal, K.; Baum, M.; Carbon, J. Centromeric DNA sequences in the pathogenic yeast *Candida albicans* are all different and unique. *Proc. Natl. Acad. Sci. USA* **2004**, *101*, 11374–11379. [CrossRef] [PubMed]
124. Mishra, P.K.; Baum, M.; Carbon, J. Centromere size and position in *Candida albicans* are evolutionarily conserved independent of DNA sequence heterogeneity. *Mol. Genet. Genom.* **2007**, *278*, 455–465. [CrossRef] [PubMed]
125. Harrington, J.J.; Van Bokkelen, G.; Mays, R.W.; Gustashaw, K.; Willard, H.F. Formation of de novo centromeres and construction of first-generation human artificial microchromosomes. *Nat. Genet.* **1997**, *15*, 345–355. [CrossRef] [PubMed]
126. Ebersole, T.A.; Ross, A.; Clark, E.; McGill, N.; Schindelhauer, D.; Cooke, H.; Grimes, B. Mammalian artificial chromosome formation from circular alphoid input DNA does not require telomere repeats. *Hum. Mol. Genet.* **2000**, *9*, 1623–1631. [CrossRef] [PubMed]
127. Doolittle, W.F.; Sapienza, C. Selfish genes, the phenotype paradigm and genome evolution. *Nature* **1980**, *284*, 601–603. [CrossRef]
128. Orgel, L.E.; Crick, F.H.C. Selfish DNA: the ultimate parasite. *Nature* **1980**, *284*, 604–607. [CrossRef]
129. Ravindran, S. Barbara McClintock and the discovery of jumping genes. *Proc. Natl. Acad. Sci. USA* **2012**. [CrossRef]
130. Wessler, S.R. Transposable elements and the evolution of eukaryotic genomes. *Proc. Natl. Acad. Sci. USA* **2006**, *103*, 17600–17601. [CrossRef]
131. Craig, N.L.; Craigie, R.; Gellert, M.; Lambowitz, A.M. *Mobile DNA II*; American Society for Microbiology Press: Washington, DC, USA, 2002.
132. Howe, K.; Clark, M.D.; Torroja, C.F.; Torrance, J.; Berthelot, C.; Muffato, M.; Collins, J.E.; Humphray, S.; McLaren, K.; Matthews, L.; et al. The zebrafish reference genome sequence and its relationship to the human genome. *Nature* **2013**, *496*, 498–503. [CrossRef] [PubMed]
133. Sotero-Caio, C.G.; Platt, R.N.; Suh, A.; Ray, D.A. Evolution and diversity of transposable elements in vertebrate genomes. *Genome Biol. Evol.* **2017**, *9*, 161–177. [CrossRef] [PubMed]
134. Schnable, P.S.; Ware, D.; Fulton, R.S.; Stein, J.C.; Wei, F.; Pasternak, S.; Liang, C.; Zhang, J.; Fulton, L.; Graves, T.A.; et al. The B73 maize genome: Complexity, diversity, and dynamics. *Science* **2009**, *326*, 1112–1115. [CrossRef] [PubMed]
135. Mills, R.E.; Bennett, E.A.; Iskow, R.C.; Devine, S.E. Which transposable elements are active in the human genome? *Trends Genet.* **2007**, *23*, 183–191. [CrossRef] [PubMed]
136. Sundaram, V.; Cheng, Y.; Ma, Z.; Li, D.; Xing, X.; Edge, P.; Snyder, M.P.; Wang, T. Widespread contribution of transposable elements to the innovation of gene regulatory networks. *Genome Res.* **2014**, *24*, 1963–1976. [CrossRef] [PubMed]
137. Wittkopp, P.J.; Kalay, G. Cis-regulatory elements: Molecular mechanisms and evolutionary processes underlying divergence. *Nat. Rev. Genet.* **2011**, *13*, 59–69. [CrossRef]
138. Jacques, P.E.; Jeyakani, J.; Bourque, G. The majority of primate-specific regulatory sequences are derived from transposable elements. *PLoS Genet.* **2013**, *9*, e1003504. [CrossRef]
139. Bourque, G.; Leong, B.; Vega, V.B.; Chen, X.; Lee, Y.L.; Srinivasan, K.G.; Chew, J.L.; Ruan, Y.; Wei, C.L.; Ng, H.H.; et al. Evolution of the mammalian transcription factor binding repertoire via transposable elements. *Genome Res.* **2008**, *18*, 1752–1762. [CrossRef]
140. Wang, T.; Zeng, J.; Lowe, C.B.; Sellers, R.G.; Salama, S.R.; Yang, M.; Burgess, S.M.; Brachmann, R.K.; Haussler, D. Species-specific endogenous retroviruses shape the transcriptional network of the human tumor suppressor protein p53. *Proc. Natl. Acad. Sci. USA* **2007**, *104*, 18613–18618. [CrossRef]
141. Roman, A.C.; Benitez, D.A.; Carvajal-Gonzalez, J.M.; Fernandez-Salguero, P.M. Genome-wide B1 retrotransposon binds the transcription factors dioxin receptor and Slug and regulates gene expression in vivo. *Proc. Natl. Acad. Sci. USA* **2008**, *105*, 1632–1637. [CrossRef]
142. Cao, Y.; Chen, G.; Wu, G.; Zhang, X.; McDermott, J.; Chen, X.; Xu, C.; Jiang, Q.; Chen, Z.; Zeng, Y.; et al. Widespread roles of enhancer-like transposable elements in cell identity and long-range genomic interactions. *Genome Res.* **2018**. [CrossRef]

143. Makarevitch, I.; Waters, A.J.; West, P.T.; Stitzer, M.; Hirsch, C.N.; Ross-Ibarra, J.; Springer, N.M. Transposable elements contribute to activation of maize genes in response to abiotic stress. *PLoS Genet.* **2015**, *11*, e1004915. [CrossRef]
144. Feschotte, C. Transposable elements and the evolution of regulatory networks. *Nat. Rev. Genet.* **2008**, *9*, 397–405. [CrossRef] [PubMed]
145. Chuong, E.B.; Elde, N.C.; Feschotte, C. Regulatory activities of transposable elements: From conflicts to benefits. *Nat. Rev. Genet.* **2017**, *18*, 71–86. [CrossRef] [PubMed]
146. Slotkin, R.K.; Martienssen, R. Transposable elements and the epigenetic regulation of the genome. *Nat. Rev. Genet.* **2007**, *8*, 272–285. [CrossRef] [PubMed]
147. Ariumi, Y. Guardian of the human genome: host defense mechanisms against LINE-1 retrotransposition. *Front. Chem.* **2016**, *4*, 28. [CrossRef] [PubMed]
148. Schueler, M.G.; Higgins, A.W.; Rudd, M.K.; Gustashaw, K.; Willard, H.F. Genomic and genetic definition of a functional human centromere. *Science* **2001**, *294*, 109–115. [CrossRef] [PubMed]
149. Kazakov, A.E.; Shepelev, V.A.; Tumeneva, I.G.; Alexandrov, A.A.; Yurov, Y.B.; Alexandrov, I.A. Interspersed repeats are found predominantly in the "old" α satellite families. *Genomics* **2003**, *82*, 619–627. [CrossRef]
150. Rosenbloom, K.R.; Armstrong, J.; Barber, G.P.; Casper, J.; Clawson, H.; Diekhans, M.; Dreszer, T.R.; Fujita, P.A.; Guruvadoo, L.; Haeussler, M.; et al. The UCSC Genome Browser database: 2015 update. *Nucleic Acids Res.* **2015**, *43*, D670–D681. [CrossRef]
151. Burrack, L.S.; Berman, J. Neocentromeres and epigenetically inherited features of centromeres. *Chromosome Res.* **2012**, *20*, 607–619. [CrossRef]
152. Chueh, A.C.; Northrop, E.L.; Brettingham-Moore, K.H.; Choo, K.H.; Wong, L.H. LINE retrotransposon RNA is an essential structural and functional epigenetic component of a core neocentromeric chromatin. *PLoS Genet.* **2009**, *5*, e1000354. [CrossRef]
153. Carbone, L.; Harris, R.A.; Mootnick, A.R.; Milosavljevic, A.; Martin, D.I.; Rocchi, M.; Capozzi, O.; Archidiacono, N.; Konkel, M.K.; Walker, J.A.; et al. Centromere remodeling in *Hoolock leuconedys* (Hylobatidae) by a new transposable element unique to the gibbons. *Genome Biol. Evol.* **2012**, *4*, 648–658. [CrossRef] [PubMed]
154. Tolomeo, D.; Capozzi, O.; Stanyon, R.R.; Archidiacono, N.; D'Addabbo, P.; Catacchio, C.R.; Purgato, S.; Perini, G.; Schempp, W.; Huddleston, J.; et al. Epigenetic origin of evolutionary novel centromeres. *Sci. Rep.* **2017**, *7*, 41980. [CrossRef]
155. Kato, M.; Takashima, K.; Kakutani, T. Epigenetic control of CACTA transposon mobility in *Arabidopsis thaliana*. *Genetics* **2004**, *168*, 961–969. [CrossRef] [PubMed]
156. Bourc'his, D.; Bestor, T.H. Meiotic catastrophe and retrotransposon reactivation in male germ cells lacking Dnmt3L. *Nature* **2004**, *431*, 96–99. [CrossRef] [PubMed]
157. Metcalfe, C.J.; Bulazel, K.V.; Ferreri, G.C.; Schroeder-Reiter, E.; Wanner, G.; Rens, W.; Obergfell, C.; Eldridge, M.D.; O'Neill, R.J. Genomic instability within centromeres of interspecific marsupial hybrids. *Genetics* **2007**, *177*, 2507–2517. [CrossRef]
158. O'Neill, R.J.; O'Neill, M.J.; Graves, J.A. Undermethylation associated with retroelement activation and chromosome remodelling in an interspecific mammalian hybrid. *Nature* **1998**, *393*, 68–72. [CrossRef] [PubMed]
159. May, B.P.; Lippman, Z.B.; Fang, Y.; Spector, D.L.; Martienssen, R.A. Differential regulation of strand-specific transcripts from Arabidopsis centromeric satellite repeats. *PLoS Genet.* **2005**, *1*, e79. [CrossRef] [PubMed]
160. Ferreri, G.C.; Brown, J.D.; Obergfell, C.; Jue, N.; Finn, C.E.; O'Neill, M.J.; O'Neill, R.J. Recent amplification of the kangaroo endogenous retrovirus, KERV, limited to the centromere. *J. Virol.* **2011**, *85*, 4761–4771. [CrossRef] [PubMed]
161. Contreras-Galindo, R.; Kaplan, M.H.; He, S.; Contreras-Galindo, A.C.; Gonzalez-Hernandez, M.J.; Kappes, F.; Dube, D.; Chan, S.M.; Robinson, D.; Meng, F.; et al. HIV infection reveals widespread expansion of novel centromeric human endogenous retroviruses. *Genome Res.* **2013**, *23*, 1505–1513. [CrossRef] [PubMed]
162. Cheng, Z.J.; Murata, M. A centromeric tandem repeat family originating from a part of Ty3/gypsy-retroelement in wheat and its relatives. *Genetics* **2003**, *164*, 665–672. [PubMed]
163. Kapitonov, V.V.; Jurka, J. Molecular paleontology of transposable elements from *Arabidopsis thaliana*. *Genetica* **1999**, *107*, 27–37. [CrossRef] [PubMed]

164. Heikkinen, E.; Launonen, V.; Muller, E.; Bachmann, L. The *pvB370 Bam*HI satellite DNA family of the *Drosophila virilis* group and its evolutionary relation to mobile dispersed genetic pDv elements. *J. Mol. Evol.* **1995**, *41*, 604–614. [CrossRef] [PubMed]
165. Kapitonov, V.V.; Holmquist, G.P.; Jurka, J. L1 repeat is a basic unit of heterochromatin satellites in cetaceans. *Mol. Biol. Evol.* **1998**, *15*, 611–612. [CrossRef] [PubMed]
166. McGurk, M.P.; Barbash, D.A. Double insertion of transposable elements provides a substrate for the evolution of satellite DNA. *Genome Res.* **2018**, *28*, 714–725. [CrossRef] [PubMed]
167. Dias, G.B.; Svartman, M.; Delprat, A.; Ruiz, A.; Kuhn, G.C. Tetris is a foldback transposon that provided the building blocks for an emerging satellite DNA of *Drosophila virilis*. *Genome Biol. Evol.* **2014**, *6*, 1302–1313. [CrossRef] [PubMed]
168. Mestrovic, N.; Mravinac, B.; Pavlek, M.; Vojvoda-Zeljko, T.; Satovic, E.; Plohl, M. Structural and functional liaisons between transposable elements and satellite DNAs. *Chromosome Res.* **2015**, *23*, 583–596. [CrossRef] [PubMed]
169. Ahmed, M.; Liang, P. Transposable elements are a significant contributor to tandem repeats in the human genome. *Comp. Funct. Genom.* **2012**, *2012*, 947089. [CrossRef]
170. Satovic, E.; Vojvoda Zeljko, T.; Luchetti, A.; Mantovani, B.; Plohl, M. Adjacent sequences disclose potential for intra-genomic dispersal of satellite DNA repeats and suggest a complex network with transposable elements. *BMC Genom.* **2016**, *17*, 997. [CrossRef]
171. McLaughlin, R.N., Jr.; Malik, H.S. Genetic conflicts: The usual suspects and beyond. *J. Exp. Biol.* **2017**, *220*, 6–17. [CrossRef]
172. Yoder, J.A.; Walsh, C.P.; Bestor, T.H. Cytosine methylation and the ecology of intragenomic parasites. *Trends Genet.* **1997**, *13*, 335–340. [CrossRef]
173. Fedoroff, N.V. Presidential address. Transposable elements, epigenetics, and genome evolution. *Science* **2012**, *338*, 758–767. [CrossRef]
174. Symer, D.E.; Connelly, C.; Szak, S.T.; Caputo, E.M.; Cost, G.J.; Parmigiani, G.; Boeke, J.D. Human l1 retrotransposition is associated with genetic instability in vivo. *Cell* **2002**, *110*, 327–338. [CrossRef]
175. Gilbert, N.; Lutz, S.; Morrish, T.A.; Moran, J.V. Multiple fates of L1 retrotransposition intermediates in cultured human cells. *Mol. Cell. Biol.* **2005**, *25*, 7780–7795. [CrossRef]
176. Kazazian, H.H., Jr.; Moran, J.V. Mobile DNA in health and disease. *N. Engl. J. Med.* **2017**, *377*, 361–370. [CrossRef]
177. Beck, C.R.; Garcia-Perez, J.L.; Badge, R.M.; Moran, J.V. LINE-1 elements in structural variation and disease. *Annu. Rev. Genom. Hum. Genet.* **2011**, *12*, 187–215. [CrossRef]
178. Divashuk, M.G.; Khuat, T.M.; Kroupin, P.Y.; Kirov, I.V.; Romanov, D.V.; Kiseleva, A.V.; Khrustaleva, L.I.; Alexeev, D.G.; Zelenin, A.S.; Klimushina, M.V.; et al. Variation in copy number of Ty3/Gypsy centromeric retrotransposons in the genomes of *Thinopyrum intermedium* and its diploid progenitors. *PLoS ONE* **2016**, *11*, e0154241. [CrossRef]
179. Han, J.; Masonbrink, R.E.; Shan, W.; Song, F.; Zhang, J.; Yu, W.; Wang, K.; Wu, Y.; Tang, H.; Wendel, J.F.; et al. Rapid proliferation and nucleolar organizer targeting centromeric retrotransposons in cotton. *Plant J.* **2016**, *88*, 992–1005. [CrossRef]
180. Yang, X.; Zhao, H.; Zhang, T.; Zeng, Z.; Zhang, P.; Zhu, B.; Han, Y.; Braz, G.T.; Casler, M.D.; Schmutz, J.; et al. Amplification and adaptation of centromeric repeats in polyploid switchgrass species. *New Phytol.* **2018**, *218*, 1645–1657. [CrossRef]
181. Pardue, M.L.; Gall, J.G. Chromosomal localization of mouse satellite DNA. *Science* **1970**, *168*, 1356–1358. [CrossRef]
182. Jones, K.W. Chromosomal and nuclear location of mouse satellite DNA in individual cells. *Nature* **1970**, *225*, 912–915. [CrossRef] [PubMed]
183. Yunis, J.J.; Yasmineh, W.G. Heterochromatin, satellite DNA, and cell function. Structural DNA of eucaryotes may support and protect genes and aid in speciation. *Science* **1971**, *174*, 1200–1209. [CrossRef] [PubMed]
184. Britten, R.J.; Kohne, D.E. Repeated sequences in DNA. Hundreds of thousands of copies of DNA sequences have been incorporated into the genomes of higher organisms. *Science* **1968**, *161*, 529–540. [CrossRef]
185. Hall, L.E.; Mitchell, S.E.; O'Neill, R.J. Pericentric and centromeric transcription: A perfect balance required. *Chromosome Res.* **2012**, *20*, 535–546. [CrossRef] [PubMed]

186. Talbert, P.B.; Henikoff, S. Transcribing centromeres: noncoding RNAs and kinetochore assembly. *Trends Genet.* **2018**, *34*, 587–599. [CrossRef] [PubMed]
187. Braselton, J.P. Ribonucleoprotein staining of *Allium cepa* kinetochores. *Cytobiologie* **1975**, *12*, 148–151.
188. Rieder, C.L. Ribonucleoprotein staining of centrioles and kinetochores in newt lung cell spindles. *J. Cell Biol.* **1979**, *80*, 1–9. [CrossRef]
189. Blower, M.D.; Sullivan, B.A.; Karpen, G.H. Conserved organization of centromeric chromatin in flies and humans. *Dev. Cell* **2002**, *2*, 319–330. [CrossRef]
190. Sullivan, B.A.; Karpen, G.H. Centromeric chromatin exhibits a histone modification pattern that is distinct from both euchromatin and heterochromatin. *Nat. Struct. Mol. Biol.* **2004**, *11*, 1076–1083. [CrossRef]
191. Bergmann, J.H.; Rodriguez, M.G.; Martins, N.M.; Kimura, H.; Kelly, D.A.; Masumoto, H.; Larionov, V.; Jansen, L.E.; Earnshaw, W.C. Epigenetic engineering shows H3K4me2 is required for HJURP targeting and CENP-A assembly on a synthetic human kinetochore. *EMBO J.* **2011**, *30*, 328–340. [CrossRef] [PubMed]
192. Choi, E.S.; Stralfors, A.; Castillo, A.G.; Durand-Dubief, M.; Ekwall, K.; Allshire, R.C. Identification of noncoding transcripts from within CENP-A chromatin at fission yeast centromeres. *J. Biol. Chem.* **2011**, *286*, 23600–23607. [CrossRef] [PubMed]
193. Chan, F.L.; Marshall, O.J.; Saffery, R.; Kim, B.W.; Earle, E.; Choo, K.H.; Wong, L.H. Active transcription and essential role of RNA polymerase II at the centromere during mitosis. *Proc. Natl. Acad. Sci. USA* **2012**, *109*, 1979–1984. [CrossRef] [PubMed]
194. Rosic, S.; Kohler, F.; Erhardt, S. Repetitive centromeric satellite RNA is essential for kinetochore formation and cell division. *J. Cell Biol.* **2014**, *207*, 335–349. [CrossRef] [PubMed]
195. Chen, C.C.; Bowers, S.; Lipinszki, Z.; Palladino, J.; Trusiak, S.; Bettini, E.; Rosin, L.; Przewloka, M.R.; Glover, D.M.; O'Neill, R.J.; et al. Establishment of centromeric chromatin by the CENP-A assembly factor CAL1 requires FACT-mediated transcription. *Dev. Cell* **2015**, *34*, 73–84. [CrossRef] [PubMed]
196. Ohkuni, K.; Kitagawa, K. Endogenous transcription at the centromere facilitates centromere activity in budding yeast. *Curr. Biol.* **2011**, *21*, 1695–1703. [CrossRef]
197. Chan, F.L.; Wong, L.H. Transcription in the maintenance of centromere chromatin identity. *Nucleic Acids Res.* **2012**, *40*, 11178–11188. [CrossRef]
198. McNulty, S.M.; Sullivan, L.L.; Sullivan, B.A. Human centromeres produce chromosome-specific and array-specific α satellite transcripts that are complexed with CENP-A and CENP-C. *Dev. Cell* **2017**, *42*, 226–240. [CrossRef]
199. Topp, C.N.; Zhong, C.X.; Dawe, R.K. Centromere-encoded RNAs are integral components of the maize kinetochore. *Proc. Natl. Acad. Sci. USA* **2004**, *101*, 15986–15991. [CrossRef]
200. Carone, D.M.; Zhang, C.; Hall, L.E.; Obergfell, C.; Carone, B.R.; O'Neill, M.J.; O'Neill, R.J. Hypermorphic expression of centromeric retroelement-encoded small RNAs impairs CENP-A loading. *Chromosome Res.* **2013**, *21*, 49–62. [CrossRef]
201. Saffery, R.; Sumer, H.; Hassan, S.; Wong, L.H.; Craig, J.M.; Todokoro, K.; Anderson, M.; Stafford, A.; Choo, K.H. Transcription within a functional human centromere. *Mol. Cell* **2003**, *12*, 509–516. [CrossRef]
202. Wong, L.H.; Brettingham-Moore, K.H.; Chan, L.; Quach, J.M.; Anderson, M.A.; Northrop, E.L.; Hannan, R.; Saffery, R.; Shaw, M.L.; Williams, E.; et al. Centromere RNA is a key component for the assembly of nucleoproteins at the nucleolus and centromere. *Genome Res.* **2007**, *17*, 1146–1160. [CrossRef]
203. Ugarkovic, D. Functional elements residing within satellite DNAs. *EMBO Rep.* **2005**, *6*, 1035–1039. [CrossRef] [PubMed]
204. Neumann, P.; Yan, H.; Jiang, J. The centromeric retrotransposons of rice are transcribed and differentially processed by RNA interference. *Genetics* **2007**, *176*, 749–761. [CrossRef] [PubMed]
205. Scott, K.C.; White, C.V.; Willard, H.F. An RNA polymerase III-dependent heterochromatin barrier at fission yeast centromere 1. *PLoS ONE* **2007**, *2*, e1099. [CrossRef]
206. Cardinale, S.; Bergmann, J.H.; Kelly, D.; Nakano, M.; Valdivia, M.M.; Kimura, H.; Masumoto, H.; Larionov, V.; Earnshaw, W.C. Hierarchical inactivation of a synthetic human kinetochore by a chromatin modifier. *Mol. Biol. Cell* **2009**, *20*, 4194–4204. [CrossRef] [PubMed]
207. Nakano, M.; Cardinale, S.; Noskov, V.N.; Gassmann, R.; Vagnarelli, P.; Kandels-Lewis, S.; Larionov, V.; Earnshaw, W.C.; Masumoto, H. Inactivation of a human kinetochore by specific targeting of chromatin modifiers. *Dev. Cell* **2008**, *14*, 507–522. [CrossRef] [PubMed]

208. Bergmann, J.H.; Jakubsche, J.N.; Martins, N.M.; Kagansky, A.; Nakano, M.; Kimura, H.; Kelly, D.A.; Turner, B.M.; Masumoto, H.; Larionov, V.; et al. Epigenetic engineering: Histone H3K9 acetylation is compatible with kinetochore structure and function. *J. Cell Sci.* **2012**, *125*, 411–421. [CrossRef] [PubMed]
209. Ting, D.T.; Lipson, D.; Paul, S.; Brannigan, B.W.; Akhavanfard, S.; Coffman, E.J.; Contino, G.; Deshpande, V.; Iafrate, A.J.; Letovsky, S.; et al. Aberrant overexpression of satellite repeats in pancreatic and other epithelial cancers. *Science* **2011**, *331*, 593–596. [CrossRef]
210. Hill, A.; Bloom, K. Genetic manipulation of centromere function. *Mol. Cell. Biol.* **1987**, *7*, 2397–2405. [CrossRef]
211. Bouzinba-Segard, H.; Guais, A.; Francastel, C. Accumulation of small murine minor satellite transcripts leads to impaired centromeric architecture and function. *Proc. Natl. Acad. Sci. USA* **2006**, *103*, 8709–8714. [CrossRef]
212. Ferri, F.; Bouzinba-Segard, H.; Velasco, G.; Hube, F.; Francastel, C. Non-coding murine centromeric transcripts associate with and potentiate Aurora B kinase. *Nucleic Acids Res.* **2009**, *37*, 5071–5080. [CrossRef] [PubMed]
213. Chan, D.Y.L.; Moralli, D.; Khoja, S.; Monaco, Z.L. Noncoding centromeric RNA expression impairs chromosome stability in human and murine stem cells. *Dis. Mark.* **2017**, *2017*, 7506976. [CrossRef] [PubMed]
214. Du, Y.; Topp, C.N.; Dawe, R.K. DNA binding of centromere protein C (CENPC) is stabilized by single-stranded RNA. *PLoS Genet.* **2010**, *6*, e1000835. [CrossRef]
215. Sandmann, M.; Talbert, P.; Demidov, D.; Kuhlmann, M.; Rutten, T.; Conrad, U.; Lermontova, I. Targeting of Arabidopsis KNL2 to centromeres depends on the conserved CENPC-k motif in its C terminus. *Plant Cell* **2017**, *29*, 144–155. [CrossRef] [PubMed]
216. Dawe, R.K. RNA interference, transposons, and the centromere. *Plant Cell* **2003**, *15*, 297–301. [CrossRef] [PubMed]
217. Wong, L.H.; Choo, K.H. Evolutionary dynamics of transposable elements at the centromere. *Trends Genet.* **2004**, *20*, 611–616. [CrossRef] [PubMed]
218. O'Neill, R.J.; Carone, D.M. The role of ncRNA in centromeres: A lesson from marsupials. *Prog. Mol. Subcell. Biol.* **2009**, *48*, 77–101. [CrossRef] [PubMed]
219. Kasinathan, S.; Henikoff, S. Non-B-form DNA is enriched at centromeres. *Mol. Biol. Evol.* **2018**, *35*, 949–962. [CrossRef]
220. Bobkov, G.O.M.; Gilbert, N.; Heun, P. Centromere transcription allows CENP-A to transit from chromatin association to stable incorporation. *J. Cell Biol.* **2018**, *217*, 1957–1972. [CrossRef]
221. Quenet, D.; Dalal, Y. A long non-coding RNA is required for targeting centromeric protein A to the human centromere. *Elife* **2014**, *3*, e03254, Erratum in **2018**, *7*, e41593. [CrossRef]
222. Chueh, A.C.; Wong, L.H.; Wong, N.; Choo, K.H. Variable and hierarchical size distribution of L1-retroelement-enriched CENP-A clusters within a functional human neocentromere. *Hum. Mol. Genet.* **2005**, *14*, 85–93. [CrossRef] [PubMed]
223. McClintock, B. The behaviour of successive nuclear divisions of a chromosome broken at meiosis. *Proc. Natl. Acad. Sci. USA* **1939**, *25*, 405–416. [CrossRef] [PubMed]
224. McClintock, B. The stability of broken ends of chromosomes in *Zea Mays*. *Genetics* **1941**, *26*, 234–282. [PubMed]
225. McClintock, B. The production of homozygous deficient tissues with mutant characteristics by means of the aberrant mitotic behavior of ring-shaped chromosomes. *Genetics* **1938**, *23*, 315–376. [PubMed]

© 2019 by the authors. Licensee MDPI, Basel, Switzerland. This article is an open access article distributed under the terms and conditions of the Creative Commons Attribution (CC BY) license (http://creativecommons.org/licenses/by/4.0/).

Article

Does the Presence of Transposable Elements Impact the Epigenetic Environment of Human Duplicated Genes?

Romain Lannes [1,†], Carène Rizzon [2] and Emmanuelle Lerat [1,*]

1. Laboratoire de Biométrie et Biologie Evolutive UMR 5558, Université de Lyon, Université Lyon 1, CNRS, F-69622 Villeurbanne, France; romain.lannes@gmail.com
2. Laboratoire de Mathématiques et Modélisation d'Evry (LaMME), Université d'Evry Val d'Essonne, UMR CNRS 8071, ENSIIE, USC INRA, 23 bvd de France, 91037 Evry CEDEX Paris, France; carene.rizzon@univ-evry.fr
* Correspondence: emmanuelle.lerat@univ-lyon1.fr; Tel.: +33-4-72-43-29-18; Fax: +33-4-72-43-13-88
† Present Address: Institut de Biologie Paris-Seine (IBPS), Sorbonne Universites, UPMC Universite Paris 06, 75005 Paris, France.

Received: 29 January 2019; Accepted: 22 March 2019; Published: 26 March 2019

Abstract: Epigenetic modifications have an important role to explain part of the intra- and inter-species variation in gene expression. They also have a role in the control of transposable elements (TEs) whose activity may have a significant impact on genome evolution by promoting various mutations, which are expected to be mostly deleterious. A change in the local epigenetic landscape associated with the presence of TEs is expected to affect the expression of neighboring genes since these modifications occurring at TE sequences can spread to neighboring sequences. In this work, we have studied how the epigenetic modifications of genes are conserved and what the role of TEs is in this conservation. For that, we have compared the conservation of the epigenome associated with human duplicated genes and the differential presence of TEs near these genes. Our results show higher epigenome conservation of duplicated genes from the same family when they share similar TE environment, suggesting a role for the differential presence of TEs in the evolutionary divergence of duplicates through variation in the epigenetic landscape.

Keywords: transposable elements; gene duplication; gene evolution; epigenetics

---

1. Introduction

Epigenetic changes can explain part of the variation in gene expression observed between tissues from the same organism [1–4], or the fate of individuals like in honeybees by affecting the differentiation between the queen and the workers [5] or in the determination of the different casts in ants [6]. These examples are likely to represent only a tiny fraction of all the possible effects of epigenetic processes. In sum, epigenetic modifications are important actors of the gene expression modulation such as variation in expression among tissues, developmental stages or in response to environmental changes [7]. Three epigenetic mechanisms have been identified that can work jointly to regulate gene expression. DNA methylation usually occurs in the context of CpG dinucleotides and is associated with transcription silencing [8–11]. RNA interference mechanism is characterized by the synthesis of small noncoding RNAs, which, when associated with a protein complex, can target messenger RNAs and trigger their degradation [12,13]. Histone modifications correspond to post-translational biochemical changes occurring at particular amino acid residues of these proteins that are at the basis of nucleosomes [11,14,15]. According to the type of histone modifications, the effect can either compact or relax the chromatin structure; both have a direct impact on the gene expression [3,16]. Due to their important role in gene regulation, epigenetic modifications can potentially cause diseases under certain circumstances when a global modification of the epigenetic landscape happens [17].

It has long been suspected that changes in gene regulation may play a role in the adaptation and evolution of organisms [18]. In particular, epigenetic divergence has been proposed to affect species divergence by conferring hybrid incompatibility like in the example of the formation of mouse subspecies, which is linked to methylation of lysine 4 from histone 3 (H3K4me) [19]. In three cell lines, variation of gene expression in primates could be associated with changes in H3K4me3 localization [20]. Similarly, changes in DNA methylation have been shown to partly explain the divergence of gene expression in the brains of humans and chimps [21]. This variation in DNA methylation could also explain the evolution of vulnerability to some diseases in humans since among the list of impacted genes, several of them have been associated with human diseases like neurodevelopmental and psychological disorders. Epigenetic conservation or divergence is also linked to the DNA sequence conservation. For example, in humans, hypomethylated CpG islands have been shown to be under selective constraints [22]. These CpG islands were also shown to be more enriched in trimethylated H3K4 and H3K36, and in acetylated H3K27 [23]. The acquisition of hypermethylated DNA in humans is coupled to a very rapid nucleotidic evolution near CpG sites [24]. In this last work, the authors showed a genome-wide conservation of DNA methylation profiles when comparing humans and various primates, with the presence of regions with human specific patterns not localized near transcription start sites. Some epigenetic modifications can be conserved between species. For example, the trimethylated H3K36 modification is conserved in exons and introns between humans and mice [25]. A wide comparison of three histone modifications among several cell types from humans and mice showed a strong association between the stability among the cell types (intraspecies) and the conservation between species of these modifications against both genetic and environmental changes [26]. Among invertebrates, gene body DNA methylation has been shown to be conserved on very long evolutionary time scales, suggesting a function of DNA methylation in the different genomes [27]. The same kind of results have been observed in plants in which a strong conservation of gene body methylation was observed that targeted slowly evolving genes, indicating that the methylation level can have evolutionary consequences [28]. At an intraspecific level, epigenetic modifications may be implicated in functional divergence by facilitating tissue-specific regulation. For example, human duplicated genes are initially highly methylated, then gradually lose DNA methylation as they age [29]. Within each pair of genes, DNA methylation divergence increases with time. Moreover, tissue-specific DNA methylation of duplicates correlates with tissue-specific expression, implying that DNA methylation could be a causative factor for functional divergence of duplicated genes [29]. However, epigenetic modifications may also play a role in the functional conservation. For example, in some plants, paralogous genes associated with trimethylated H3K27 showed the highest coding sequence divergence but the highest similarity in expression patterns and in regulatory regions when compared to paralogous genes in which only one gene was the target of this histone modification [30]. In this case, the histone modification could be responsible for the conservation of gene expression. By comparing segmental duplications regions in humans, a widespread conservation of DNA methylation and some histone modifications was observed when considering recently duplicated regions [31]. For the regions displaying divergence in DNA methylation and chromatin states, particular DNA motifs were detected.

Eukaryotic genomes are formed from a variety of elements among which protein-coding genes are a minority. In the human genome, for example, the protein-coding genes represent only a very small fraction (<2% of the genome), whereas repetitive sequences represent more than half [32]. While the non-coding part was first thought to have no function [33], it is now known to be composed of a mixture of repetitive DNA and non-functional sequences interspersed with non-coding RNA genes and regions that are crucial for transcriptional and post-transcriptional regulation [34,35]. The greater part of repeated DNA is classified as transposable elements (TEs), with several millions of them inserted throughout the human genome. Because of their presence in genomes, TEs have a significant impact on genome evolution and on gene functioning [36,37]. For example, a bias in the distribution of TEs in and near genes has been observed, showing that TEs are found to be under represented inside genes, which indicates that they are counter selected in these regions [34,38]. Moreover, TEs have been shown to be associated with the evolution of duplicated genes [39,40]. To counteract their deleterious effects, TEs are regulated by the host genome via epigenetic mechanisms to suppress or silence their activity [41,42]. In normal mammalian cells, TEs are usually methylated, therefore transcriptionally silenced [41]. In some abnormal cells where DNA methylation is abolished, TEs can be mobilized, resulting in a potential impact on the integrity of the cell [43,44]. A change in the local epigenetic landscape associated with the presence of TE sequences is expected to affect the expression of the neighboring genes since these modifications occurring at TE sequences can spread to neighboring sequences, as has been observed in mice, in plants or in fungi [45–51]. In humans, the recent insertion of an Alu element was identified as the cause of increasing levels of DNA methylation in its surrounding genomic area, which inactivated the neighboring gene expression [52]. When comparing histone modification of genes between normal and cancer conditions in humans, we found that the presence of TEs near genes was associated with more changes in histone enrichment [53]. In primates, some TEs have been identified as a source of novelty in gene regulation, in association with changes in histone modifications [54]. Alu elements were observed to be enriched around methylated sites of discordant paralogous regions corresponding to segmental duplications in human [31]. Differentially methylated regions between humans and primates were shown to be enriched in endogenous retroviruses in hypomethylated human specific regions [24]. Thus, the presence of TEs in a genome may have a direct influence on the epigenetic variations directed on the host genes, potentially influencing their fate and functioning.

In this work, we have explored how the epigenetic modifications of genes are conserved and what the role of TEs is in this conservation. For that, we have studied the conservation of the epigenome at an intraspecific level in humans. By measuring, in different cell types, the divergence of epigenetic modifications associated with duplicated genes and linked to the presence of TEs near the genes, we have determined the impact of TEs on epigenetic changes and expression divergence associated with the time since duplication. Our results show that the presence of TEs is associated with variation in histone modification enrichment and methylation level of neighboring genes but also that a similar TE environment near duplicated genes is related to higher conservation of epigenetic modification and expression.

2. Material and Methods

2.1. Duplicated Genes

Gene families were retrieved from the HOGENOM database [55], which contains functional proteins from 1400 organisms grouped by sequence homology coming from various nucleotide sequence collections. Among the 10,064 gene families for which we were able to identify Ensembl gene access numbers in the human genome version GRCh38, 1420 families contain two functional human genes (list provided as Supplementary data S1). We determined for each of these pairs the divergence between the two genes by aligning the protein sequences and subsequently the nucleotidic sequences to keep the codon alignments. The sequence divergence estimates between duplicated genes of a given family were computed using the YN00 module of paml [56] to obtain the synonymous substitution rate (dS), the non synonymous substitution rate (dN) and the omega ratio (dN/dS).

2.2. Epigenetic Modification and Expression Data

This study makes use of data generated by the BLUEPRINT Consortium (www.blueprint-epigenome.eu). We have retrieved epigenetics and expression data from four normal cell types extracted from cord blood of female individuals and corresponding to two precursor cell types (cd14+cd16− (access number C005PS) and erythroblast (access number S002S3)) and to two differentiated cell types (macrophage (access number S00BHQ) and cd8T (access number C0066P)). Methylation, histone modification and expression data have been generated by the alignment of BS-seq, ChIP-seq and RNA-seq reads on the human genome (version GRCh38) using the mapper bwa [57] with a random location assignment for multiple hits [58]. We thus used the methylation status (hypomethylated, hypermethylated, and standard), and the histone enrichment for six histone modifications (H3K27me3, H3K9me3, H3K27ac, H3K4me1, H3K4me3 and H3K36me3) of genomic regions, and the expression level of annotated genes (FPKM) as provided by the BLUEPRINT Consortium. The mean histone enrichment was computed for each gene and corresponds to the average fold enrichment of the given histone modification for the positions covered by the gene, normalized by the gene size [53]. We have determined the mean level of methylation of each gene from the identified hyper- and hypo-methylated regions covering the gene. Hyper-methylated regions correspond to regions with an average methylation level of >0.75 and hypo-methylated regions have an average methylation level of <0.25. These values correspond to the ratio of reads with an unconverted cytosine (i.e., C) over the sum of all reads containing either an unconverted cytosine or a converted cytosine (i.e., T). We thus have considered a gene globally hypo- or hyper-methylated when the average methylation ratio covering its position was <0.25 or >0.75 respectively. Its level of methylation was considered as standard otherwise. For the expression analysis, a gene was considered as expressed if it has an FPKM value of at least 0.5 [59]. As recommended [60], the expression data of a given gene i in each cell type were converted from FPKM to TPM using the formula $TPM_i = \frac{FPKM_i}{\sum FPKM} \cdot 10^6$ to normalize the values in each cell type allowing direct comparisons. The divergence of expression between the two genes g1 and g2 from a given family was estimated by the Manhattan distance d_m across the four samples according to the formula:

$$d_m = \frac{1}{2} \sum_{k=1}^{4} \left| \frac{g_{1,k}}{\sum_{k=1}^{4} g_{1,k}} - \frac{g_{2,k}}{\sum_{k=1}^{4} g_{2,k}} \right|$$

2.3. Transposable Elements Neighborhood

The TE annotation from the latest version of the human genome assembly was obtained by parsing the repeat-masker output file available on the website of the University of California, Santa Cruz (http://hgdownload.cse.ucsc.edu/goldenPath/hg38/bigZips/) using the program one-code-to-find-them-all [61] with the –*strict* option to avoid false positive identification. This program assembles each TE copy and determine their positions in the genome. Although polymorphic TE insertions are present when comparing different individuals and may locally have an important impact on health, they represent only thousand of insertions, which is fare less than the millions of fixed ones [62]. In this work, we are investigating the influence of fixed TE insertions for normal conditions. For each human coding gene, we computed the TE density and the TE coverage using a 2kb-flanking region upstream and downstream the gene as proposed by Grégoire et al. [53] to cover the promoter region of the genes in addition to the entire gene. The density estimates the number of TEs in a given region normalized by the size of the region and the coverage measures the proportion of nucleotides belonging to an TE in the considered region. We have considered in our approach all types of TEs globally, without differentiating the classes. It is known that epigenetic modifications may differ according to the type of TEs [63]; however, it would be impossible to have a large enough sample size of duplicated genes if considering only those with just one type of TE in their vicinity, the unique condition to really analyze the TE type contribution without any confounding factors due to the presence of other TEs.

2.4. Gene Classification

All human coding genes (18,938 genes) were clustered according to their level of density and coverage of TEs using the K-medoids algorithm as implemented in the pam() function of the R package [64], which allows an unsupervised classification in a defined number of classes. We thus defined five gene categories from TE-free genes (genes with no TE in their neighborhood) to TE-very-rich genes (genes with numerous TE in their neighborhood). The genes with density and coverage of 0 were defined as TE-free genes. The remaining genes were clustered using both density and coverage values to discriminate between the TE-very-poor (mean density of 0.0003 insertions/pb and mean coverage of 0.086), TE-poor (mean density of 0.0007 insertions/pb and mean coverage of 0.196), TE-rich (mean density of 0.0012 insertions/pb and mean coverage of 0.304), and TE-very-rich genes (mean density of 0.0025 insertions/pb and mean coverage of 0.419).

We determined three age classes (young, middle-age and old) of gene families based on the intra family synonymous substitution rate (dS) values with young families corresponding to gene pairs with dS < 1, middle-age families corresponding to gene pairs with $1 \leq dS < 2$, and old families corresponding to gene pairs with dS > 2 [29].

2.5. Statistical Tests

All statistical analyses were performed using R version 3.2.3 [64]. The Kolmogorov-Smirnov test was used to compare the distribution of two samples, the Kruskall-Wallis test was used to determine whether samples originated from the same distribution, and the Spearman test was used to determine if the correlations between the compared data were significantly not null. The Pearson's chi-squared goodness of fit test was used to determine whether there was a significant difference between the expected and the observed frequencies in one or more categories of possible associations of TE context for duplicated gene pairs. It is designed to test the null hypothesis that an observed frequency distribution is consistent with a hypothesized theoretical distribution. *P*-values were computed by Monte Carlo simulations with 2000 replicates. In this test, simulations are done by random sampling from the discrete distribution specified by the given theoretical distribution, each sample being of size $n = sum(x)$, with x the numeric vector of absolute observed frequencies (see help of R for more details). To account for multiple testing, we used the *Benjamini–Hochberg* procedure to compute *q*-values.

3. RESULTS

3.1. Duplicated Genes in the Human Genome Are Mainly Located on Different Chromosomes, Represent Old Events and Display Similar TE Environment

Among the 10,064 homologous families present in the HOGENOM database grouping 16,144 human proteins, about 75% of the families contain single copy genes (7576 families). The 25% remaining families contain from 2 to 345 human genes. We decided to focus our analyses on gene families with two copies (that we will refer to in this manuscript as "duplicated genes"), which represent 14.53% of all gene families (1462 families containing 2924 proteins). Among the 2924 proteins, we were able to find the corresponding gene ids in Ensembl for 2840 genes.

These duplicated genes are quite old as confirmed by the elevate mean synonymous rate we obtained when comparing the gene pairs (mean dS = 3.136). Indeed, this rate increases with the time since the duplication event [29]. Only 48 pairs of duplicated genes displayed dS values less than 0.25, which indicates that they represent very recent duplicates, among the 99 pairs of duplicated genes that we qualified as young families. Among the others, 189 pairs were considered as middle-age families and 1132 pairs were considered as old families. We determined the physical distances between the duplicated genes. The vast majority (2464 over 2840 genes representing 86.76% of all duplicated genes) is located on different chromosomes. Among the remaining 376, 26 duplicated genes are overlapping and the global distance between the other 350 duplicated genes is quite high since the median distance is about 81 kb (72kb when considering only young families and 109 kb when considering only middle-age and old families). When we looked at the position of genes according to the age of the family (young, middle-age and old), 72% and 94% of middle-age and old families, respectively, had their genes on different chromosomes, whereas only 37% of the young families had their genes on different chromosomes. We examined the level of sequence divergence by estimating the omega ratio (corresponding to the dN/dS ratio) for all duplicate pairs (Figure 1A). The ratios centered around a median at 0.129 with only four families with omega > 1. This indicates rather slow rates of protein evolution, suggesting that the genes of all these families are evolving under purifying selection.

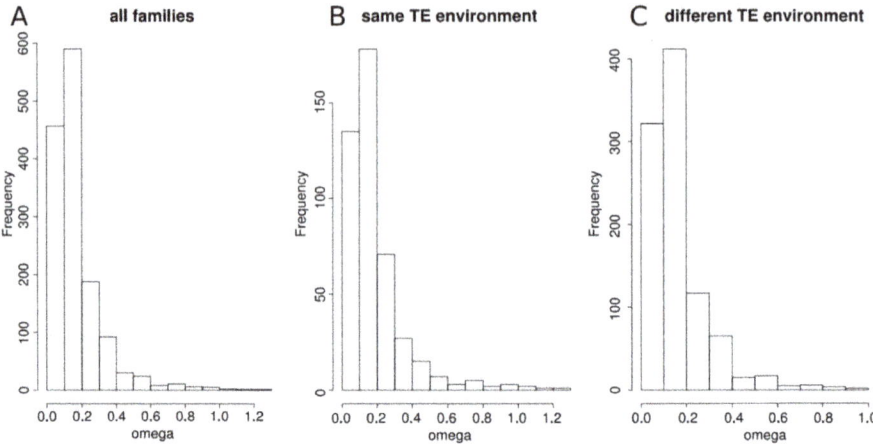

Figure 1. Distribution of the omega (dN/dS) ratio computed between duplicated genes from a same family, (**A**) for all families, (**B**) for families whose two genes have the same transposable elements (TE) environment, (**C**) for families whose two genes have different TE environment.

We then explored the TE environment of the duplicated genes. All coding genes from the human genome were clustered according to their TE environment (see method) and we then considered only the duplicated genes. The distributions corresponding to the number of genes according to their TE neighborhood category between all genes in the genome and the duplicated genes are not different (Table 1; X-squared = 2.4439, df = 4, p-value = 0.6547).

Table 1. Number of genes according to their TE neighborhood category.

TE Category	All Protein Coding Genes	Duplicated Genes
TE-free	773 (4.08%)	109 (3.84%)
TE-very-poor	4830 (25.50%)	713 (25.10%)
TE-poor	5885 (31.08%)	915 (32.22%)
TE-rich	4848 (25.60%)	729 (25.67%)
TE-very-rich	2602 (13.74%)	374 (13.17%)

In both cases, TE-free genes are the less abundant category since they represent less than 5% of all genes. The TE-very-rich genes are also less frequent (less than 14%). Both TE-very-poor and TE-rich genes represent the same proportion in the genome (>25%). The most represented category concerned the TE-poor genes (>30%). We then explored for each gene family if the two duplicated genes have similar TE environment. We observed that in a large proportion of the families (31.9%—453 families), the two genes are assigned to the same TE neighborhood cluster. This is significantly higher than when grouping randomly two genes (24%; X-squared = 35.584, df = 1, p-value = 2.443×10^{-9}) and this remains significant when considering only families whose genes are located on different chromosomes. When considering all possible associations of TE context for duplicated gene pairs, their observed occurrences are significantly different than expected according to the frequencies of TE categories in the entire genome (Table 2; X-squared = 226.52, p-value estimated according to 2000 replicates using Monte Carlo test = 0.0004998).

Table 2. Number of gene families according to the TE neighborhood category of each duplicated gene.

		TE Environment of the Second Gene				
		TE-Free	TE-Very-Poor	TE-Poor	TE-Rich	TE-Very-Rich
TE Environment of the First Gene	TE-free	**20 (1.41%)**	/	/	/	/
	TE-very-poor	36 (2.53%)	**121 (8.52%)**	/	/	/
	TE-poor	*18 (1.27%)*	220 (15.49%)	**169 (11.90%)**	/	/
	TE-rich	*13 (0.91%)*	143 (10.07%)	229 (16.13%)	**110 (7.75%)**	/
	TE-very-rich	*2 (0.14%)*	72 (5.07%)	110 (7.75%)	124 (8.73%)	**33 (2.321%)**

In bold: excess; italic: depletion; the percentages of gene families are indicated in parenthesis.

Moreover, the results indicate that there is an excess of families whose genes are either in the same or in close categories (Table 2). This observation remains true when considering each class of age independently (Supplementary Table S1), although the comparison between the three age classes indicates that when the families are recent, the proportion of genes with the same TE environment is larger than in older families (X-squared = 8.65, df = 2, *p*-value = 0.01323). We looked at the omega ratio of gene families, taking into account the TE environment of their genes. For that, we separated families in which both genes had a similar TE environment and those in which genes had a different TE environment (Figure 1B,C). The distributions of the omega ratio are not different between the two groups (Two-sample Kolmogorov-Smirnov test D = 0.074374, *p*-value = 0.06847), indicating similar evolutionary constraints on the families, irrespective of their TE environment. This remains true whatever the age of the family.

3.2. Duplicated Genes Have Similar Histone Modification Enrichment Especially If They Share a Similar TE Environment

We determined the histone enrichment of each duplicated gene according to their TE neighborhood in four cell types. Figure 2 displays the normalized average histone enrichment in each cell type and inside each gene category related to their TE neighborhood (from TE-free to TE-very-rich genes).

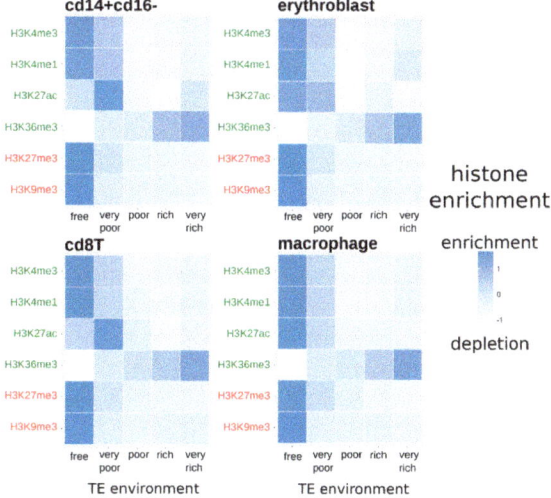

Figure 2. Normalized average histone enrichment of duplicated genes in each cell type according to the category related to their TE neighborhood (from TE-free to TE-very-rich genes). White color indicates a depletion in the considered histone modification and dark blue indicates an enrichment of the histone modification (Kruskal-Wallis tests, *q*-values < 0.05—Supplementary Table S2). Activating and repressive histone modifications are represented, respectively, in green and red.

Inside each cell type and for each histone modification, there are significant differences between genes according to their TE neighborhood (Kruskal-Wallis tests, q-values < 0.05—Supplementary Table S2). In particular, for all histone modifications, there is a decrease in the histone enrichment of genes associated with an increase in the presence of TEs in their neighborhood, excepted for H3K36me3 for which it is the contrary with more enrichment when genes have a neighborhood richer in TEs.

We then wanted to determine if genes from the same family could have similar histone enrichment and if this could be linked with any similarity in the amount of TEs found nearby. We thus tested the correlations inside each family of the histone enrichment of genes (Table 3).

The results showed an effect of the gene family since for all cell types and for all histone modifications, there are significant positive correlations between the histone enrichment of genes from the same family. According to the histone modification considered, the positive correlations are more or less pronounced. For example, the genes have a higher positive correlation for their enrichment in H3K27me3 (0.31 in CD14+CD16−, 0.34 in macrophages, 0.32 in CD8T and in erythroblasts) than in H3K9me3 (0.18 in CD14+CD16−, 0.17 in macrophages, 0.10 in CD8T, and 0.15 in erythroblasts). In order to determine if these positive correlations may be only due to the fact that genes are from the same gene family or if their respective TE environment may be involved, we tested the same correlations between duplicated genes having similar TE environment on one hand and between duplicated genes with different TE environment on the other hand. The second case is expected to underline any correlations due only to the belonging of the same family. In that last case, we observed positive correlations but they were weaker than when considering all genes (Table 3). For the histone modification H3K9me3, the correlations even disappeared in CD8T and was barely significant in erythroblasts. However, when considering only gene families for which both genes have similar TE environment, the positive correlations observed before were stronger, especially for the H3K27me3 modification (0.41 in CD14+CD16−, 0.40 in macrophages, 0.43 in CD8T and 0.44 in erythroblasts). In Figure 3 (Supplementary Table S3), we displayed the correlations for each histone modification and according to the TE neighborhood, for all cell types taken together.

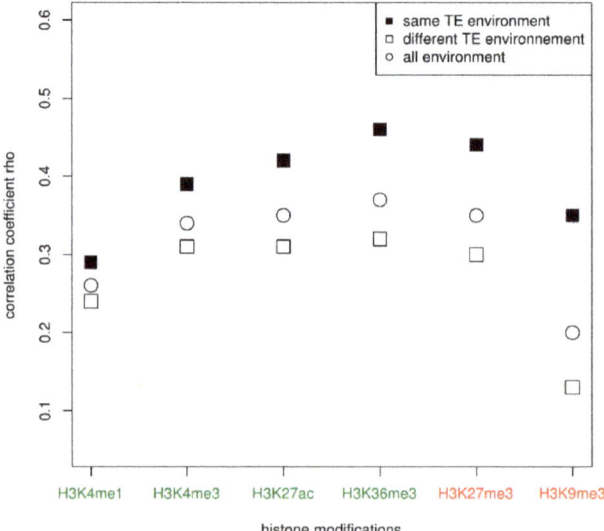

Figure 3. Correlations of the histone enrichment between paired genes according to the similarity of their TE neighborhood in the four cell types. Activating and repressive histone modifications are represented, respectively, in green and red.

Table 3. Correlations of the histone enrichment between paired genes.

		Repressive Modifications						Activating Modifications					
		H3K27me3		H3K9me3		H3K36me3		H3K27ac		H3K4me1		H3K4me3	
		Spearman Rho	q Value	Spearman Rho	q Value	Spearman Rho	q Value	Spearman Rho	q Value	Spearman Rho	q Value	Spearman Rho	q Value
Duplicated genes	CD14+CD16−	0.31 *	4.525714×10^{-16}	0.18 *	2.767059×10^{-11}	0.29 *	4.525714×10^{-16}	0.31 *	4.525714×10^{-16}	0.19 *	1.496681×10^{-13}	0.33 *	4.525714×10^{-16}
	erythroblast	0.32 *	4.525714×10^{-16}	0.15 *	2.088000×10^{-8}	0.30 *	4.525714×10^{-16}	0.31 *	4.525714×10^{-16}	0.27 *	4.525714×10^{-16}	0.39 *	4.525714×10^{-16}
	CD8T	0.32 *	4.525714×10^{-16}	0.10 *	1.878261×10^{-4}	0.29 *	4.525714×10^{-16}	0.16 *	4.189091×10^{-9}	0.24 *	4.525714×10^{-16}	0.30 *	4.525714×10^{-16}
	macrophage	0.34 *	4.525714×10^{-16}	0.17 *	1.369385×10^{-10}	0.30 *	4.525714×10^{-16}	0.23 *	4.525714×10^{-16}	0.20 *	1.224000×10^{-14}	0.33 *	4.525714×10^{-16}
Duplicated genes with same TE environment	CD14+CD16−	0.41 *	4.525714×10^{-16}	0.25 *	4.865806×10^{-8}	0.36 *	8.623256×10^{-15}	0.38 *	4.525714×10^{-16}	0.21 *	7.111385×10^{-6}	0.35 *	1.936000×10^{-14}
	erythroblast	0.44 *	4.525714×10^{-16}	0.29 *	4.727547×10^{-10}	0.38 *	4.525714×10^{-16}	0.38 *	4.525714×10^{-16}	0.27 *	4.189091×10^{-9}	0.42 *	4.525714×10^{-16}
	CD8T	0.43 *	4.525714×10^{-16}	0.22 *	1.714286×10^{-6}	0.32 *	4.838400×10^{-12}	0.26 *	3.234098×10^{-8}	0.26 *	2.088000×10^{-8}	0.36 *	5.356098×10^{-15}
	macrophage	0.40 *	4.525714×10^{-16}	0.26 *	1.743158×10^{-8}	0.44 *	4.525714×10^{-16}	0.19 *	3.299104×10^{-5}	0.22 *	1.890000×10^{-6}	0.36 *	5.356098×10^{-15}
Duplicated genes with different TE environment	CD14+CD16−	0.25 *	2.368421×10^{-15}	0.14 *	1.963636×10^{-5}	0.25 *	7.645714×10^{-15}	0.27 *	4.525714×10^{-16}	0.18 *	1.998621×10^{-8}	0.31 *	4.525714×10^{-16}
	erythroblast	0.26 *	7.297297×10^{-16}	0.08 *	1.454197×10^{-2}	0.26 *	4.525714×10^{-16}	0.27 *	4.525714×10^{-16}	0.26 *	4.940000×10^{-16}	0.36 *	4.525714×10^{-16}
	CD8T	0.25 *	4.098462×10^{-15}	0.03	2.789000×10^{-1}	0.28 *	4.525714×10^{-16}	0.11 *	5.142857×10^{-4}	0.22 *	4.599184×10^{-12}	0.26 *	4.525714×10^{-16}
	macrophage	0.31 *	4.525714×10^{-16}	0.12 *	1.058824×10^{-4}	0.22 *	3.750000×10^{-12}	0.24 *	8.405217×10^{-14}	0.18 *	7.315714×10^{-9}	0.31 *	4.525714×10^{-16}

bold * statistically significant correlations (q values < 0.05).

The figure clearly shows that genes of a given family that display a similar TE environment have a stronger positive correlation for their histone modification enrichment compared to genes that have a different TE environment. For example, the Spearman correlation rho of the duplicated genes for their enrichment in H3K36me3 is 0.37 when it increases to 0.46 when considering only families whose duplicates share the same TE environment (Supplementary Table S3). To determine whether this correlation may be linked to the age of the family, we computed the correlations in the three age groups. The previous observation remains globally true in some cell types, especially for middle-age families (Supplementary Table S4). Indeed, irrespective of the cell type and the histone modification, duplicated genes having the same TE environment usually displayed a positive correlation for their histone enrichment that is stronger than duplicated genes with different TE environment when they belong to middle-age families. For the histone modifications H3K36me3 and H3K27me3, this positive correlation is also observed between genes from young families. However, the correlation is generally less strong or at least the same between genes from old families, independently of the TE environment with few exceptions. For example, there is a higher positive correlation for H3K36me3 enrichment in macrophage and erythroblast among duplicated genes with the same TE environment in old families when compared to genes with a different TE environment, which is not the case in CD14+CD16 and CD8T in which the positive correlation is the same irrespective of the TE environment.

3.3. Duplicated Genes Have a Similar Methylation Level That Is Linked to Both the TE Environment Conservation and the Age of the Gene Family

We looked at the methylation level of each duplicated gene linked with its richness in TEs in the four cell types (Figure 4).

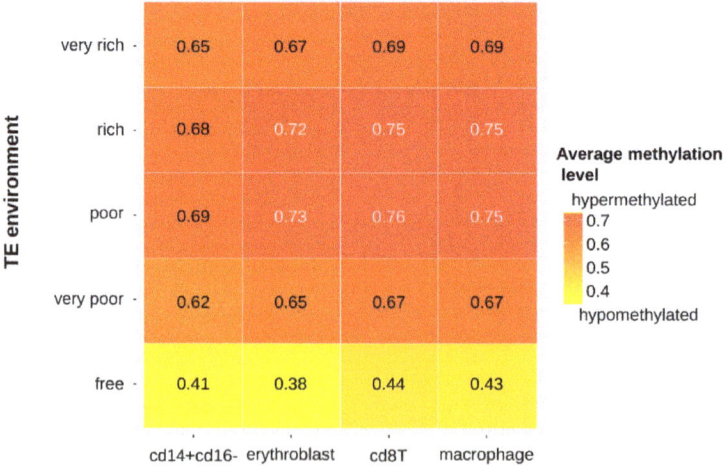

Figure 4. Average methylation level of the duplicated genes according to their neighborhood in TEs in the four cell types. (Kruskal-Wallis tests, p-values $< 2.2 \times 10^{-16}$).

In all cell types, the methylation level of the genes is associated with the TE category of each gene (Kruskal-Wallis tests, p-values $< 2.2 \times 10^{-16}$). In particular, TE-free genes are systematically less methylated than the other genes. Interestingly, the genes categorized as TE-poor and TE-rich displayed the highest methylation levels when we could have expected TE-very-rich genes to behave this way if the presence of TEs was mainly responsible for the methylation level of the genes.

We then compared inside each gene family the methylation level of the duplicated genes (Table 4).

Table 4. Correlations of the methylation level between duplicated genes.

	Duplicated Genes		Duplicated Genes with the Same TE Environment		Duplicated Genes with a Different TE Environment	
	Spearman Rho	q Value	Spearman Rho	q Value	Spearman Rho	q Value
CD14+CD16−	0.14 *	1.911360×10^{-6}	0.17 *	3.354545×10^{-4}	0.11 *	9.217000×10^{-4}
erythroblast	0.16 *	9.492000×10^{-9}	0.21 *	2.280000×10^{-5}	0.13 *	3.475500×10^{-5}
CD8T	0.15 *	1.333800×10^{-8}	0.22 *	9.620000×10^{-6}	0.12 *	1.370400×10^{-4}
macrophage	0.18 *	6.968400×10^{-11}	0.27 *	1.333800×10^{-8}	0.13 *	4.400000×10^{-5}

bold * statistically significant correlations (q values < 0.05).

We observed that there is a positive correlation in the methylation level between the genes belonging to the same family. For example, in the macrophage cell type, the Spearman correlation rho is 0.18 (q value = 6.97×10^{-11}), indicating a weak but significant positive correlation. To investigate the implication of the quantity of nearby TEs around the genes, we compared the genes only from families whose two genes had similar TE environment. In that case, we observed a stronger positive correlation (Table 4). For example, in the macrophage cell type, the Spearman rho is 0.27 (q value = 1.33×10^{-8}). On the contrary, when considering families in which the two genes have a different TE environment, although there is still a positive correlation, it is weaker (for example, r = 0.13, q value = 4.40×10^{-5} in the macrophage cell type).

We investigated whether the age of the family may be implicated in the observed correlations. There is a positive correlation in the methylation level between the duplicated genes for young and middle age families but this correlation is either absent or not significant for old families (Supplementary Table S5). When taking into account the TE environment around the genes, the conservation of the methylation level is higher in young and middle-age families for families whose genes have a similar TE environment, except for young families with respect to erythroblasts (Supplementary Table S5). This is true also for old families for three cell types (erythroblast, macrophage, and CD8T), although the correlation values are weak. It thus seems that in the case of methylation conservation, the age of the family plays an important role, in addition to the TE environment.

3.4. The Duplicated Genes with Very Low Expression Divergence Display a High Conservation of Epigenetic Modifications and TE Environment

We found that the majority of the genes are either expressed in the four cell types (n = 1677—59.05%) or in none of them (n = 729—25.67%). Of the 1420 families, genes from 267 of them presented no expression in all cell types for both members of the pair. These gene families were not considered in the remaining analyses.

To determine the divergence in expression between duplicates of a given family, we computed the normalized Manhattan distance d_m to compare differences in the relative abundance of the two genes across the four cell types. We looked at the average Manhattan distance inside each class of ages (Figure 5A).

As could be expected, the divergence of expression between the genes is associated with the age of the family, genes from old families having an expression divergence higher than genes from young families (Kruskal-Wallis chi-squared = 15.789, df = 2, p-value = 0.0003727). We then tested whether the TE environment around the genes could be associated with the observed expression divergence. For that, we separated the gene families according to the TE environment of the duplicated genes (similar or different). The results are presented on the Figure 5B,C. We observed the same tendency as when all gene families were considered together, irrespective of the TE environment, to have a difference in gene expression divergence according to the age of the families (different TE environment, Kruskal-Wallis chi-squared = 7.8622, df = 2, p-value = 0.01962; same TE environment Kruskal-Wallis chi-squared = 10.421, df = 2, p-value = 0.00546). However, the expression divergence is the same for a given class of age independently of the TE environment

(Two-sample Kolmogorov-Smirnov tests D = 0.29895, p-value = 0.1048, D = 0.21009, p-value = 0.1433, and D = 0.03746, p-value = 0.9661 for young, middle-age, and old families, respectively).

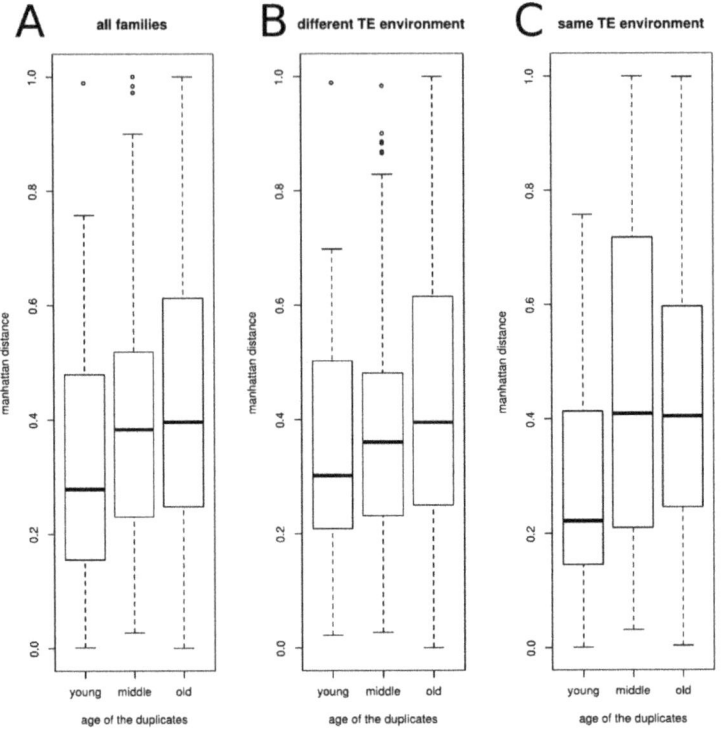

Figure 5. Average Manhattan distance of the duplicated gene expression level inside each class of ages, (**A**) for all families (Kruskal-Wallis chi-squared = 15.789, df = 2, p-value = 0.0003727), (**B**) for families whose two genes have different TE environment (Kruskal-Wallis chi-squared = 7.8622, df = 2, p-value = 0.01962), (**C**) for families whose two genes have the same TE environment (Kruskal-Wallis chi-squared = 10.421, df = 2, p-value = 0.00546).

We then studied the association between the different types of epigenetic modifications of the duplicated genes, the expression divergence and the TE environment. We separated the gene families according to the level of expression divergence in four classes (very low d_m [0–0.25], low d_m [0.25–0.5], medium d_m [0.5–0.75], and high d_m [0.75–1]). The results showed that a strong positive correlation in the DNA methylation level can be observed only for the families whose duplicates share similar TE environment and have a very low expression divergence (Spearman correlation rho = 0.33, q value = 6.88×10^{-3}; Supplementary Table S6). We performed the same kind of analysis considering the mean histone modification enrichment between the duplicated genes of each family (Supplementary Table S6). As previously, we separated the gene families according to the level of gene expression divergence. The results showed also a strong positive correlation of the histone enrichment between paired genes when they have similar TE environment and a very low expression divergence (from rho = 0.31 q value = 1.28×10^{-2} for H3K9me3 to rho = 0.57, q value = 1.77×10^{-7} for H3K4me3 Supplementary Table S6). A less strong positive correlation is also observed for three histone modifications (H3K4me3, H3K36me3 and H3K9me3) for genes with a low expression divergence and a same TE environment (rho = 0.29 q value = 8.16×10^{-3}, rho = 0.29 q value = 8.16×10^{-3}, and rho = 0.26 q value = 2.09×10^{-2}, respectively). There is also a positive correlation for families whose genes have a different TE environment and a very low expression divergence for H3K4me3, H3K27ac, H3K36me3, and H3K27me3, these correlations being less strong than for genes having the same TE environment (rho = 0.33 q value = 5.23 × 10^{-5}, rho = 0.29 q value = 4.65 × 10^{-4}, rho = 0.25 q value = 3.16 × 10^{-3}, and rho = 0.19 q value = 3.10 × 10^{-2}, respectively). The conservation of epigenetic modifications and of the TE environment around genes is associated with a very low expression divergence between duplicated genes.

4. Discussion

The maintenance and evolution of duplicated genes have been proposed to be linked to variation in epigenetic modifications [65]. For example, it has been shown in zebrafish that epigenetic divergence of duplicated genes affects both their expression and their functional divergence [66]. In humans, duplicated genes display highly consistent patterns of DNA methylation divergence across multiple tissues due to different frequencies of sequence motifs, which allowed the proposal that DNA methylation could be a causative factor for functional divergence of duplicated genes [29]. In *Arabidopsis*, the presence of H3K27me3 correlates with a slower rate of function evolution in duplicated gene families [30]. These various examples indicate that epigenetic modifications can have an evolutionary importance in the fate of duplicated genes. To gain insight into this question, we have analyzed in this work different histone modifications enrichment and DNA methylation level between pairs of genes from a same family, in different cell types, taking into account the presence of TEs near the genes, to evaluate their impact in any potential conservation or divergence of these epigenetic modifications.

We have focused our interest on gene families of size two. They represent the majority of the multigenic families in the human genome. We have observed that on average, these families are quite old. This is consistent with the hypothesis of the two rounds of whole genome duplication that occurred early in the evolution of vertebrates [67,68]. It was indeed predicted that we should have an excess of two and four size gene families in the human genome due to extensive gene losses that occurred later [69]. Even if the average age is quite high for those families, there remains a substantial number of families that appeared recently, via other mechanisms. Their synonymous substitution rate distribution is consistent with what was observed for all duplicated genes with Ks values less than 1 [70], indicating that the young families of size two are representative of all young duplicated families. When we analyzed the TE environment around these duplicated genes, we found that the proportion of TEs around them is not different than when considering all human genes, with TE-free genes being the less abundant category of genes, followed by TE-rich genes. Interestingly, we observed that genes from the same family tend to globally conserve the same type of TE neighborhood. It could be expected to observe this tendency only for young gene families, whose genes did not have time to differently accumulate new TE insertions. Young families indeed present an excess of similar TE environments in both genes. However, it is also true for older gene families, even if the proportion decreases. Although this indicates a link between the conservation of TE neighborhood between duplicated genes and the age of the duplication, it is not the only explanation since we can still observe this effect in old families. In old families, we could also hypothesize that some selective pressures to conserve the gene environment are at work that could explain the similar TE environment. The duplicated genes displayed a similar level of selection acting on them, indicating that almost all genes in our dataset evolve under purifying selection. These selective pressures could thus explain why genes from the same family tend to conserve the same TE environment. Selective constraints acting on genes have already been shown to be associated with the presence of TE insertions near the genes [71,72]. In particular, TE-free genes were shown to be subjected to a stronger purifying selection when compared to TE-rich genes [72]. However, the same selective pressure is also acting on gene pairs for which the TE environment is different. Then, the purifying selection that could act against TE insertions is not enough to explain why the members of some gene families conserve the same TE environment. Another possibility to explain the conservation of the TE environment could be linked with the gene function of duplicated genes. However, we did not detect any functional bias among the duplicated genes with the same TE environment when compared to all duplicated genes in the human genome. To go deeper to explore this question would be to focus more specifically on larger gene families for which more data are available concerning their function.

We have shown in this work that according to the proportion of TEs inserted near genes, there are variations in the level of methylation and the enrichment in histone modification of genes. In particular, TE-free genes are depleted in H3K36me3 whereas TE-very-rich genes are on the contrary enriched for this modification. This modification has usually been described to be associated with active chromatin but it has also been shown to be implicated in various other mechanisms like transcriptional repression, alternative splicing or DNA methylation [73]. Interestingly, this modification can promote repressive chromatin within actively transcribed genes, preventing spurious transcription initiation from cryptic promoters or TE remnants [74]. The histone modifications H3K4me1, H3K4me3, H3K27ac, H3K27me3 and H3K9me3 were found to be more present in TE-free genes rather than in genes with TEs in their surroundings. This could be expected for the modifications H3K4me1, H3K4me3, and H3K27ac, which have been shown to be associated with actively transcribed regions, if we consider that TEs are rather associated with repressive modifications [63,75]. This could be more surprising concerning the repressive modifications H3K27me3 and H3K9me3, which have been shown to be associated with TE repression in various cell types and organisms [43,63,75–78]. Since histone modifications can spread at TE insertions [46], it could be expected that genomic regions with numerous TE insertions would be impacted by repressive modifications originating in TEs. However, in this work, we are considering TEs that are found near or in genes, rather than intergenic insertions. We are thus considering TE insertions among which some could potentially have a role in the regulation of gene expression and some could just be neutral with no particular effect. Indeed, it has been observed that SINE elements are depleted in H3K9me3, especially when they are close to genes, supporting a potential role of these elements in the gene regulation [76]. Moreover, we already observed these results in other work [53], that could be explained by the "exaptation hypothesis" [77], considering that epigenetic modifications associated with specific TE insertions could be adaptive. This would imply that among all TE insertions in a genome, not all of them will have the same impact on gene expression. We also observed that TE-free genes displayed the lowest level of methylation when compared to genes with TEs in their surrounding. This is what could be expected if the presence of TEs in or near genes triggers DNA methylation, since this epigenetic modification has been largely associated with TE silencing, especially in mammals [79]. Interestingly, the proportion of TEs does not seem to impact the level of methylation since even TE-very-poor genes displayed as much DNA methylation than TE-very-rich genes. This could indicate that the methylation level does not increase with the number of TEs but as soon as even a few TEs are present, they are susceptible to trigger a significant amount of methylation.

We compared the histone enrichment and methylation level between both members of the same gene family in four different cell types to determine whether duplicated genes tend to conserve their epigenetic environment. As we could expect, there is a positive correlation of the epigenetic modification between genes from the same family, especially when the families are young. This is consistent with what was previously observed concerning the DNA methylation divergence of duplicated genes, with young duplicates displaying similar levels of methylation compared to older duplicates [29,31]. This could be explained by the fact that young duplicates are likely to be in a similar genomic environment. Indeed, when we considered only young duplicates (99 pairs of genes), there is a strong positive correlation for the histone enrichment irrespective of the TE environment, when the two genes are on the same chromosome (62 pairs of genes) (Supplementary Table S7). However, the duplicated genes we analyzed are on average very far away and sometimes even on different chromosomes. The conservation of epigenetic modification is in contradiction to the results presented by a study on segmental duplications in which an asymmetry was observed in the methylation level and in the histone acetylation that could be linked to pseudogenization [80]. Although in this last study, the genes considered may not all be pseudogenes, the discrepancy could be explained by the fact that in our work, we focused only on duplicated genes that are both functional. It was proposed that when a gene is in a different genomic environment, this could trigger changes in epigenetic modifications that could allow new duplicates to be submitted to new selective pressures preventing their pseudogenization [81]. The correlation we observed is stronger when the TE environment of both genes is similar. This could be a byproduct of selective pressure acting on those genes that would have the consequence to conserve the same proportion of TE insertions by removing any new insertions. However, when duplicated genes with different TE environments are submitted to the same selective constraints, then the selective pressure acting on duplicated genes is not enough to explain this observation. In *Arabidopsis*, there is an association between the conservation of H36K27me3 of paralogs and conserved noncoding sequences (CNS) [82]. The same mechanism could be at work in this case. However, we did not find much overlap between the duplicated genes and CNS previously identified in humans [83]. Only 17% ($n = 484$) of the duplicated genes from our analysis were overlapping with at least one CNS. This overlapping concerned the two duplicated genes of only 49 gene families. The presence of TEs could thus be implicated in both the maintenance and the divergence of epigenetic modifications.

In conclusion, our results point out the possibility for TE insertions to participate in the modulation of epigenetic variation of genes, especially inside duplicated gene families. New TE insertions could help trigger new epigenetic modifications that could have an impact in the functional divergence of the duplicated genes, whereas ancestral insertions would on the contrary have an effect of conservation. This hypothesis is supported by the fact that we observed a strong positive correlation in epigenetic modification between both duplicates when they display very low expression divergence and the same TE environment, irrespective of the age of the family. Perspectives on this work will require to work at the individual TE insertion level in order to identify, without any ambiguity, epigenetic modifications associated with them to clearly identify their effect on gene regulation.

Supplementary Materials: The following are available online at http://www.mdpi.com/2073-4425/10/3/249/s1. Supplementary data S1: List and positions of the duplicated genes used in this work. Supplementary Table S1: Number of gene families according to the TE neighborhood category of each duplicated gene and the age of the families. Supplementary Table S2: mean histone enrichment of genes for each tissue type and according to their TE neighborhood. Supplementary Table S3: correlations of histone enrichment between duplicated genes of each family across all cell types. Supplementary Table S4: correlation of the histone enrichment between genes from the same family, according to the TE neighborhood and the age of the family. Supplementary Table S5: correlations of the methylation level between duplicated genes from a same family, according to the age of the family and the TE neighborhood. Supplementary Table S6: correlations of the methylation level or histone enrichment of the duplicated genes according to the level of expression divergence between the two genes across all tissues. Supplementary Table S7: correlation of the histone enrichment between genes from the same young family, according to the TE neighborhood and the position on the chromosome.

Author Contributions: E.L. conceived the analysis. R.L., C.R. and E.L. performed the analyses and interpreted the results. E.L. wrote the first version of the manuscript, all authors reviewed and edited the different versions.

Funding: This work was funded by the CNRS, the University Lyon 1 and the Laboratory "Biométrie et Biologie Evolutive".

Acknowledgments: This work was performed using the computing facilities of the CC LBBE/PRABI. This study makes use of data generated by the Blueprint Consortium. A full list of the investigators who contributed to the generation of the data is available from www.blueprint-epigenome.eu. This work was supported by the CNRS, the University Lyon 1 and the Laboratory "Biométrie et Biologie Evolutive". This work is dedicated to the memory of my long term partner Uhuru.

Conflicts of Interest: The authors declare no conflict of interest.

References

1. Straussman, R.; Nejman, D.; Roberts, D.; Steinfeld, I.; Blum, B.; Benvenisty, N.; Simon, I.; Yakhini, Z.; Cedar, H. Developmental programming of CpG island methylation profiles in the human genome. *Nat. Struct. Mol. Biol.* **2009**, *16*, 564–571. [CrossRef] [PubMed]
2. Varley, K.E.; Gertz, J.; Bowling, K.M.; Parker, S.L.; Reddy, T.E.; Pauli-Behn, F.; Cross, M.K.; Williams, B.A.; Stamatoyannopoulos, J.A.; Crawford, G.E.; et al. Dynamic DNA methylation across diverse human cell lines and tissues. *Genome Res.* **2013**, *23*, 555–567. [CrossRef]

3. Ha, M.; Ng, D.W.-K.; Li, W.-H.; Chen, Z.J. Coordinated histone modifications are associated with gene expression variation within and between species. *Genome Res.* **2011**, *21*, 590–598. [CrossRef]
4. Ghosh, D.; Qin, Z.S. Statistical issues in the analysis of ChIP-seq and RNA-seq data. *Genes* **2010**, *1*, 317–334. [CrossRef]
5. Kucharski, R.; Maleszka, J.; Foret, S.; Maleszka, R. Nutritional control of reproductive status in honeybees via DNA methylation. *Science* **2008**, *319*, 1827–1830. [CrossRef] [PubMed]
6. Chittka, A.; Wurm, Y.; Chittka, L. Epigenetics: The Making of Ant Castes. *Curr. Biol.* **2012**, *22*, R835–R838. [CrossRef] [PubMed]
7. Jaenisch, R.; Bird, A. Epigenetic regulation of gene expression: how the genome integrates intrinsic and environmental signals. *Nat. Genet.* **2003**, *33*, 245–254. [CrossRef]
8. Bird, A. DNA methylation patterns and epigenetic memory. *Genes Dev.* **2002**, *16*, 6–21. [CrossRef] [PubMed]
9. Weber, M.; Schübeler, D. Genomic patterns of DNA methylation: targets and function of an epigenetic mark. *Curr. Opin. Cell Biol.* **2007**, *19*, 273–280. [CrossRef]
10. Jones, P.A.; Liang, G. Rethinking how DNA methylation patterns are maintained. *Nat. Rev. Genet.* **2009**, *10*, 805–811. [CrossRef]
11. Bernstein, B.E.; Meissner, A.; Lander, E.S. The Mammalian Epigenome. *Cell* **2007**, *128*, 669–681. [CrossRef]
12. Carthew, R.W.; Sontheimer, E.J. Origins and Mechanisms of miRNAs and siRNAs. *Cell* **2009**, *136*, 642–655. [CrossRef] [PubMed]
13. Ghildiyal, M.; Zamore, P.D. Small silencing RNAs: an expanding universe. *Nat. Rev. Genet.* **2009**, *10*, 94–108. [CrossRef]
14. Grant, P.A. A tale of histone modifications. *Genome Biol.* **2001**, *2*, REVIEWS0003. [CrossRef] [PubMed]
15. Peterson, C.L.; Laniel, M.-A. Histones and histone modifications. *Curr. Biol.* **2004**, *14*, R546–551. [CrossRef]
16. Li, B.; Carey, M.; Workman, J.L. The role of chromatin during transcription. *Cell* **2007**, *128*, 707–719. [CrossRef] [PubMed]
17. Esteller, M. Cancer epigenomics: DNA methylomes and histone-modification maps. *Nat. Rev. Genet.* **2007**, *8*, 286–298. [CrossRef]
18. Britten, R.J.; Davidson, E.H. Gene regulation for higher cells: A theory. *Science* **1969**, *165*, 349–357. [CrossRef] [PubMed]
19. Mihola, O.; Trachtulec, Z.; Vlcek, C.; Schimenti, J.C.; Forejt, J. A mouse speciation gene encodes a meiotic histone H3 methyltransferase. *Science* **2009**, *323*, 373–375. [CrossRef] [PubMed]
20. Cain, C.E.; Blekhman, R.; Marioni, J.C.; Gilad, Y. Gene expression differences among primates are associated with changes in a histone epigenetic modification. *Genetics* **2011**, *187*, 1225–1234. [CrossRef]
21. Zeng, J.; Konopka, G.; Hunt, B.G.G.; Preuss, T.M.M.; Geschwind, D.; Yi, S.V.V. Divergent whole-genome methylation maps of human and chimpanzee brains reveal epigenetic basis of human regulatory evolution. *Am. J. Hum. Genet.* **2012**, *91*, 455–465. [CrossRef] [PubMed]
22. Cocozza, S.; Akhtar, M.M.; Miele, G.; Monticelli, A. CpG islands undermethylation in human genomic regions under selective pressure. *PLoS ONE* **2011**, *6*, e23156. [CrossRef]
23. Akhtar, M.M.; Scala, G.; Cocozza, S.; Miele, G.; Monticelli, A. (2013) CpG islands under selective pressure are enriched with H3K4me3, H3K27ac and H3K36me3 histone modifications. *BMC Evol. Biol.* **2013**, *13*, 145. [CrossRef]
24. Hernando-Herraez, I.; Heyn, H.; Fernandez-Callejo, M.; Vidal, E.; Fernandez-Bellon, H.; Prado-Martinez, J.; Sharp, A.J.; Esteller, M.; Marques-Bonet, T. The interplay between DNA methylation and sequence divergence in recent human evolution. *Nucleic Acids Res.* **2015**, *43*, 8204–8214. [CrossRef]
25. Kolasinska-Zwierz, P.; Down, T.; Latorre, I.; Liu, T.; Liu, X.S.; Ahringer, J. Differential chromatin marking of introns and expressed exons by H3K36me3. *Nat. Genet.* **2009**, *41*, 376–381. [CrossRef] [PubMed]
26. Woo, Y.H.; Li, W.H. Evolutionary conservation of histone modifications in mammals. *Mol. Biol. Evol.* **2012**, *29*, 1757–1767. [CrossRef]
27. Sarda, S.; Zeng, J.; Hunt, B.G.; Yi, S.V. The evolution of invertebrate gene body methylation. *Mol. Biol. Evol.* **2012**, *29*, 1907–1916. [CrossRef] [PubMed]
28. Takuno, S.; Gaut, B.S. Gene body methylation is conserved between plant orthologs and is of evolutionary consequence. *Proc. Natl. Acad. Sci. USA* **2013**, *110*, 1797–1802. [CrossRef] [PubMed]
29. Keller, T.E.; Yi, S.V. DNA methylation and evolution of duplicate genes. *Proc. Natl. Acad. Sci. USA* **2014**, *111*, 5932–5937. [CrossRef]

30. Berke, L.; Sanchez-Perez, G.F.; Snel, B. Contribution of the epigenetic mark H3K27me3 to functional divergence after whole genome duplication in Arabidopsis. *Genome Biol.* **2012**, *13*, R94. [CrossRef]
31. Prendergast, J.G.D.; Chambers, E.V.; Semple, C.A.M. Sequence-Level Mechanisms of Human Epigenome Evolution. *Genome Biol. Evol.* **2014**, *6*, 1758–1771. [CrossRef] [PubMed]
32. Consortium International Human Genome Sequencing. Finishing the euchromatic sequence of the human genome. *Nature* **2004**, *431*, 931–945. [CrossRef] [PubMed]
33. Lander, E.S.; Linton, L.M.; Birren, B.; Nusbaum, C.; Zody, M.C.; Baldwin, J.; Devon, K.; Dewar, K.; Doyle, M.; FitzHugh, W.; et al. Initial sequencing and analysis of the human genome. *Nature* **2001**, *409*, 860–921. [PubMed]
34. Cordaux, R.; Batzer, M.A. The impact of retrotransposons on human genome evolution. *Nat. Rev. Genet.* **2009**, *10*, 691–703. [CrossRef] [PubMed]
35. Ludwig, M. Functional evolution of noncoding DNA. *Curr. Opin. Genet. Dev.* **2002**, *12*, 634–639. [CrossRef]
36. Kidwell, M.G.; Lisch, D.R. Transposable elements and host genome evolution. *Trends Ecol. Evol.* **2000**, *15*, 95–99. [CrossRef]
37. Biémont, C.; Vieira, C. Genetics: Junk DNA as an evolutionary force. *Nature* **2006**, *443*, 521–524. [CrossRef] [PubMed]
38. Medstrand, P.; van de Lagemaat, L.N.; Mager, D.L. Retroelement distributions in the human genome: Variations associated with age and proximity to genes. *Genome Res.* **2002**, *12*, 1483–1495. [CrossRef]
39. Hoen, D.R.; Park, K.C.; Elrouby, N.; Yu, Z.; Mohabir, N.; Cowan, R.K.; Bureau, T.E. Transposon-mediated expansion and diversification of a family of ULP-like genes. *Mol. Biol Evol.* **2006**, *2006 23*, 1254–1268. [CrossRef]
40. Juretic, N.; Hoen, D.R.; Huynh, M.L.; Harrison, P.M.; Bureau, T.E. The evolutionary fate of MULE-mediated duplications of host gene fragments in rice. *Genome Res.* **2005**, *15*, 1292–1297. [CrossRef]
41. Slotkin, R.K.; Martienssen, R. Landscape of Somatic Retrotransposition in Human Cancers. *Nat. Rev. Genet.* **2007**, *8*, 272–285. [CrossRef]
42. Huda, A.; Jordan, I.K. Epigenetic regulation of mammalian genomes by transposable elements. *Ann. N. Y. Acad. Sci.* **2009**, *1178*, 276–284. [CrossRef]
43. Kulis, M.; Esteller, M. DNA methylation and cancer. *Adv. Genet.* **2010**, *70*, 27–56.
44. Ross, J.P.; Rand, K.N.; Molloy, P.L. Hypomethylation of repeated DNA sequences in cancer. *Epigenomics* **2010**, *2*, 245–269. [CrossRef]
45. Morgan, H.D.; Sutherland, H.G.; Martin, D.I.; Whitelaw, E. Epigenetic inheritance at the agouti locus in the mouse. *Nat. Genet.* **1999**, *23*, 314–318. [CrossRef]
46. Rebollo, R.; Karimi, M.M.; Bilenky, M.; Gagnier, L.; Miceli-Royer, K.; Zhang, Y.; Goyal, P.; Keane, T.M.; Jones, S.; Hirst, M.; et al. Retrotransposon-induced heterochromatin spreading in the mouse revealed by insertional polymorphisms. *PLoS Genet.* **2011**, *7*, e1002301. [CrossRef] [PubMed]
47. Eichten, S.R.; Ellis, N.A.; Makarevitch, I.; Yeh, C.T.; Gent, J.I.; Guo, L.; McGinnis, K.M.; Zhang, X.; Schnable, P.S.; Vaughn, M.W.; et al. Spreading of Heterochromatin Is Limited to Specific Families of Maize Retrotransposons. *PLoS Genet.* **2012**, *8*, e1003127. [CrossRef] [PubMed]
48. Gendrel, A.-V.; Lippman, Z.; Yordan, C.; Colot, V.; Martienssen, R.A. Dependence of heterochromatic histone H3 methylation patterns on the *Arabidopsis* gene *DDM1*. *Science* **2002**, *297*, 1871–1873. [CrossRef] [PubMed]
49. Volpe, T.A.; Kidner, C.; Hall, I.M.; Teng, G.; Grewal, S.I.S.; Martienssen, R.A. Regulation of heterochromatic silencing and histone H3 lysine-9 methylation by RNAi. *Science* **2002**, *297*, 1833–1837. [CrossRef] [PubMed]
50. Lippman, Z.; Gendrel, A.-V.; Black, M.; Vaughn, M.W.; Dedhia, N.; McCombie, W.R.; Lavine, K.; Mittal, V.; May, B.; Kasschau, K.D.; et al. Role of transposable elements in heterochromatin and epigenetic control. *Nature* **2004**, *430*, 471–476. [CrossRef]
51. Mirouze, M.; Vitte, C. Transposable elements, a treasure trove to decipher epigenetic variation: Insights from *Arabidopsis* and crop epigenomes. *J. Exp. Bot.* **2014**, *65*, 2801–2812. [CrossRef] [PubMed]
52. Byun, H.-M.; Heo, K.; Mitchell, K.J.; Yang, A.S. Mono-allelic retrotransposon insertion addresses epigenetic transcriptional repression in human genome. *J. Biomed. Sci.* **2012**, *19*, 13. [CrossRef]
53. Grégoire, L.; Haudry, A.; Lerat, E. The transposable element environment of human genes is associated with histone and expression changes in cancer. *BMC Genomics* **2016**, *17*, 588. [CrossRef]

54. Trizzino, M.; Park, Y.; Holsbach-Beltrame, M.; Aracena, K.; Mika, K.; Caliskan, M.; Perry, G.H.; Lynch, V.J.; Brown, C.D. Transposable elements are the primary source of novelty in primate gene regulation. *Genome Res.* **2017**, *27*, 1623–1633. [CrossRef]
55. Penel, S.; Arigon, A.-M.; Dufayard, J.-F.; Sertier, A.-S.; Daubin, V.; Duret, L.; Gouy, M.; Perrière, G. Databases of homologous gene families for comparative genomics. *BMC Bioinform.* **2009**, *10* (Suppl 6), S3. [CrossRef] [PubMed]
56. Yang, Z. PAML 4: Phylogenetic analysis by maximum likelihood. *Mol. Biol. Evol.* **2007**, *24*, 1586–1591. [CrossRef] [PubMed]
57. Li, H.; Durbin, R. Fast and accurate short read alignment with Burrows-Wheeler transform. *Bioinformatics* **2009**, *25*, 1754–1760. [CrossRef] [PubMed]
58. Stunnenberg, H.G.; Hirst, M.; de Almeida, M.; Altucci, L.; Amin, V.; Amit, I.; Antonarakis, S.E.; Aparicio, S.; Arima, T.; International Human Epigenome Consortium; et al. The International Human Epigenome Consortium: A Blueprint for Scientific Collaboration and Discovery. *Cell* **2016**, *167*, 1145–1149. [CrossRef]
59. Hart, T.; Komori, H.K.; LaMere, S.; Podshivalova, K.; Salomon, D.R. Finding the active genes in deep RNA-seq gene expression studies. *BMC Genomics* **2013**, *14*, 778. [CrossRef]
60. Conesa, A.; Madrigal, P.; Tarazona, S.; Gomez-Cabrero, D.; Cervera, A.; McPherson, A.; Szcześniak, M.W.; Gaffney, D.J.; Elo, L.L.; Zhang, X.; et al. A survey of best practices for RNA-seq data analysis. *Genome Biol.* **2016**, *17*, 13. [CrossRef]
61. Bailly-Bechet, M.; Haudry, A.; Lerat, E. 'One code to find them all': A perl tool to conveniently parse RepeatMasker output files. *Mob. DNA* **2014**, *5*, 13. [CrossRef]
62. Stewart, C.; Kural, D.; Strömberg, M.P.; Walker, J.A.; Konkel, M.K.; Stütz, A.M.; Urban, A.E.; Grubert, F.; Lam, H.Y.; Lee, W.P.; et al. A comprehensive map of mobile element insertion polymorphisms in humans. *PLoS Genet.* **2011**, *7*, e1002236. [CrossRef]
63. Walter, M.; Teissandier, A.; Pérez-Palacios, R.; Bourc'his, D. An epigenetic switch ensures transposon repression upon dynamic loss of DNA methylation in embryonic stem cells. *Elife* **2016**, *5*, e11418. [CrossRef]
64. R Core Team. *R: A Language and Environment for Statistical Computing*; R Foundation for Statistical Computing: Vienna, Austria, 2018. Available online: https://www.R-project.org/ (accessed on 29 January 2019).
65. Rodin, S.N.; Riggs, A.D. Epigenetic Silencing May Aid Evolution by Gene Duplication. *J. Mol. Evol.* **2003**, *56*, 718–729. [CrossRef] [PubMed]
66. Zhong, Z.; Du, K.; Yu, Q.; Zhang, Y.E.; He, S. Divergent DNA Methylation Provides Insights into the Evolution of Duplicate Genes in Zebrafish. *G3* **2016**, *6*, 3581–3591. [CrossRef] [PubMed]
67. McLysaght, A.; Hokamp, K.; Wolfe, K.H. Extensive genomic duplication during early chordate evolution. *Nat. Genet.* **2002**, *31*, 200–204. [CrossRef]
68. Dehal, P.; Boore, J.L. Two rounds of whole genome duplication in the ancestral vertebrate. *PLoS Biol.* **2005**, *3*, e314. [CrossRef] [PubMed]
69. Mendivil Ramos, O.; Ferrier, D.E.K. Mechanisms of Gene Duplication and Translocation and Progress towards Understanding Their Relative Contributions to Animal Genome Evolution. *Int. J. Evol. Biol.* **2012**, *2012*, 1–10. [CrossRef]
70. Zhang, P.; Min, W.; Li, W.-H. Different age distribution patterns of human, nematode, and Arabidopsis duplicate genes. *Gene* **2004**, *342*, 263–268. [CrossRef]
71. Mortada, H.; Vieira, C.; Lerat, E. Genes devoid of full-length transposable element insertions are involved in development and in the regulation of transcription in human and closely related species. *J. Mol. Evol.* **2010**, *71*, 180–191. [CrossRef]
72. Nellåker, C.; Keane, T.M.; Yalcin, B.; Wong, K.; Agam, A.; Belgard, T.G.; Flint, J.; Adams, D.J.; Frankel, W.N.; Ponting, C.P. The genomic landscape shaped by selection on transposable elements across 18 mouse strains. *Genome Biol.* **2012**, *13*, R45. [CrossRef]
73. Wagner, E.J.; Carpenter, P.B. Understanding the language of Lys36 methylation at histone H3. *Nat. Rev. Mol. Cell Biol.* **2012**, *13*, 115–126. [CrossRef]
74. Teissandier, A.; Bourc'his, D. Gene body DNA methylation conspires with H3K36me3 to preclude aberrant transcription. *EMBO J.* **2017**, *36*, 1471–1473. [CrossRef] [PubMed]
75. Kondo, Y.; Issa, J.-P.J. Enrichment for histone H3 lysine 9 methylation at Alu repeats in human cells. *J. Biol. Chem.* **2003**, *278*, 27658–27662. [CrossRef] [PubMed]

76. Martens, J.H.A.; O'Sullivan, R.J.; Braunschweig, U.; Opravil, S.; Radolf, M.; Steinlein, P.; Jenuwein, T. The profile of repeat-associated histone lysine methylation states in the mouse epigenome. *EMBO J.* **2005**, *24*, 800–812. [CrossRef]
77. Huda, A.; Mariño-Ramírez, L.; Jordan, I.K. Epigenetic histone modifications of human transposable elements: genome defense versus exaptation. *Mob. DNA* **2010**, *1*, 2. [CrossRef] [PubMed]
78. Pauler, F.M.; Sloane, M.A.; Huang, R.; Regha, K.; Koerner, M.V.; Tamir, I.; Sommer, A.; Aszodi, A.; Jenuwein, T.; Barlow, D.P. H3K27me3 forms BLOCs over silent genes and intergenic regions and specifies a histone banding pattern on a mouse autosomal chromosome. *Genome Res.* **2009**, *19*, 221–233. [CrossRef] [PubMed]
79. Li, E.; Zhang, Y. DNA methylation in mammals. *Cold Spring Harb. Perspect. Biol.* **2014**, *6*, a019133. [CrossRef]
80. Zheng, D. Gene duplication in the epigenomic era. *Epigenetics* **2008**, *3*, 250–253. [CrossRef]
81. Rodin, S.N.; Parkhomchuk, D.V.; Rodin, A.S.; Holmquist, G.P.; Riggs, A.D. Repositioning-dependent fate of duplicate genes. *DNA Cell Biol.* **2005**, *24*, 529–542. [CrossRef]
82. Berke, L.; Snel, B. The Histone Modification H3K27me3 Is Retained after Gene Duplication and Correlates with Conserved Noncoding Sequences in Arabidopsis. *Genome Biol. Evol.* **2014**, *6*, 572–579. [CrossRef] [PubMed]
83. Babarinde, I.A.; Saitou, N. Genomic Locations of Conserved Noncoding Sequences and Their Proximal Protein-Coding Genes in Mammalian Expression Dynamics. *Mol. Biol. Evol.* **2016**, *33*, 1807–1817. [CrossRef] [PubMed]

© 2019 by the authors. Licensee MDPI, Basel, Switzerland. This article is an open access article distributed under the terms and conditions of the Creative Commons Attribution (CC BY) license (http://creativecommons.org/licenses/by/4.0/).

Review

Centromeric Satellite DNAs: Hidden Sequence Variation in the Human Population

Karen H. Miga

UC Santa Cruz Genomics Institute, University of California, Santa Cruz, California, CA 95064, USA; khmiga@soe.ucsc.edu; Tel.: +1-831-459-5232

Received: 2 April 2019; Accepted: 3 May 2019; Published: 8 May 2019

Abstract: The central goal of medical genomics is to understand the inherited basis of sequence variation that underlies human physiology, evolution, and disease. Functional association studies currently ignore millions of bases that span each centromeric region and acrocentric short arm. These regions are enriched in long arrays of tandem repeats, or satellite DNAs, that are known to vary extensively in copy number and repeat structure in the human population. Satellite sequence variation in the human genome is often so large that it is detected cytogenetically, yet due to the lack of a reference assembly and informatics tools to measure this variability, contemporary high-resolution disease association studies are unable to detect causal variants in these regions. Nevertheless, recently uncovered associations between satellite DNA variation and human disease support that these regions present a substantial and biologically important fraction of human sequence variation. Therefore, there is a pressing and unmet need to detect and incorporate this uncharacterized sequence variation into broad studies of human evolution and medical genomics. Here I discuss the current knowledge of satellite DNA variation in the human genome, focusing on centromeric satellites and their potential implications for disease.

Keywords: satellite DNA; centromere; sequence variation; structural variation; repeat; alpha satellite; human satellites; genome assembly

1. Introduction

Genome-scale initiatives, such as the Human Genome Project and the 1000 Genome (1KG) consortium [1–3], have provided a wealth of genomic information that have greatly advanced basic and biomedical research. However, in light of this progress, the millions of bases that span each human centromeric region remain largely disconnected from contemporary genetic and genomic analyses. This has historically been due to the challenge of generating and validating linear assemblies of tandemly-repeated DNA (e.g., thousands of copies of a repeat with a limited number of sequence variants to guide overlap-consensus derived assemblies), which are known to span each centromeric region [4]. Our understanding of the sequence content and organization of human centromeres improved dramatically with the release of the GRCh38 reference genome, and recent efforts to generate true linear assemblies using "ultra-long" sequencing (i.e., reads that span hundreds of kilobases [5]), wherein the centromere-assigned gaps on each chromosome assembly were updated with sequence information [6,7]. Thus, we are entering a new era in genomics where centromeric DNAs are available for detailed study, either within a single karyotype or across human populations, that will drive research aimed to understand repeat variation that contributes to genome stability, population variation, and disease.

Centromeric satellite DNA arrays are known to vary extensively in the human population, yet few genomic tools have been developed to study the full extent of this sequence variation, thereby ignoring a fraction of the human genome expected to contribute directly to cancer and human disease [8–10]. The extent of variation has been documented at the cytogenetic level, and gross estimates of rearrangement and/or repeat expansion have been associated with cancer and infertility [11–13]. Additionally, the epigenetic regulation of satellite DNAs, as well as anomalous methylation and altered transcription of satellite DNAs, have been associated with human diseases [9,14]. However, these early observations are challenged by inconsistencies in association [15], small sample size and perhaps an incomplete (or often, low-resolution) understanding of underlying genomic structure and array variant composition.

Acknowledging the differences between the satellite arrays, there is limited utility in restricting studies to the use of a single genomic map. Rather, it is important to extend our survey of satellite DNA genomics to large panels of diverse individuals, thus enabling high-resolution maps of human sequence diversity in these regions. Such sampling efforts would be the foundation for a modern era of satellite DNA genomics: establishing allelic frequencies for satellite variants necessary to expand disease-association studies. Here I discuss our current understanding of satellite DNA variation as determined from whole-genome sequencing projects, with a focus on the largest families in the human genome and their association with disease.

2. What Proportion of the Human Genome is Defined by Peri/Centromeric Satellite DNAs?

The largest arrays of satellite DNAs in the human genome are organized within centromeric and pericentromeric regions [2,3,16]. Although several distinct satellite DNA families are known to contribute to pericentromeric regions (e.g., gamma, beta, and subtelomeric satellites [17–19]), this review is focused on alpha satellite and human satellites 2, 3, which are the most abundant in the human genome and most commonly associated with human disease [8]. The alpha satellite DNA family is defined by a group of related, highly divergent AT-rich repeats or 'monomers', each approximately 171 bp in length, which are found in every normal human centromere [20–22]. Previous genome-wide estimates of alpha satellite have observed that this family represents ~2.6% of the human genome [23], which roughly aligns with early hybridization-based estimates [16]. Additionally, previous physical maps of centromeric regions, and pulse-field gel electrophoresis (PFGE) southern-based estimates of chromosome-assigned satellite arrays, revealed an average (~3 Mbps) amount of alpha satellite per centromeric region [24–26]. Therefore, we can assign a very rough estimate of 72 Mbps across all 22 autosomes and two sex chromosomes (i.e., 2.4%), which remains in agreement with all previous genome-wide estimates. Human satellites 2, 3 (HSat2,3), are collectively defined by enrichment of a pentameric repeat, (CATTC)n, and represent the largest heterochromatin blocks (documented as at least 10 Mbps in length) in human pericentromeric regions; notably, on chromosomes 1, 9, 16, and Y [27–32]. In total, HSat2,3 are observed to be less abundant than alpha satellite (~1.5% of the genome) [31], yet early estimates of array lengths on the DYZ1 array suggest that abundance estimates may vary considerably in the human population [31,33–35]. Therefore, efforts to better understand the true extent of satellite subfamily overall variation would benefit from surveying a much larger panel of diverse individuals. In doing so, one can define the lower and upper bounds of satellite DNA content in the genome. For example, can one individual have 3% alpha satellite and another individual have closer to 10%? What defines these bounds? Further, do these fluctuations in overall satellite composition contribute to our understanding of chromosome segregation, genome stability, and disease?

Genome-wide estimates of alpha satellite and HSat2,3 content have relied on constructing comprehensive sequence databases using raw read data from each satellite DNA family, thereby avoiding underestimates due to assembly collapse of identical/near-identical repeats [23,31]. Alpha satellite and HSat2,3 exhibit considerable sequence heterogeneity [20,21,32], as observed most readily in the ability to hybridize specifically to divergent repeat sequences within chromosome-specific arrays [20,36–38]. Therefore, efforts to construct genome-wide libraries from short-read datasets rely on methods that are comprehensive and inclusive with respect to the potential heterogeneity within satellite families. One approach is to reformat published satellite sequence libraries [7,23,31] into catalogs of short oligonucleotide sequences (24 bps, representing sequences in both orientations) that are specific to a given satellite family; that is, each oligo is only observed within a respective satellite database and never observed to have an exact match anywhere else in the genome. It is then possible to survey existing low-coverage, publicly available, population datasets, such as 1KG data [1], to identify what percentage of reads (as determined by exact matches with oligo libraries) are assigned to each respective satellite family (Figure 1a,b). In a study of low-coverage 1KG sequence data representing 14 diverse populations (400 male individuals and 414 female individuals) [1,7], alpha satellite has a median of 3.1% genome-wide estimate, with a range between 1% and 5% (Figure 1a). These initial HSat2,3 estimates reveal that although this satellite family is typically less abundant than alpha (median, 2.1%, range ~1–7%), it is observed in many cases to match, or surpass, alpha satellite abundance [31] (Figure 1b).

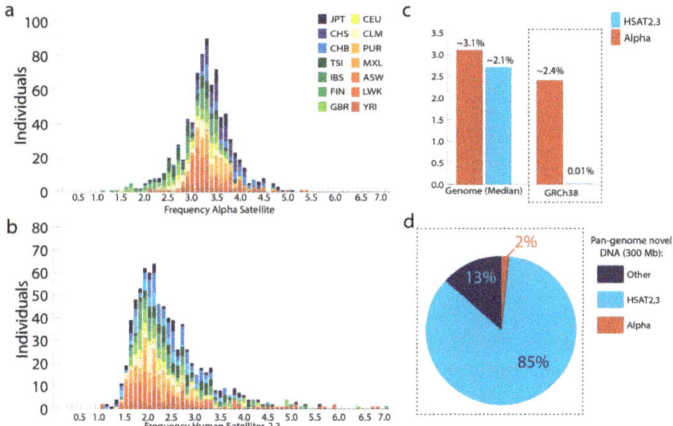

Figure 1. Proportion of alpha satellite and human satellites 2,3 in the human population. Using 1KG [1] data representing 14 diverse populations (400 male individuals and 414 female individuals) (a) the frequency of 24-mers that have an exact match with alpha satellite [23], (b) the frequency of 24-mers that have an exact match with human satellite 2,3 [31]. (c) Median frequencies (from panel (a) alpha and (b) HSat2,3) are listed relative to the observed frequency in the human reference genome assembly (GRCh38; GCA_000001405.15). (d) Evaluation of 300 Mb of DNA from the collective genomes of 910 people of African descent, previously determined to be missing or unaligned to GRCh38 [39]. Key for human subpopulations: CHB: Han Chinese in Beijing, China; JPT: Japanese in Tokyo, Japan; CHS: Southern Han Chinese; CEU: Utah Residents (CEPH) with Northern and Western European Ancestry; TSI: Toscani in Italia; FIN: Finnish in Finland; GBR: British in England and Scotland; IBS: Iberian Population in Spain; YRI: Yoruba in Ibadan, Nigeria; LWK: Luhya in Webuye, Kenya; GWD: Gambian in Western Divisions in the Gambia; ASW: Americans of African Ancestry in SW USA; MXL: Mexican Ancestry from Los Angeles USA; PUR: Puerto Ricans from Puerto Rico; CLM: Colombians from Medellin, Colombia.

Proper representation of satellite DNAs in human reference assemblies will be critical to ensure faithful short read mapping and accurate assessments of satellite family variability in the future. Notably, the addition of millions of bases of alpha satellite "reference models" with the release of the GRCh38 reference genome has provided initial short read mapping targets [7,40]. This has proven to be useful in decreasing off-target alignments and has enabled high-resolution studies aimed to study the epigenetic structure in centromeres [41,42]. In contrast, HSat2,3 are woefully underrepresented in all human assemblies, with only ~0.01% representation in GRCh38 (Figure 1c). This can lead to pronounced differences in the way we annotate and study human variation. For example, a recent study of 910 individuals of African descent identified roughly 300 Mbp of sequences not present in the human reference (GRCh38) [39], of which the largest proportion have exact oligo matches with HSat2,3 (Figure 1d). This demonstrates the importance of proper satellite DNA representation in the reference assembly in shaping our interpretation of novel sequences in the population. Notably, alpha satellite, a satellite family with great representation in the current reference assembly, still has candidate sequences that are not aligned to the reference models derived from the HuRef genome (Figure 1d, red). This result emphasizes a second important point: because satellites are expected to vary between genomes, the use of a single individual's genome as a reference, in this case HuRef [43], is not sufficient to capture the sequence diversity in the human population. Expanding the representation of human sequence diversity has been previously shown to improve mapping of variants [44,45], highlighting the need for a 'pan-human genome reference' to improve mapping efficiency and satellite variation studies in the future.

3. What is the Nature of Sequence Variation within a Single Satellite Array?

The variation of satellite DNAs genome-wide abundance is driven by repeat expansion and contraction, commonly attributed to mechanisms of non-homologous crossover and/or conversion [46,47]. Genomic-based studies of satellite DNA evolution have greatly benefited from the advancement of software designed to study tandem repeat variation in unassembled reads (reviewed [48]). The advancement of such high-resolution studies across a broad number of species is expected to dramatically advance our knowledge of the rates and mechanisms driving satellite array evolution. Previous studies of comprehensive studies of satellite DNA classes. Efforts in the past to study satellite repeat variation have focused on shorter microsatellites and tandem repeat classes that are amenable to complete assembly using long read technologies. However, recent efforts to study tandem repeat variation in human rDNA arrays revealed a high level of heterogeneity (i.e., an average rate of 7.5 variants per kb). Each rDNA unit is 45 kb with roughly 500 copies per diploid cell, and much like the satellite arrays, rDNA array length can vary significantly in size from just a few units to >100 between individuals [49,50].

The relationship between repeat units from different alpha satellite arrays would suggest that the rate of intra-chromosomal exchange (i.e., sister chromatid exchange) is higher than inter-homologue exchange [51]. As a result, most satellite arrays in the human genome can be defined by highly homogeneous arrays that can be often typified by chromosome-specific multi-monomeric repeat units, or higher-order repeats (HOR) [20,52]. Although the chromosome-assignment of HORs is largely invariant between individuals, as demonstrated by the effectiveness of commercially available satellite Fluorescence in situ hybridization (FISH) markers for chromosome labeling in clinical cytogenetics, the thousands of copies of the HOR that comprise a single array are expected to represent a mixture of expansion/contraction of repeat variants, shifts in orientation, and mobile element insertions (Figure 2a) [7,20,37,53].

This ever-changing genomic landscape guides our understanding of centromere function and chromosome stability. For example, the repeat structure and array length are expected to change the frequency of and spacing of the 17-bp centromere protein B binding motif, or CENP-B box. The functional role of CENP-B at human centromeres is not yet fully understood [54,55], yet recent studies suggest that the periodicity may contribute to kinetochore function and centromere fidelity [56,57]. Rearrangement in canonical HOR units, i.e., insertions and/or deletions presumably due to unequal crossing-over events, are observed at different frequencies between spatially distinct arrays. The Chr17-specific alpha satellite HOR (D17Z1) is characterized by arrays containing approximately 1000 repeat units that range in length from 11–16 monomers [36,56,58]. The frequency and ordering of these variants have been shown to influence the centromere location on human chromosomes with metastable epialleles [59,60]. Ultimately, sequence composition within each satellite array is thought to influence expression of the repeats [10,61,62], transcription factor binding [63,64], and replication efficiency [65–67]. Therefore, the high-resolution and comprehensive study of array sequence composition and structure is key to our understanding of how these specialized loci function.

Previous methods have used unassembled reads from whole-genome sequencing projects to evaluate chromosome-specific satellite overall abundance, or copy number, and the frequency of variants (e.g., HOR rearrangements, inversions, transposition, and single nucleotide variants (SNVs)) within the array [7,31,68]. Specifically, the centromeric regions on the X and Y chromosomes in male genomes offer a unique opportunity to study the variation in haploid array length within the human population. Altemose et al. [31], estimated the array size using low-coverage, short-read sequencing data from 396 male 1KG individuals [1], showing that the DYZ1 array varies over an order of magnitude (7–98 Mbps, with a mean of 24 Mb), consistent with previous experimental observations of Y-chromosome length variability [34,69,70]. Similarly, 1KG read-depth-based estimates of alpha satellite array lengths on the X and Y chromosomes (DXZ1 and DYZ3) agree with prior PFGE Southern experiments [7,25,71]. Although the X array has been predicted to have a 10-fold size range (800 kb to 8 Mbps), the medians of predicted X array lengths per human population, are observed to fall within experimentally validated lengths of 2.2–3.7 Mbps (mean 3010 kb) [7,25] (Figure 2b). This further corroborates the accuracy of short-read-based array length estimates applied to diverse groups of people.

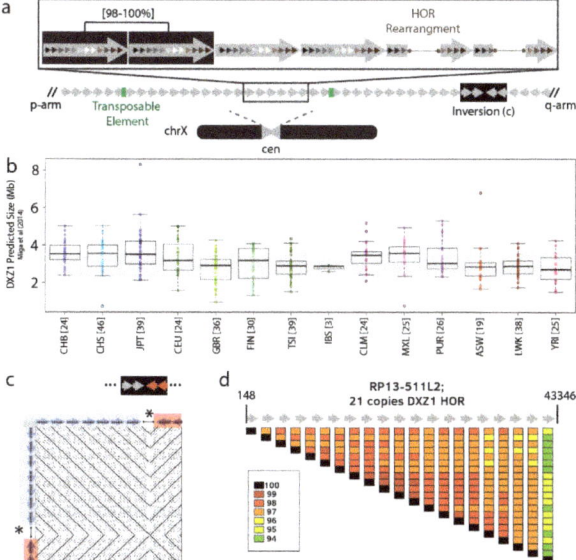

Figure 2. Intra-array satellite sequence variation. (a) All normal human centromeric regions contain at least one alpha satellite array, shown in grey, which is tandemly organized in a head-to-tail orientation with occasionally transposable element interruptions (green) and shifts in directionality (black box). The fundamental alpha satellite repeat unit, or ~171 bp monomer, is shown in a variation of shaded colors to illustrate the heterogeneity of the sequencing identity. Multi-monomer repeat units, or 'higher-order repeats (HORs), are shown by the larger grey arrows that encompass the collection of smaller repeats. In contrast to the individual monomers, these repeats are shown to be identical, or near-identical (98–100%). In addition to single nucleotide differences between the HORs, larger rearrangements (shown as a deletion of five monomers) are observed to occur and expand and contract within the array. (b) Satellite array length predictions on the X chromosome (DXZ1) [7], grey shading marks the previously observed PFGE Southern length range [25]. (c) Inversion detected using error-corrected PacBio reads [68]. (d) RP13-511L2 is an X-specific BAC that represents the transition from core alpha satellite to the edge of the array. HOR pair-wise repeat identity (muscle alignment [72]) showing increased divergence approaching the chromosome arm (43,346 bp), as typically observed at the edge of the array.

Use of error-corrected long reads (e.g., Pacific Biosciences, PacBio) prior to assembly provide an automated method to identify larger structural variation (SV) in satellite arrays, such as: HOR rearrangement (insertion and/or deletion), inversions, and interruption by transposons [68]. In addition to changes in the HOR structure, one can monitor precise sites where shifts in orientation or inversions take place within the array (Figure 2c). Further, when tracking sites of transposable element insertion, LINE1 is documented to be the most prevalent, consistent with the literature of alpha satellite DNA [73]. In addition to advancing our understanding of sequence organization and centromere function, such low-copy sequence variants that interrupt the uniformity of the satellite array are also expected to also guide linear assembly efforts [4,6]. Likewise, low-copy SNVs have been shown to be useful in overlap-consensus assembly, but they depend on high sequencing accuracy often obtained from Illumina reads and/or high-coverage of long-read data [6]. Satellite DNA studies using bacterial artificial chromosome (BAC) data provide a snapshot of local SNV spacing, where increased divergence is expected at the edges of the array (closest to the transition with the chromosome arms) with sparse, and infrequent informative sites within the array (Figure 2d) [73]. Ultimately, efforts to construct robust databases of satellite-associated SVs and SNVs will benefit from additional high-coverage, long read (PacBio or nanopore sequencing) datasets from diverse individuals. Such databases would provide allele-frequency data needed to guide future disease associations of variants.

4. Centromeric Regions Span Variants Associated with Disease.

Entire multi-megabase-sized centromeric regions, including the heterochromatic regions in the pericentromere, suppress meiotic recombination and are commonly observed as a single haplotype block, or 'cenhap' (Figure 3a) [74]. Little is known about the unique evolution and regulatory properties of those sequences that are associated with these highly specialized regions. Position effect variegation (PEV), or the mosaic pattern of gene expression when placed within or near heterochromatic environments, has been observed in organisms from yeast to humans [71]. The extensive range of satellite array sizes observed within the human population may contribute to studies of PEV variability and gene regulation in the human genome. Sequences directly adjacent to the centromeric satellite arrays have been documented as hypermutable, with a speculation that the increased mutation rate may be attributed to centromere activity [73,74]. Further, genes that are largely excluded by recombination may influence the efficacy of selection and create a 'protected' environment for gene mutations, inheritance, and disease.

These immense linkage blocks encompass satellite DNAs, segmental duplications [75], and a collection of well-annotated genes [76], many of which have been previously attributed to human clinical and disease phenotypes. Although the functional implications of gene-level associations are difficult to infer due to the large region of linkage disequilibrium, it may be useful for studies to recognize and bin these centromere-associated genomic regions as it is likely that they are share a compartment of the genome with specialized inheritance and evolution. The Xq cenhap region contains eight genes that are documented in the Online Mendelian Inheritance in Man (OMIM) noting the potential for allelic variants to represent disease-causing mutations (Figure 3b) [77,78]. Additionally, genome-wide association studies (GWAS) have identified SNPs in cenhap regions as associated with human disease (as shown in Figure 3b for NHGRI-EBI GWAS data, each selected with p-values $< 1.0 \times 10-5$), many of which do not overlap with genes or annotated sequence features [79,80]. Studies of variants directly adjacent to centromeric regions have been associated with chromosome instability and disease. For example, multiple independent signals associated with chromosome X loss around the centromere of chromosome X have been reported in a study of mosaic chromosomal alterations in clonal hematopoiesis [81], with a strong association ($P = 6.6 \times 10-27$, with an observed 1.9:1 bias in the lost haplotype) near the centromere array (DXZ1, Xp11.1). Further examples of centromere-adjacent or associated SNPs have been used to predict a significant association with multiple sclerosis risk around the chromosome 1 (lod = 4.9; with initial scan of 484 cases and 1043 controls; genotyped at 1082 SNPs) [82]. It is likely that many other disease association loci exist in these centromere-proximal regions, as association with centromeric SNPs (defined as within 2 Mbps of an alpha satellite reference model in GRCh38 that do not overlap with a known gene or segmental duplication) have been observed for a variety of clinical studies, including various cancers, neurodegenerative disorders and cardiovascular diseases (Table 1) [80,83].

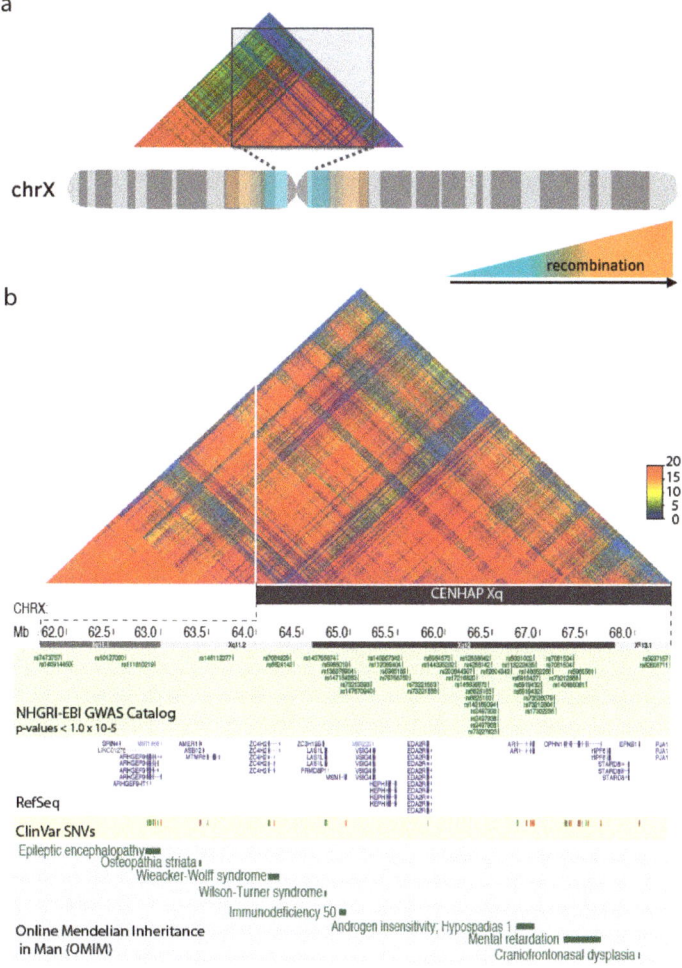

Figure 3. Disease-associated variants in centromere-associated haplotypes. (a) Centromeres act as the primary constriction of chromosomes, and are historically defined by the reduction of meiotic recombination (indicated by blue). Therefore, sequences in these regions are commonly inherited in large linkage blocks, or cenhaps (shown in the linkage disequilibrium heat map) [70]. (b) Study of disease and clinically associated single nucleotide variants (GWAS Catalog, green), ClinVar SNVs (yellow) in the Xq cenhap region (with Linkage Disequilibrium heat map from (a) enlarged) and a collection of annotated genes (RefSeq, white), of which variation have been attributed to a human disease (OMIM data, grey).

In addition to studies that involve the cenhap associated regions, extensive sequence variation within the satellite array is expected to contribute to our understanding of centromere instability and disease. Cytogenetic staining has revealed the constitutive heterochromatin in human centromeric regions has a highly heteromorphic structure. Given the critical importance of centromeres in ensuring proper chromosome segregation, such genomic variation is hypothesized to drive genome stability, and have been linked with human disease and cancers (reviewed [84]). Nevertheless, in the case of cancers where cells are expected to present increased genomic rearrangements, altered regulation and localization of kinetochore proteins, it will be important to estimate the rate of neocentromere formation with respect to native centromere sequence stability to test functionally relevant satellite DNA variants. We are only beginning to understand the sequence organization, allelic frequency, and evolution of satellite DNAs in the human population. Indeed, an analysis of optical genome maps of 154 individuals from 26 populations provided evidence for a large proportion of structural variants in satellite DNAs [85]. Such high-resolution diversity maps are expected to guide studies aimed to characterize satellite array structures that are associated with disease from those that have little functional consequence.

Table 1. Description of centromere-adjacent single nucleotide polymorphisms (SNPs) identified by published Genome-Wide Association Studies (GWAS), collected in the NHGRI-EBI GWAS Catalog published jointly by the National Human Genome Research Institute (NHGRI) and the European Bioinformatics Institute (EMBL-EBI) [80]. SNPs are included if found within a two-megabase window of an alpha satellite reference model (GRCh38) and do not overlap with annotated genes or segmental duplication).

Trait	SNPs	CEN adjacent (2Mb) Regions	Citation
Cancer	rs930395, rs2241024, rs142427110, rs35951924, rs199501877, rs11146838, rs6490525, rs2050203, rs7278690, rs35505947	4p12; 5p12; 5q11; 10p11; 13q12; 18p11; 19q11; 20p11; 21q11	[86–89]
Cardiovascular disease	rs10132760, rs12186641, rs9367716, rs71566846, rs223290, rs144961578, rs3813127, rs1657346, rs1254531, rs10793514	5q11.2; 6p11.2; 6q11.1; 10q11.21; 14q11.2; 18q11.2	[90–92]
Neurodegenerative diseases	rs11826064, rs13168838, rs62365447, rs140996952, rs1480597, rs10783624, rs7989524, rs6822736, rs13110633, rs2424635	4p11; 4q12; 5p12; 5q11.1; 6q11.1; 10q11; 11p11; 12q12; 13q12; 20p11	[93–100]
Scoliosis/Bone Density (Spine)	rs8111296, rs11652527, rs1436931, rs6061081, rs17599071, rs10136383, rs9288898, rs10772040, rs4562194, rs810967, rs6050182, rs6511621, rs11229654, rs6551418, rs1006899	3p11.1; 3q11.2; 6q12; 7q11.21; 10q11.21; 11q11; 12p11.21; 14q11.2; 17q11.2; 19p12; 20p11.21; 21q11.2	[101,102]
Digestive system disease	rs4243971, rs2342002, rs4800353, rs6058869, rs6087990	6q11.1; 18q11.2; 20q11.21	[103,104]

5. Concluding Remarks

In conclusion, human satellite DNAs provide a new, largely uncharted source of sequence variation in the human population. Chromosome-specific satellite arrays are expected to vary considerably in the human population, and measuring the overall range in the abundance and frequencies of repeat variants will contribute to ongoing studies of centromere biology and genome instability. Efforts to identify and study these variants will rely on improved, comprehensive genomic methods capable of mapping the full extent of satellite sequence heterogeneity that cannot be captured using a single reference genome. Such maps are necessary to direct future biomedical research to variants that are associated with disease, rather than natural sequence variation, which may have little or no clinical consequence.

Funding: The research was made possible by the generous financial support of the W.M. Keck Foundation.

Acknowledgments: I would like to thank Nicolas Altemose, Charles Langley, and Sasha Langley for valuable comments on the manuscript. The results presented herein were obtained at the Genomics Institute at the University of California, Santa Cruz.

Conflicts of Interest: The author declares no conflict of interest.

References

1. 1000 Genomes Project Consortium; Auton, A.; Brooks, L.D.; Durbin, R.M.; Garrison, E.P.; Kang, H.M.; Korbel, J.O.; Marchini, J.L.; McCarthy, S.; McVean, G.A.; et al. A global reference for human genetic variation. *Nature* **2015**, *526*, 68–74.
2. Lander, E.S.; Linton, L.M.; Birren, B.; Nusbaum, C.; Zody, M.C.; Baldwin, J.; Devon, K.; Dewar, K.; Doyle, M.; FitzHugh, W.; et al. Initial sequencing and analysis of the human genome. *Nature* **2001**, *409*, 860–921. [PubMed]
3. Venter, J.C.; Adams, M.D.; Myers, E.W.; Li, P.W.; Mural, R.J.; Sutton, G.G.; Smith, H.O.; Yandell, M.; Evans, C.A.; Holt, R.A.; et al. The sequence of the human genome. *Science* **2001**, *291*, 1304–1351. [CrossRef]
4. Miga, K.H. Completing the human genome: The progress and challenge of satellite DNA assembly. *Chromosome Res.* **2015**, *23*, 421–426. [CrossRef]
5. Jain, M.; Koren, S.; Miga, K.H.; Quick, J.; Rand, A.C.; Sasani, T.A.; Tyson, J.R.; Beggs, A.D.; Dilthey, A.T.; Fiddes, I.T.; et al. Nanopore sequencing and assembly of a human genome with ultra-long reads. *Nat. Biotechnol.* **2018**, *36*, 338. [CrossRef]
6. Jain, M.; Olsen, H.E.; Turner, D.; Stoddart, D.; Paten, B.; Haussler, D.; Willard, H.F.; Akeson, M.; Miga, K.H. Linear assembly of a human centromere on the Y chromosome. *Nat. Biotechnol.* **2018**, *36*, 321–323. [CrossRef] [PubMed]
7. Miga, K.H.; Newton, Y.; Jain, M.; Altemose, N.; Willard, H.F.; Kent, W.J. Centromere reference models for human chromosomes X and Y satellite arrays. *Genome Res.* **2014**, *24*, 697–707. [CrossRef] [PubMed]
8. Black, E.M.; Giunta, S. Repetitive Fragile Sites: Centromere Satellite DNA As a Source of Genome Instability in Human Diseases. *Genes* **2018**, *9*, 615. [CrossRef] [PubMed]
9. Ferreira, D.; Meles, S.; Escudeiro, A.; Mendes-da-Silva, A.; Adega, F.; Chaves, R. Satellite non-coding RNAs: The emerging players in cells, cellular pathways and cancer. *Chromosome Res.* **2015**, *23*, 479–493. [CrossRef]
10. Enukashvily, N.I.; Donev, R.; Waisertreiger, I.S.-R.; Podgornaya, O.I. Human chromosome 1 satellite 3 DNA is decondensed, demethylated and transcribed in senescent cells and in A431 epithelial carcinoma cells. *Cytogenet. Genome Res.* **2007**, *118*, 42–54. [CrossRef] [PubMed]
11. Atkin, N.B.; Brito-Babapulle, V. Heterochromatin polymorphism and human cancer. *Cancer Genet. Cytogenet.* **1981**, *3*, 261–272. [CrossRef]
12. Berger, R.; Bernheim, A.; Kristoffersson, U.; Mitelman, F.; Olsson, H. C-band heteromorphism in breast cancer patients. *Cancer Genet. Cytogenet.* **1985**, *18*, 37–42. [CrossRef]
13. Sahin, F.I.; Yilmaz, Z.; Yuregir, O.O.; Bulakbasi, T.; Ozer, O.; Zeyneloglu, H.B. Chromosome heteromorphisms: An impact on infertility. *J. Assist. Reprod. Genet.* **2008**, *25*, 191–195. [CrossRef]
14. Ting, D.T.; Lipson, D.; Paul, S.; Brannigan, B.W.; Akhavanfard, S.; Coffman, E.J.; Contino, G.; Deshpande, V.; Iafrate, A.J.; Letovsky, S.; et al. Aberrant overexpression of satellite repeats in pancreatic and other epithelial cancers. *Science* **2011**, *331*, 593–596. [CrossRef]
15. Atkin, N.B.; Brito-Babapulle, V. Chromosome 1 heterochromatin variants and cancer: A reassessment. *Cancer Genet. Cytogenet.* **1985**, *18*, 325–331. [CrossRef]
16. Wu, J.C.; Manuelidis, L. Sequence definition and organization of a human repeated DNA. *J. Mol. Biol.* **1980**, *142*, 363–386. [CrossRef]
17. Lee, C.; Wevrick, R.; Fisher, R.B.; Ferguson-Smith, M.A.; Lin, C.C. Human centromeric DNAs. *Hum. Genet.* **1997**, *100*, 291–304. [CrossRef]
18. Rudd, M.K.; Willard, H.F. Analysis of the centromeric regions of the human genome assembly. *Trends Genet.* **2004**, *20*, 529–533. [CrossRef]
19. Eichler, E.E.; Clark, R.A.; She, X. An assessment of the sequence gaps: Unfinished business in a finished human genome. *Nat. Rev. Genet.* **2004**, *5*, 345–354. [CrossRef]
20. Willard, H.F. Chromosome-specific organization of human alpha satellite DNA. *Am. J. Hum. Genet.* **1985**, *37*, 524–532.
21. Waye, J.S.; Willard, H.F. Nucleotide sequence heterogeneity of alpha satellite repetitive DNA: A survey of alphoid sequences from different human chromosomes. *Nucleic Acids Res.* **1987**, *15*, 7549–7569. [CrossRef]
22. Manuelidis, L.; Wu, J.C. Homology between human and simian repeated DNA. *Nature* **1978**, *276*, 92–94. [CrossRef]

23. Hayden, K.E.; Strome, E.D.; Merrett, S.L.; Lee, H.-R.; Rudd, M.K.; Willard, H.F. Sequences associated with centromere competency in the human genome. *Mol. Cell. Biol.* **2013**, *33*, 763–772. [CrossRef]
24. Wevrick, R.; Willard, H.F. Long-range organization of tandem arrays of alpha satellite DNA at the centromeres of human chromosomes: High-frequency array-length polymorphism and meiotic stability. *Proc. Natl. Acad. Sci. USA* **1989**, *86*, 9394–9398. [CrossRef]
25. Mahtani, M.M.; Willard, H.F. Pulsed-field gel analysis of alpha-satellite DNA at the human X chromosome centromere: High-frequency polymorphisms and array size estimate. *Genomics* **1990**, *7*, 607–613. [CrossRef]
26. Marçais, B.; Bellis, M.; Gérard, A.; Pagès, M.; Boublik, Y.; Roizès, G. Structural organization and polymorphism of the alpha satellite DNA sequences of chromosomes 13 and 21 as revealed by pulse field gel electrophoresis. *Hum. Genet.* **1991**, *86*, 311–316. [CrossRef]
27. Jones, K.W.; Prosser, J.; Corneo, G.; Ginelli, E. The chromosomal location of human satellite DNA III. *Chromosoma* **1973**, *42*, 445–451. [CrossRef]
28. Jones, K.W.; Corneo, G. Location of satellite and homogeneous DNA sequences on human chromosomes. *Nat. New Biol.* **1971**, *233*, 268–271. [CrossRef]
29. Gosden, J.R.; Mitchell, A.R.; Buckland, R.A.; Clayton, R.P.; Evans, H.J. The location of four human satellite DNAs on human chromosomes. *Exp. Cell Res.* **1975**, *92*, 148–158. [CrossRef]
30. Tagarro, I.; Fernández-Peralta, A.M.; González-Aguilera, J.J. Chromosomal localization of human satellites 2 and 3 by a FISH method using oligonucleotides as probes. *Hum. Genet.* **1994**, *93*, 383–388. [CrossRef]
31. Altemose, N.; Miga, K.H.; Maggioni, M.; Willard, H.F. Genomic characterization of large heterochromatic gaps in the human genome assembly. *PLoS Comput. Biol.* **2014**, *10*, e1003628. [CrossRef]
32. Prosser, J.; Frommer, M.; Paul, C.; Vincent, P.C. Sequence relationships of three human satellite DNAs. *J. Mol. Biol.* **1986**, *187*, 145–155. [CrossRef]
33. Cooke, H. Repeated sequence specific to human males. *Nature* **1976**, *262*, 182–186. [CrossRef]
34. Kunkel, L.M.; Smith, K.D.; Boyer, S.H.; Borgaonkar, D.S.; Wachtel, S.S.; Miller, O.J.; Breg, W.R.; Jones, H.W., Jr.; Rary, J.M. Analysis of human Y-chromosome-specific reiterated DNA in chromosome variants. *Proc. Natl. Acad. Sci. USA* **1977**, *74*, 1245–1249. [CrossRef]
35. Nakahori, Y.; Mitani, K.; Yamada, M.; Nakagome, Y. A human Y-chromosome specific repeated DNA family (DYZ1) consists of a tandem array of pentanucleotides. *Nucleic Acids Res.* **1986**, *14*, 7569–7580. [CrossRef]
36. Willard, H.F.; Waye, J.S. Chromosome-specific subsets of human alpha satellite DNA: Analysis of sequence divergence within and between chromosomal subsets and evidence for an ancestral pentameric repeat. *J. Mol. Evol.* **1987**, *25*, 207–214. [CrossRef]
37. Alexandrov, I.; Kazakov, A.; Tumeneva, I.; Shepelev, V.; Yurov, Y. Alpha-satellite DNA of primates: Old and new families. *Chromosoma* **2001**, *110*, 253–266. [CrossRef]
38. Jeanpierre, M.; Weil, D.; Gallano, P.; Creau-Goldberg, N.; Junien, C. The organization of two related subfamilies of a human tandemly repeated DNA is chromosome specific. *Hum. Genet.* **1985**, *70*, 302–310. [CrossRef]
39. Sherman, R.M.; Forman, J.; Antonescu, V.; Puiu, D.; Daya, M.; Rafaels, N.; Boorgula, M.P.; Chavan, S.; Vergara, C.; Ortega, V.E.; et al. Assembly of a pan-genome from deep sequencing of 910 humans of African descent. *Nat. Genet.* **2019**, *51*, 30. [CrossRef]
40. Schneider, V.A.; Graves-Lindsay, T.; Howe, K.; Bouk, N.; Chen, H.-C.; Kitts, P.A.; Murphy, T.D.; Pruitt, K.D.; Thibaud-Nissen, F.; Albracht, D.; et al. Evaluation of GRCh38 and de novo haploid genome assemblies demonstrates the enduring quality of the reference assembly. *Genome Res.* **2017**, *27*, 849–864. [CrossRef]
41. Miga, K.H.; Eisenhart, C.; Kent, W.J. Utilizing mapping targets of sequences underrepresented in the reference assembly to reduce false positive alignments. *Nucleic Acids Res.* **2015**, *43*, e133. [CrossRef]
42. Nechemia-Arbely, Y.; Fachinetti, D.; Miga, K.H.; Sekulic, N.; Soni, G.V.; Kim, D.H.; Wong, A.K.; Lee, A.Y.; Nguyen, K.; Dekker, C.; et al. Human centromeric CENP-A chromatin is a homotypic, octameric nucleosome at all cell cycle points. *J. Cell Biol.* **2017**, *216*, 607–621. [CrossRef]
43. Levy, S.; Sutton, G.; Ng, P.C.; Feuk, L.; Halpern, A.L.; Walenz, B.P.; Axelrod, N.; Huang, J.; Kirkness, E.F.; Denisov, G.; et al. The diploid genome sequence of an individual human. *PLoS Biol.* **2007**, *5*, e254. [CrossRef]
44. Audano, P.A.; Sulovari, A.; Graves-Lindsay, T.A.; Cantsilieris, S.; Sorensen, M.; Welch, A.E.; Dougherty, M.L.; Nelson, B.J.; Shah, A.; Dutcher, S.K.; et al. Characterizing the Major Structural Variant Alleles of the Human Genome. *Cell* **2019**, *176*, 663–675.e19. [CrossRef]

45. Chaisson, M.J.P.; Wilson, R.K.; Eichler, E.E. Genetic variation and the de novo assembly of human genomes. *Nat. Rev. Genet.* **2015**, *16*, 627–640. [CrossRef] [PubMed]
46. Marçais, B.; Charlieu, J.P.; Allain, B.; Brun, E.; Bellis, M.; Roizès, G. On the mode of evolution of alpha satellite DNA in human populations. *J. Mol. Evol.* **1991**, *33*, 42–48. [CrossRef] [PubMed]
47. Smith, G.P. Evolution of repeated DNA sequences by unequal crossover. *Science* **1976**, *191*, 528–535. [CrossRef]
48. Lower, S.S.; McGurk, M.P.; Clark, A.G.; Barbash, D.A. Satellite DNA evolution: old ideas, new approaches. *Curr. Opin. Genet. Dev.* **2018**, *49*, 70–78. [CrossRef] [PubMed]
49. Stults, D.M.; Killen, M.W.; Pierce, H.H.; Pierce, A.J. Genomic architecture and inheritance of human ribosomal RNA gene clusters. *Genome Res.* **2008**, *18*, 13–18. [CrossRef] [PubMed]
50. Kim, J.-H.; Dilthey, A.T.; Nagaraja, R.; Lee, H.-S.; Koren, S.; Dudekula, D.; Wood, W.H., III; Piao, Y.; Ogurtsov, A.Y.; Utani, K.; et al. Variation in human chromosome 21 ribosomal RNA genes characterized by TAR cloning and long-read sequencing. *Nucleic Acids Res.* **2018**, *46*, 6712–6725. [CrossRef]
51. Warburton, P.E.; Willard, H.F. Interhomologue sequence variation of alpha satellite DNA from human chromosome 17: Evidence for concerted evolution along haplotypic lineages. *J. Mol. Evol.* **1995**, *41*, 1006–1015. [CrossRef]
52. Willard, H.F.; Waye, J.S. Hierarchical order in chromosome-specific human alpha satellite DNA. *Trends Genet.* **1987**, *3*, 192–198. [CrossRef]
53. Hayden, K.E. Human centromere genomics: Now it's personal. *Chromosome Res.* **2012**, *20*, 621–633. [CrossRef] [PubMed]
54. Pluta, A.F.; Saitoh, N.; Goldberg, I.; Earnshaw, W.C. Identification of a subdomain of CENP-B that is necessary and sufficient for localization to the human centromere. *J. Cell Biol.* **1992**, *116*, 1081–1093. [CrossRef]
55. Hudson, D.F.; Fowler, K.J.; Earle, E.; Saffery, R.; Kalitsis, P.; Trowell, H.; Hill, J.; Wreford, N.G.; de Kretser, D.M.; Cancilla, M.R.; et al. Centromere protein B null mice are mitotically and meiotically normal but have lower body and testis weights. *J. Cell Biol.* **1998**, *141*, 309–319. [CrossRef]
56. Warburton, P.E.; Waye, J.S.; Willard, H.F. Nonrandom localization of recombination events in human alpha satellite repeat unit variants: Implications for higher-order structural characteristics within centromeric heterochromatin. *Mol. Cell. Biol.* **1993**, *13*, 6520–6529. [CrossRef] [PubMed]
57. Fachinetti, D.; Han, J.S.; McMahon, M.A.; Ly, P.; Abdullah, A.; Wong, A.J.; Cleveland, D.W. DNA Sequence-Specific Binding of CENP-B Enhances the Fidelity of Human Centromere Function. *Dev. Cell* **2015**, *33*, 314–327. [CrossRef] [PubMed]
58. Waye, J.S.; Willard, H.F. Structure, organization, and sequence of alpha satellite DNA from human chromosome 17: Evidence for evolution by unequal crossing-over and an ancestral pentamer repeat shared with the human X chromosome. *Mol. Cell. Biol.* **1986**, *6*, 3156–3165. [CrossRef] [PubMed]
59. Maloney, K.A.; Sullivan, L.L.; Matheny, J.E.; Strome, E.D.; Merrett, S.L.; Ferris, A.; Sullivan, B.A. Functional epialleles at an endogenous human centromere. *Proc. Natl. Acad. Sci. USA* **2012**, *109*, 13704–13709. [CrossRef]
60. Aldrup-MacDonald, M.E.; Kuo, M.E.; Sullivan, L.L.; Chew, K.; Sullivan, B.A. Genomic variation within alpha satellite DNA influences centromere location on human chromosomes with metastable epialleles. *Genome Res.* **2016**, *26*, 1301–1311. [CrossRef]
61. McNulty, S.M.; Sullivan, L.L.; Sullivan, B.A. Human Centromeres Produce Chromosome-Specific and Array-Specific Alpha Satellite Transcripts that Are Complexed with CENP-A and CENP-C. *Dev. Cell* **2017**, *42*, 226–240.e6. [CrossRef]
62. Hall, L.L.; Byron, M.; Carone, D.M.; Whitfield, T.W.; Pouliot, G.P.; Fischer, A.; Jones, P.; Lawrence, J.B. Demethylated HSATII DNA and HSATII RNA Foci Sequester PRC1 and MeCP2 into Cancer-Specific Nuclear Bodies. *Cell Rep.* **2017**, *18*, 2943–2956. [CrossRef]
63. Cobb, B.S.; Morales-Alcelay, S.; Kleiger, G.; Brown, K.E.; Fisher, A.G.; Smale, S.T. Targeting of Ikaros to pericentromeric heterochromatin by direct DNA binding. *Genes Dev.* **2000**, *14*, 2146–2160. [CrossRef]
64. Nishibuchi, G.; Déjardin, J. The molecular basis of the organization of repetitive DNA-containing constitutive heterochromatin in mammals. *Chromosome Res.* **2017**, *25*, 77–87. [CrossRef]
65. Delpu, Y.; McNamara, T.F.; Griffin, P.; Kaleem, S.; Narayan, S.; Schildkraut, C.; Miga, K.H.; Tahiliani, M. Chromosomal rearrangements at hypomethylated Satellite 2 sequences are associated with impaired replication efficiency and increased fork stalling. *bioRxiv* **2019**. [CrossRef]

66. Erliandri, I.; Fu, H.; Nakano, M.; Kim, J.-H.; Miga, K.H.; Liskovykh, M.; Earnshaw, W.C.; Masumoto, H.; Kouprina, N.; Aladjem, M.I.; et al. Replication of alpha-satellite DNA arrays in endogenous human centromeric regions and in human artificial chromosome. *Nucleic Acids Res.* **2014**, *42*, 11502–11516. [CrossRef]
67. Bersani, F.; Lee, E.; Kharchenko, P.V.; Xu, A.W.; Liu, M.; Xega, K.; MacKenzie, O.C.; Brannigan, B.W.; Wittner, B.S.; Jung, H.; et al. Pericentromeric satellite repeat expansions through RNA-derived DNA intermediates in cancer. *Proc. Natl. Acad. Sci. USA* **2015**, *112*, 15148–15153. [CrossRef]
68. Sevim, V.; Bashir, A.; Chin, C.-S.; Miga, K.H. Alpha-CENTAURI: Assessing novel centromeric repeat sequence variation with long read sequencing. *Bioinformatics* **2016**, *32*, 1921–1924. [CrossRef]
69. Pathak, D.; Premi, S.; Srivastava, J.; Chandy, S.P.; Ali, S. Genomic instability of the DYZ1 repeat in patients with Y chromosome anomalies and males exposed to natural background radiation. *DNA Res.* **2006**, *13*, 103–109. [CrossRef]
70. Rahman, M.M.; Bashamboo, A.; Prasad, A.; Pathak, D.; Ali, S. Organizational variation of DYZ1 repeat sequences on the human Y chromosome and its diagnostic potentials. *DNA Cell Biol.* **2004**, *23*, 561–571. [CrossRef]
71. Oakey, R.; Tyler-Smith, C. Y chromosome DNA haplotyping suggests that most European and Asian men are descended from one of two males. *Genomics* **1990**, *7*, 325–330. [CrossRef]
72. Edgar, R.C. MUSCLE: Multiple sequence alignment with high accuracy and high throughput. *Nucleic Acids Res.* **2004**, *32*, 1792–1797. [CrossRef]
73. Schueler, M.G.; Higgins, A.W.; Rudd, M.K.; Gustashaw, K.; Willard, H.F. Genomic and genetic definition of a functional human centromere. *Science* **2001**, *294*, 109–115. [CrossRef]
74. Langley, S.A.; Miga, K.; Karpen, G.H.; Langley, C.H. Haplotypes spanning centromeric regions reveal persistence of large blocks of archaic DNA. *BioRxiv* **2018**. [CrossRef]
75. She, X.; Horvath, J.E.; Jiang, Z.; Liu, G.; Furey, T.S.; Christ, L.; Clark, R.; Graves, T.; Gulden, C.L.; Alkan, C.; et al. The structure and evolution of centromeric transition regions within the human genome. *Nature* **2004**, *430*, 857–864. [CrossRef]
76. Pruitt, K.D.; Tatusova, T.; Maglott, D.R. NCBI Reference Sequence (RefSeq): A curated non-redundant sequence database of genomes, transcripts and proteins. *Nucleic Acids Res.* **2005**, *33*, D501–D504. [CrossRef]
77. Amberger, J.; Bocchini, C.A.; Scott, A.F.; Hamosh, A. McKusick's Online Mendelian Inheritance in Man (OMIM®). *Nucleic Acids Res.* **2009**, *37*, D793–D796. [CrossRef]
78. Hamosh, A.; Scott, A.F.; Amberger, J.S.; Bocchini, C.A.; McKusick, V.A. Online Mendelian Inheritance in Man (OMIM), a knowledgebase of human genes and genetic disorders. *Nucleic Acids Res.* **2005**, *33*, D514–D517. [CrossRef]
79. Landrum, M.J.; Lee, J.M.; Benson, M.; Brown, G.; Chao, C.; Chitipiralla, S.; Gu, B.; Hart, J.; Hoffman, D.; Hoover, J.; et al. ClinVar: Public archive of interpretations of clinically relevant variants. *Nucleic Acids Res.* **2016**, *44*, D862–D868. [CrossRef]
80. Hindorff, L.A.; Sethupathy, P.; Junkins, H.A.; Ramos, E.M.; Mehta, J.P.; Collins, F.S.; Manolio, T.A. Potential etiologic and functional implications of genome-wide association loci for human diseases and traits. *Proc. Natl. Acad. Sci. USA* **2009**, *106*, 9362–9367. [CrossRef]
81. Loh, P.-R.; Genovese, G.; Handsaker, R.E.; Finucane, H.K.; Reshef, Y.A.; Palamara, P.F.; Birmann, B.M.; Talkowski, M.E.; Bakhoum, S.F.; McCarroll, S.A.; et al. Insights into clonal haematopoiesis from 8,342 mosaic chromosomal alterations. *Nature* **2018**, *559*, 350–355. [CrossRef]
82. Reich, D.; Patterson, N.; De Jager, P.L.; McDonald, G.J.; Waliszewska, A.; Tandon, A.; Lincoln, R.R.; DeLoa, C.; Fruhan, S.A.; Cabre, P.; et al. A whole-genome admixture scan finds a candidate locus for multiple sclerosis susceptibility. *Nat. Genet.* **2005**, *37*, 1113–1118. [CrossRef] [PubMed]
83. Karolchik, D.; Hinrichs, A.S.; Furey, T.S.; Roskin, K.M.; Sugnet, C.W.; Haussler, D.; Kent, W.J. The UCSC Table Browser data retrieval tool. *Nucleic Acids Res.* **2004**, *32*, D493–D496. [CrossRef]
84. Barra, V.; Fachinetti, D. The dark side of centromeres: types, causes and consequences of structural abnormalities implicating centromeric DNA. *Nat. Commun.* **2018**, *9*, 4340. [CrossRef]
85. Levy-Sakin, M.; Pastor, S.; Mostovoy, Y.; Li, L.; Leung, A.K.Y.; McCaffrey, J.; Young, E.; Lam, E.T.; Hastie, A.R.; Wong, K.H.Y.; et al. Genome maps across 26 human populations reveal population-specific patterns of structural variation. *Nat. Commun.* **2019**, *10*, 1025. [CrossRef]

86. Michailidou, K.; Lindström, S.; Dennis, J.; Beesley, J.; Hui, S.; Kar, S.; Lemaçon, A.; Soucy, P.; Glubb, D.; Rostamianfar, A.; et al. Association analysis identifies 65 new breast cancer risk loci. *Nature* **2017**, *551*, 92–94. [CrossRef]
87. O'Donnell, P.H.; Stark, A.L.; Gamazon, E.R.; Wheeler, H.E.; McIlwee, B.E.; Gorsic, L.; Im, H.K.; Huang, R.S.; Cox, N.J.; Dolan, M.E. Identification of novel germline polymorphisms governing capecitabine sensitivity. *Cancer* **2012**, *118*, 4063–4073. [CrossRef] [PubMed]
88. Moore, K.N.; Tritchler, D.; Kaufman, K.M.; Lankes, H.; Quinn, M.C.J.; Ovarian Cancer Association Consortium; Van Le, L.; Berchuck, A.; Backes, F.J.; Tewari, K.S.; et al. Genome-wide association study evaluating single-nucleotide polymorphisms and outcomes in patients with advanced stage serous ovarian or primary peritoneal cancer: An NRG Oncology/Gynecologic Oncology Group study. *Gynecol. Oncol.* **2017**, *147*, 396–401. [CrossRef] [PubMed]
89. Hofer, P.; Hagmann, M.; Brezina, S.; Dolejsi, E.; Mach, K.; Leeb, G.; Baierl, A.; Buch, S.; Sutterlüty-Fall, H.; Karner-Hanusch, J.; et al. Bayesian and frequentist analysis of an Austrian genome-wide association study of colorectal cancer and advanced adenomas. *Oncotarget* **2017**, *8*, 98623–98634. [CrossRef]
90. Deng, X.; Sabino, E.C.; Cunha-Neto, E.; Ribeiro, A.L.; Ianni, B.; Mady, C.; Busch, M.P.; Seielstad, M. REDSII Chagas Study Group from the NHLBI Retrovirus Epidemiology Donor Study-II Component International Genome wide association study (GWAS) of Chagas cardiomyopathy in Trypanosoma cruzi seropositive subjects. *PLoS ONE* **2013**, *8*, e79629. [CrossRef] [PubMed]
91. Cordell, H.J.; Bentham, J.; Topf, A.; Zelenika, D.; Heath, S.; Mamasoula, C.; Cosgrove, C.; Blue, G.; Granados-Riveron, J.; Setchfield, K.; et al. Genome-wide association study of multiple congenital heart disease phenotypes identifies a susceptibility locus for atrial septal defect at chromosome 4p16. *Nat. Genet.* **2013**, *45*, 822–824. [CrossRef]
92. van der Harst, P.; Verweij, N. Identification of 64 Novel Genetic Loci Provides an Expanded View on the Genetic Architecture of Coronary Artery Disease. *Circ. Res.* **2018**, *122*, 433–443. [CrossRef]
93. Nagel, M.; Jansen, P.R.; Stringer, S.; Watanabe, K.; de Leeuw, C.A.; Bryois, J.; Savage, J.E.; Hammerschlag, A.R.; Skene, N.G.; Muñoz-Manchado, A.B.; et al. Meta-analysis of genome-wide association studies for neuroticism in 449,484 individuals identifies novel genetic loci and pathways. *Nat. Genet.* **2018**, *50*, 920–927. [CrossRef]
94. Turley, P.; Walters, R.K.; Maghzian, O.; Okbay, A.; Lee, J.J.; Fontana, M.A.; Nguyen-Viet, T.A.; Wedow, R.; Zacher, M.; Furlotte, N.A.; et al. Multi-trait analysis of genome-wide association summary statistics using MTAG. *Nat. Genet.* **2018**, *50*, 229–237. [CrossRef]
95. Herold, C.; Hooli, B.V.; Mullin, K.; Liu, T.; Roehr, J.T.; Mattheisen, M.; Parrado, A.R.; Bertram, L.; Lange, C.; Tanzi, R.E. Family-based association analyses of imputed genotypes reveal genome-wide significant association of Alzheimer's disease with OSBPL6, PTPRG, and PDCL3. *Mol. Psychiatry* **2016**, *21*, 1608–1612. [CrossRef]
96. Fung, H.-C.; Scholz, S.; Matarin, M.; Simón-Sánchez, J.; Hernandez, D.; Britton, A.; Gibbs, J.R.; Langefeld, C.; Stiegert, M.L.; Schymick, J.; et al. Genome-wide genotyping in Parkinson's disease and neurologically normal controls: First stage analysis and public release of data. *Lancet Neurol.* **2006**, *5*, 911–916. [CrossRef]
97. Goes, F.S.; McGrath, J.; Avramopoulos, D.; Wolyniec, P.; Pirooznia, M.; Ruczinski, I.; Nestadt, G.; Kenny, E.E.; Vacic, V.; Peters, I.; et al. Genome-wide association study of schizophrenia in Ashkenazi Jews. *Am. J. Med. Genet. B Neuropsychiatr. Genet.* **2015**, *168*, 649–659. [CrossRef]
98. Li, Z.; Chen, J.; Yu, H.; He, L.; Xu, Y.; Zhang, D.; Yi, Q.; Li, C.; Li, X.; Shen, J.; et al. Genome-wide association analysis identifies 30 new susceptibility loci for schizophrenia. *Nat. Genet.* **2017**, *49*, 1576–1583. [CrossRef]
99. Beecham, G.W.; Hamilton, K.; Naj, A.C.; Martin, E.R.; Huentelman, M.; Myers, A.J.; Corneveaux, J.J.; Hardy, J.; Vonsattel, J.-P.; Younkin, S.G.; et al. Genome-wide association meta-analysis of neuropathologic features of Alzheimer's disease and related dementias. *PLoS Genet.* **2014**, *10*, e1004606. [CrossRef]
100. Wang, K.-S.; Liu, X.-F.; Aragam, N. A genome-wide meta-analysis identifies novel loci associated with schizophrenia and bipolar disorder. *Schizophr. Res.* **2010**, *124*, 192–199. [CrossRef]
101. Styrkarsdottir, U.; Halldorsson, B.V.; Gretarsdottir, S.; Gudbjartsson, D.F.; Walters, G.B.; Ingvarsson, T.; Jonsdottir, T.; Saemundsdottir, J.; Snorradóttir, S.; Center, J.R.; et al. New sequence variants associated with bone mineral density. *Nat. Genet.* **2009**, *41*, 15–17. [CrossRef] [PubMed]

102. Liu, J.; Zhou, Y.; Liu, S.; Song, X.; Yang, X.-Z.; Fan, Y.; Chen, W.; Akdemir, Z.C.; Yan, Z.; Zuo, Y.; et al. The coexistence of copy number variations (CNVs) and single nucleotide polymorphisms (SNPs) at a locus can result in distorted calculations of the significance in associating SNPs to disease. *Hum. Genet.* **2018**, *137*, 553–567. [CrossRef] [PubMed]
103. Liu, J.Z.; van Sommeren, S.; Huang, H.; Ng, S.C.; Alberts, R.; Takahashi, A.; Ripke, S.; Lee, J.C.; Jostins, L.; Shah, T.; et al. Association analyses identify 38 susceptibility loci for inflammatory bowel disease and highlight shared genetic risk across populations. *Nat. Genet.* **2015**, *47*, 979–986. [CrossRef] [PubMed]
104. Levine, D.M.; Ek, W.E.; Zhang, R.; Liu, X.; Onstad, L.; Sather, C.; Lao-Sirieix, P.; Gammon, M.D.; Corley, D.A.; Shaheen, N.J.; et al. A genome-wide association study identifies new susceptibility loci for esophageal adenocarcinoma and Barrett's esophagus. *Nat. Genet.* **2013**, *45*, 1487–1493. [CrossRef] [PubMed]

© 2019 by the author. Licensee MDPI, Basel, Switzerland. This article is an open access article distributed under the terms and conditions of the Creative Commons Attribution (CC BY) license (http://creativecommons.org/licenses/by/4.0/).

Article

Whole-Genome Analysis of Domestic Chicken Selection Lines Suggests Segregating Variation in ERV Makeups

Mats E. Pettersson and Patric Jern *

Science for Life Laboratory, Department of Medical Biochemistry and Microbiology, Uppsala University, Box 582, SE-75123 Uppsala, Sweden; Mats.Pettersson@imbim.uu.se
* Correspondence: Patric.Jern@imbim.uu.se; Tel.: +46-18-471-4593

Received: 21 January 2019; Accepted: 15 February 2019; Published: 20 February 2019

Abstract: Retroviruses have invaded vertebrate hosts for millions of years and left an extensive endogenous retrovirus (ERV) record in the host genomes, which provides a remarkable source for an evolutionary perspective on retrovirus-host associations. Here we identified ERV variation across whole-genomes from two chicken lines, derived from a common founder population subjected to 50 years of bi-directional selection on body weight, and a distantly related domestic chicken line as a comparison outgroup. Candidate ERV loci, where at least one of the chicken lines indicated distinct differences, were analyzed for adjacent host genomic landscapes, selective sweeps, and compared by sequence associations to reference assembly ERVs in phylogenetic analyses. Current data does not support selection acting on specific ERV loci in the domestic chicken lines, as determined by presence inside selective sweeps or composition of adjacent host genes. The varying ERV records among the domestic chicken lines associated broadly across the assembly ERV phylogeny, indicating that the observed insertion differences result from pre-existing and segregating ERV loci in the host populations. Thus, data suggest that the observed differences between the host lineages are best explained by substantial standing ERV variation within host populations, and indicates that even truncated, presumably old, ERVs have not yet become fixed in the host population.

Keywords: endogenous retrovirus; host genome; evolution; segregation

1. Introduction

Retroviruses have infiltrated vertebrate germline for millions of years by integrating as proviruses in host DNA, which have then passed down to the offspring through generations as inherited endogenous retroviruses (ERVs). The genomic ERV record represents retroviruses that were replicating at the time of integration and constitutes large fractions of contemporary vertebrate genomes, for example about 7–8% of human DNA [1,2] and about 3% of the chicken genome [3]. The genomic ERV record thus presents a remarkable source for an evolutionary perspective on the biology and interactions among retroviruses and their hosts.

Diverse sets of ERVs can be identified across all studied vertebrate genome assemblies [4] by screening for structural hallmarks including long terminal repeats (LTRs), which flank the ERV *gag*, *pol*, and *env* genes [5]. Over time, ERV loci may become fixed in the host population, either due to genetic drift of those loci that are least harmful or due to selection on beneficial insertions [2]. ERV contributions to the host genome structure and function include providing a substrate for genomic recombination, and effects on the host transcriptome resulting from their integration and expression with diverse effects on host genome function and evolution. Among positive effects are the expression of viral gene products as useful new genes in the host [6], modification of chromosomal gene expression by ERVs including promoter, enhancer, and insulator functions, as well as alternative splice signals from ERV integrations in host transcription units or adjacent to chromosomal genes [2,7]. On the other hand, is the potential for host gene disruption, as well as the potential for somatic spread of replicating retroviruses leading to pathogenic consequences for the host [2,8,9]. ERV-mediated genomic recombination can further contribute to the organization and plasticity of the host's genome [10–13]. Overall, it is plausible that ERVs have had considerable effects on host genome function and evolution across the entire vertebrate lineage, by shuffling genomic regions, exons, and regulatory genetic sequences into new contexts and thereby altering the dynamic functions of the host DNA.

It is desirable to identify orthologous ERV loci across the compared host lineages in order to evaluate potential effects of retroviruses and ERVs on host biology because it allows for connecting ERV integrations to host phenotypic differences and evolutionary history. ERV studies have benefited from recent advancements in sequencing technology and a growing catalogue of reference host genome assemblies, where much focus has been placed on comparing ERV records across related host species reference genome assemblies, an approach that suffers from undersampling of the diversity within vertebrate species, and thus presents challenges for reaching a better understanding of potential factors that contribute to the long-term retrovirus-host associations [4]. More recently, studies utilizing re-sequencing data to target searches for integration differences among selected ERVs within a host population to explain activities during recent evolution have made efforts to address this issue for specific virus types in host populations [14,15].

In an attempt to further explore ERV-host associations in a hitherto un-examined system, we make use of an artificial selection system where selection lines of domestic chicken that have been undergoing strong bi-directional selection on body weight at eight weeks of age for more than 50 years [16]. This selection-scheme, from a single founder population, has generated extreme phenotypes with more than 10-fold difference in average weights between the two chicken lines. We utilized whole-genome re-sequencing data from these chicken lines, as well as an outgroup commercial chicken line, to investigate ERV insertion variation and potential evolutionary contributions from inherited ERVs on host genome function.

Domestic animals provide rare possibilities, currently not feasible in human biomedicine, to study connections between genes, phenotypes, and biological function [17]. Crossbreeding of domestic animals is also a useful tool to determine genomic differences, making it possible to apply genetic analyses to reveal loci controlling phenotypic traits that have been selected during domestication [18].

The rationale for utilizing chicken as a model dates back more than 100 years to pioneering studies of retroviruses and ERVs, reviewed in [19,20], and the availability of sequence data from the chicken selection pedigree established in 1957 (see above), which, measured by the response in phenotypic traits and single nucleotide polymorphism (SNP) allele-frequency divergence, has accumulated changes that have been estimated by Johansson et al. to require about 5000 years to evolve in natural populations [21]. Overall, the chicken selection lines present a promising model for identifying ERV divergence and interpreting observations in the context of previously known results in this system, thereby estimating ERV contributions to dynamics of complex genetic traits of their hosts.

Here, we identify ERV insertion differences across available re-sequenced genomes derived from the two bi-directionally growth-selected chicken lines and compare candidate ERV loci with a commercial layer chicken outgroup. We map insertions and deletions to establish their positions relative to the adjacent host genomic landscape and compare candidate loci associated by sequence similarity to ERVs identified in the Red junglefowl reference assembly (version galGal3) along with reference retroviral sequences within a phylogenetic framework.

2. Materials and Methods

2.1. Domestic Chicken Selection Lines

Whole-genome re-sequenced DNA from domestic chicken selection lines was analyzed for differences in ERV makeups as potential markers for effects of ERVs on host genome function and evolution, see Table 1. For this purpose, we utilized a well-studied model system established in 1957, where two growth-selected lines of chicken were developed from a single founder population by bi-directional selection of body weight at 56 days of age for more than 40 generations and kept as closed populations [16]. The average body weights of individuals from respective selection lines (H: high-growth; L: low-growth) differ more than 10-fold today, and, in addition to the response to the selection of body weight, the lines have accumulated significant differences in feeding behavior and food consumption [22]. The chicken selection lines (H and L) were analyzed together with the commercial white leghorn chicken (W) for whole-genome comparisons as previously described in a study of domestication sweeps [18]. Mate-pair sequence read libraries, with read length 50 nt times two and approximately 3 to 4 kb insert sizes were produced using high throughput SOLiD sequencing technology, see Table 1, and mapped to the Red junglefowl reference genome assembly (version galGal3, accessed from the UCSC Genome Browser, http://genome.ucsc.edu) by Rubin et al. [18].

Table 1. Re-sequenced chicken selection lines and endogenous retrovirus (ERV) associated reads.

Pooled Genomes [1]	Short Name	Library [2]	n [3]	Coverage [4]	ERV Assoc. Reads	ERV-Host Read Pairs
High-growth line	H	ugc_208	11	5.53x	233,621	82,719
Low-growth line	L	ugc_209	11	5.19x	239,189	98,942
White leghorn	W	ugc_254	11	3.37x	191,873	90,759

[1] High throughput SOLiD sequencing mate-pair libraries as previously described [18]. White leghorn chicken was used as an outgroup in comparisons. [2] Approximately 3500 nt mate-pair library gap lengths mapped to the chicken genome assembly (version galGal3) [18]. [3] Numbers of pooled individuals in SOLiD sequencing. [4] Sequencing coverage as previously described [18].

2.2. Endogenous Retrovirus Mapping

Briefly, to allow identification of reference as well as non-reference assembly ERVs using mate-pair short reads sequencing technology, we applied a strategy where reads were mapped to an independent ERV library and then located along host chromosomes by anchoring their mate-pair reads to positions in the flanking host DNA. The RetroTector software [5] was used to mine the Red junglefowl reference genome assembly (version galGal3) for ERV sequences to construct an independent reference library for mapping ERV-associated SOLiD sequencing short reads for each (H, L, and W, see above) chicken selection line [18] using the SHRiMP2 software [23]. To identify ERV-host DNA junctions, SHRiMP2 ERV-associated reads scoring $\geq$400 were paired with reads that target unique chromosomal flanking sequences [18]. The number of expected loci including full-length ERVs, truncated ERVs, and solo-LTRs [24], which are the results from homologous recombination between the two provirus LTRs, can be estimated to be about 20 times more frequent than the number of full-length ERVs based on previous evaluations [1], which serves as a conservative starting point for analyses. As our RetroTector analyses of the galGal3 reference assembly identified 532 high-quality ERVs scoring $\geq$300 (as previously discussed [5]), a putative target of around 10,000 candidate ERV loci was used to determine conditions for reads clustering at ERV-host DNA junctions. Conservative (top) scores were used for chromosomal DNA positions mapped by Rubin et al. [18] and SHRiMP2 ERV-associated read scores (ranging 400–493) were used at $\geq$425. The up- and downstream ERV-host DNA junctions were clustered separately considering mate-pair reads insert size of about 3500 nt, which reflects the maximum chromosomal flanking distance to the ERV integration, and requiring short reads to cover at least 2% of that length. Up- and downstream ERV-host DNA junctions were paired given shared orientation, read associations to target reference ERV sequences, indications of both 5'- and 3'-ERV flanking sequences, and separation by less than about 20 kb to accommodate expected ERV lengths of around 7–11 kb and potential secondary transposable element integrations into the candidate ERV. Together, these clustering conditions indicated 12,709 candidate ERV loci and additional ERV-associated reads with relaxed mapping scores (≥400) were appended to the identified ERV-host DNA junctions.

2.3. Endogenous Retrovirus Integration Variation

Candidate ERV loci were tested for read mapping differences across the re-sequenced chicken lines (H, L, and W, see above) using Fisher's exact test if the minimum observed read counts at the locus were fewer than 15 and otherwise by comparison to the Chi distribution. Loci where short reads were missing in one or two of the three chicken lines were kept for further analyses if p-values for ERV-associated read counts passed the conservative threshold ($p < 4 \times 10^{-6}$) after Bonferroni correction for multiple testing.

2.4. Endogenous Retrovirus Integration Landscape

The Red junglefowl (galGal3) reference gene dataset was downloaded from the UCSC genome browser (http://genome.ucsc.edu) and intersected with positions for candidate ERV loci in order to explore biological significance of genes located adjacent to ERVs and their potential associations with the chicken selection line phenotypes. Associations among chromosomal genes and ERVs, intragenic as well as intergenic positions covering 150 kb up- and downstream of reference gene transcription start sites were analyzed. Candidate ERV loci were intersected with sweep regions previously determined for the H and L chicken selection lines [18,21]. Chromosomal reference genes identified adjacent to ERV loci were included in searches at the database for annotation, visualization, and integrated discovery (DAVID at https://david.ncifcrf.gov/) to explore biological impact of differences in ERV integrations across the H and L chicken selection lines.

2.5. Phylogenetic Framework

Phylogenetic analyses of ERVs identified in the reference assembly (version galGal3) together with reference retrovirus sequences were performed as previously described [4,25,26]. Briefly, high-quality ERVs (RetroTector score ≥ 300) identified in the reference chicken assembly (galGal3) were split according to conserved motifs and phylogenetically informative segments across the ERV *gag* and *pol* genes for multiple sequence alignments that were concatenated for phylogenetic analysis using FastTree2 [27]. The resulting phylogenetic tree was rooted using the *Caenorhabditis elegans* retrotransposon Cer1 (GenBank accession no. U15406), and visualized using FigTree v1.4.2 (http://tree.bio.ed.ac.uk/software/figtree/).

3. Results

Whole-genomes from domestic High-Growth (H), Low-Growth (L), and White Leghorn (W) chicken selection lines were previously sequenced using high throughput SOLiD technology and mapped to the chicken reference assembly (version galGal3) by Rubin et al. [18], see Table 1. Here, we utilized the RetroTector software [5] to identify ERVs in the Red junglefowl (version galGal3) assembly, which were used as an independent sequence library to map ERV-associated reads from the re-sequenced chicken lines that could then be mated with their respective chromosomal mapping reads for locating ERV-host DNA insertion junctions, even in cases where the insertion was absent from the reference assembly. Up- and downstream ERV-host junctions were clustered and paired using stringency criteria tuned for identifying about 10,000 loci, the expected number of ERV and solo LTR loci based on RetroTector results and a previously estimated 1:20 ratio between complete ERVs and solo LTRs, which are generated by homologous recombination between the two proviral LTRs [1]. The clustering of paired sequence reads identified 12,709 candidate ERV loci, of which 8340 candidate loci indicated distinct differences, measured as absence or near-absence of ERV-associated reads in at least one of the three compared chicken lines. Bonferroni correction for multiple testing left 369 differentiated candidate ERV loci. Among these candidate loci, 115 ERVs were adjacent to, or located within, 229 host genes considering 150 kb distances up- and downstream of the candidate loci, see Table 2 and Supplementary Information Table S1.

Table 2. Candidate ERV loci.

Chicken [1]	ERV Candidate Loci	ERV Loci (Corrected) [2]	ERVs Adjacent to Genes [3]	Genes Adjacent to ERVs [3]
H••	874	38	11	29
•L•	1109	52	19	38
••W	616	117	46	95
HL•	2966	103	19	36
H•W	1227	30	9	11
•LW	1548	29	11	20
HLW	4369	nd	nd	nd

[1] Re-sequenced chicken selection lines selection lines: H (High-Growth line), L (Low-Growth line) and W (White Leghorn). [2] Candidate ERV loci after Bonferroni correction ($p < 4 \times 10^{-6}$) where ERV-associated reads indicate distinct differences (present/missing reads) in one of the chicken lines compared to the others. [3] Within 150 kb.

The bi-directionally growth-selected chicken lines (H and L) diverged from a single broiler founder population about 60 years ago and were separated more than 100 years ago from the branch leading to the comparison outgroup represented here by the commercial White Leghorn (layer) chicken. For reference, the compared chicken lines share a relatively recent common ancestry, compared to the reference genome assembly, generated from the Red junglefowl, *Gallus gallus*, which was separated from the investigated chicken lines about 8000 years ago when chicken was first domesticated, see Figure 1. However, even this split is recent compared to datasets that have been the subject of previous studies [4,14,15,26], and thus the use of ERV loci comparison in a small host pedigree, such as the domestic chicken lines, relies on that integration differences may be observed as a result from selection during domestication that could require many thousands of years to become fixed in wild host populations [21]. The observed branch-specific ERV loci differences across the domestic chicken selection lines broadly reflects the time scale after divergence as the growth-selected broiler chicken lines (H and L) were separated about half of the time since the layer outgroup (W) separated from the domesticated broiler chickens, see Figure 1, indicating that differences in ERV makeups may provide potential traceable markers for host evolution, see Figure 1, Table 2.

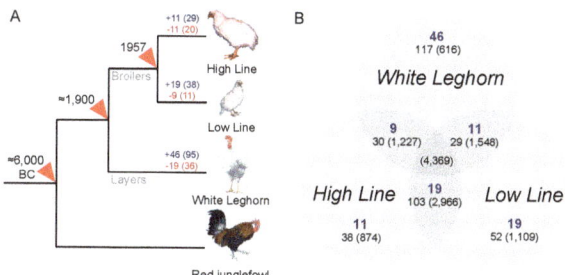

Figure 1. Chicken selection lines and ERV variation. (**A**) Phylogeny of analyzed fowl modified from Rubin et al. [18]. Blue numbers above branches indicate candidate branch-specific ERV insertions and red numbers indicate candidate branch-specific missing ERVs at the analyzed loci, see Table 2. Numbers in brackets indicate ERV-associated host genes. (**B**) Venn diagram showing distribution of identified candidate ERV differences across the domestic chicken selection lines. Blue numbers indicate counts of ERVs only found in the respective chicken line adjacent to genes. The numbers below represent the number of candidate loci after correction for multiple testing, see Table 2, and numbers within brackets show the corresponding number of candidate loci before correction.

To explore potential connections between the observed divergent ERV loci across the domestic chicken lines, we intersected chromosomal positions with the reference assembly host genes (version galGal3, downloaded from the UCSC genome browser, http://genome.ucsc.edu) and previously determined selective sweeps for the H and L chicken selection lines [18,21]. Although some ERVs overlapped with domestication sweep regions, see Supplementary Information Table S1, the observed overlap did not deviate significantly from the expectation ($p = 0.1$, binomial test), given the size of the sweep areas and the number of detected ERV insertions elsewhere in the host genomes. In addition, gene ontology searches were inconclusive and could not establish links between ERVs and adjacent host genes that could help explain the distinct phenotypes. We, therefore, analyzed candidate ERV insertion orientations and distances relative to host genes. Candidate ERV loci within host gene transcripts show a clear bias in antisense orientation relative to the host gene transcript, which could be explained by purifying selection due to potential splice interference from canonical splice signals as previously discussed [28]. Intergenic ERV orientations relative to host genes fluctuate up- and downstream and a bias pattern is not clear given the limited data, see Figure 2. It thus appears that intergenic ERV insertions may not influence host genome function to the same extent as intragenic ERVs.

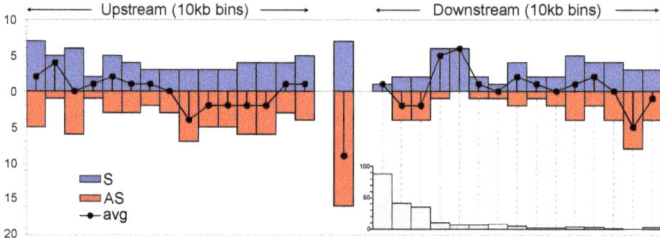

Figure 2. Chicken ERV integration landscape relative host genes (upstream, intragenic, and downstream) and orientation. Blue and red histograms show the number of identified host genes at various distances to the ERV in sense (blue) and antisense (red) relative orientations, and average points are indicated for each bin. The grey histogram insert indicates the number of analyzed genes at downstream distances, split into 10 kb bins, with respect to nearby ERVs.

To investigate relationships between the observed ERV loci varying across the analyzed chicken lines, we constructed a phylogenetic tree based on ERVs identified by the RetroTector software [5] in the Red junglefowl reference assembly (version galGal3) and appended reference retroviral sequences for comparisons as previously described [4,25,26]. Since the insert sizes and read lengths of ERV-associated mate-pair reads only allow limited coverage into the candidate ERV loci, it is useful to align reads to reference assembly ERVs that could build a phylogenetic framework, and from which the best ERV match for candidate ERV loci by can be determined, see Figure 3.

In agreement with the observed lack of significant associations between candidate ERV loci and adjacent host genes (see above), divergent ERV loci in the domestic H, L, and W chicken lines located across the phylogenetic tree that was rooted on a distant outgroup, rather than being found inside any specific retroviral clade, which is what could be expected if variation was due to retroviral expansion after the last common ancestor. Instead, the result indicates that the observed candidate ERV insertion differences do not result from recent retrovirus replication and integrations as ERVs in one or two of the chicken lineages, but rather it is consistent with standing variation of segregating ERV loci present at the onset of the bi-directional selection experiment as well as during breed formation since the domestication of chicken. Multiple radiations involving candidate ERV loci associated with assembly ERVs showing short terminal branch lengths indicate relatively recent expansions occurring within several retroviral genera across the phylogeny. It seems plausible that these radiations have generated a substantial number of segregating ERV insertions in the domestic chicken lines, thus providing the standing variation that explains the observed differences in ERV makeups, and that the number of divergent ERV loci is largely a product of the accelerated genomic divergence caused by the strong selection imposed on the H and L lines specifically, as well as directed selection of host features during domestication, which has affected all three studied (H, L, and W) chicken lines.

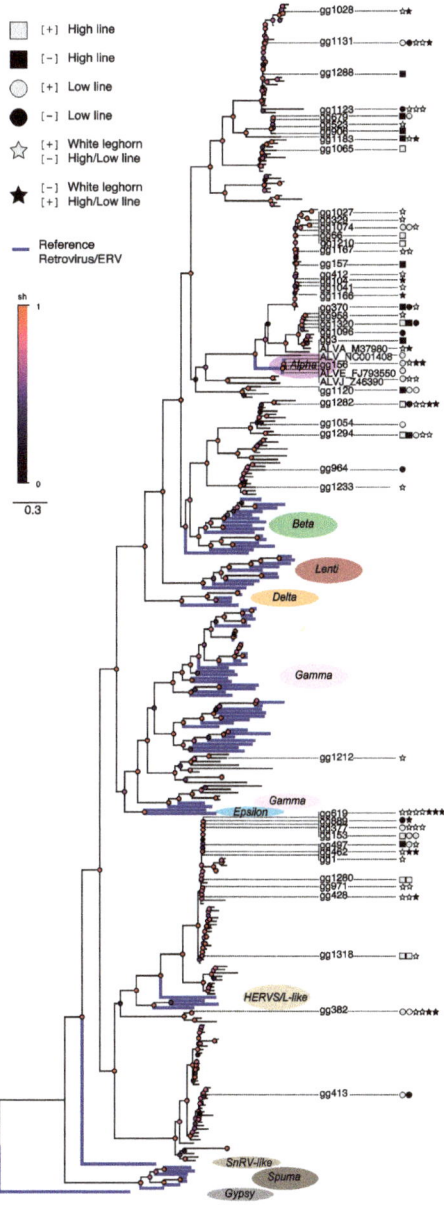

Figure 3. Chicken ERV phylogenetic tree based on Gag and Pol motifs as previously described [4,25]. Thick blue lines indicate reference retroviral and ERV sequences. Candidate ERV-associated loci differences between the H, L, and W chicken are indicated by symbols next to the ERV IDs referring to the RetroTector analysis of the chicken (galGal3) assembly (see FASTA sequences and loci annotations in Supplementary Information Data S1). The complete phylogenetic tree is presented in nexus file format (Supplementary Information Data S2) for rendering in a tree drawing software such as FigTree.

4. Discussion

The known breeding history and well-studied phenotypic traits among domestic animals make them first-rate model organisms to identify potential ERV contributions to biological functions and dynamics of complex genetic traits. Rare genomic changes resulting in host phenotypes that would require thousands of years to establish, or become lost, in wild host populations may be selected for in domestic settings during fewer generations [17,21,29]. This type of genomic data presents an excellent chance to study differences in genomic ERV makeups across many chicken selection lines.

Here, we utilized whole-genome sequences from two bi-directionally growth-selected domestic broiler chicken lines and a distantly related domestic layer chicken line [18] for identifying and comparing candidate ERV insertion differences. Using an independent ERV search library, it is possible to identify non-reference assembly ERV in the different host lineages. We show that the domestic chicken carries a large number of segregating ERVs, evidenced by the observation that 65% of detected loci display a nominal difference in frequencies and more than 350 insertions are significantly differentiated. Standing variation has previously been shown to contribute to the majority of the alleles under selection in the H and L lines [21,22,29] and here we show that segregating ERVs are a part of this variation, and thus they form part of the potential substrate for selection in these, and other, chicken lines.

As high-throughput parallel sequencing technologies generate short reads and limited coverage into the ERV loci depending on mate-pair insert sizes for reliable chromosomal anchoring, we utilized ERVs identified in the Red junglefowl reference assembly to generate an independent ERV search library and anchored loci to chromosomal positions by ERV-associated short reads mate-pair mapping. Despite limited ERV sequence coverage, it is thus possible to associate the best fit for candidate ERV loci reads with assembly ERV sequences, which could be used to construct phylogenetic frameworks and to determine associations between ERV loci and host genomic landscapes.

The whole-genome sequences were generated from pooled individual DNA, see Table 1, which complicates assessments of ERV presence/absence and we, therefore, used a conservative approach by considering loci where one or two of the three domestic chicken lines indicated missing or present ERV-associated reads. Given the large phenotypic differences between the high- and low-growth chicken lines, the domestic animal model presents a promising system to determine potential influences from ERVs on host genome structure and function.

However, although gene ontology searches for genes adjacent to divergent candidate ERV loci could not explain host phenotypic variation, which could be due to the known highly polygenic nature of the trait under selection [21,29], intersection of ERV loci positions with previously determined domestication sweep signals [18,21,29] showed only a weak association. Together, the results suggest that pre-existing ERV variation derived from a common host ancestor segregate in the domestic chicken selection lines today. This notion is supported by the estimated age range of loci that are divergent loci between the domestic chicken lines, which include loci that are presumably old, based on sequence similarities to re-constructed reference assembly ERVs, as well as newer reference assembly ERV insertions, indicating that the divergence represents a frequency shift among ERV loci that segregate in the host population.

Similar observations have also recently been made in other vertebrate host populations [26], demonstrating limitations when assessing historic retrovirus activities from the genomic ERV record using reference assemblies compared to host population data [30], due to the severe sampling effect and associated loss of diversity introduced by reducing a species population down to a single reference genome.

While it can be informative to study ERV variation from host species assemblies covering multiple species over long evolutionary time scales [4,25], analyses along single species phylogenies provide additional information regarding ERV variation and expression [31]. To analyze more recent ERV activities, it has also been successful to employ targeted analysis of specific ERVs in single host species population data [14,15]. However, sampling constraints complicate identification of standing variations in ERV makeups in these systems, and the broad searches in controlled genome groups that the domestic animal selection pedigrees provide are not easily achieved under such conditions. By narrowing the time scale using domestic and wild animal pedigrees, it has been possible to estimate segregating ERV variation for a broad range of ERV clades in host populations [26]. Use of PCR to investigate polymorphisms and incomplete lineage sorting was recently demonstrated for young ERVs [32], and further refining these types of studies by analyzing the recently diverged growth-selected chicken pedigree in this study, we conclude that standing ERV variation is a common feature in contemporary vertebrate populations.

In summary, it appears increasingly important to employ careful experimental design to control the occurrence of artifacts and incorrect inferences due to unbalanced sampling in analyses aimed at evaluating host species ERV makeups. In order to obtain valid comparisons from population and distantly related genomes, it is valuable to focus on well-known pedigrees like those offered by domestic animal selection lines, where the prior knowledge makes it possible to compare observed patterns with expectations that are based on the evolutionary context of the specific case with higher precision than is generally achievable in natural populations. Sequencing and analyses of domestic animal populations and single genomes from known selection pedigrees facilitated by improved sequencing technologies that provide depth and coverage over long insertion sizes together with newly developed and fine-tuned analysis methods will facilitate mapping of ERVs previously not feasible and thereby generate new knowledge about contributions from retroviruses and ERVs to host genome function and evolution.

Supplementary Materials: The following are available online at http://www.mdpi.com/2073-4425/10/2/162/s1, Data S1: galGalERV FASTA sequences, Data S2: galGalERV phylogeny in nexus format, Table S1: galGalERV loci variation and genomic landscape.

Author Contributions: P.J. conceived and designed the study. P.J. and M.E.P. analyzed data. P.J and M.E.P. wrote the manuscript.

Funding: This research was funded by the Swedish Research Council FORMAS, grant numbers 2010-474 and 2018-01008 (to P.J.) and by the Swedish Research Council VR, grant numbers 2015-02429 and 2018-03017 (P.J.).

Acknowledgments: We are grateful to Leif Andersson and his team for valuable discussions and data access during the start of the project. We also thank Carl-Johan Rubin and for helpful discussions. Computer resources were provided by the Uppsala Multidisciplinary Center for Advanced Computational Science (www.uppmax.uu.se), Uppsala University.

Conflicts of Interest: The authors declare no conflict of interest. The funders had no role in the design of the study; in the collection, analyses, or interpretation of data; in the writing of the manuscript, or in the decision to publish the results.

References

1. Stoye, J.P. Endogenous retroviruses: Still active after all these years? *Curr. Biol.* **2001**, *11*, R914–R916. [CrossRef]
2. Jern, P.; Coffin, J.M. Effects of retroviruses on host genome function. *Annu. Rev. Genet.* **2008**, *42*, 709–732. [CrossRef] [PubMed]
3. Mason, A.S.; Fulton, J.E.; Hocking, P.M.; Burt, D.W. A new look at the ltr retrotransposon content of the chicken genome. *BMC Genom.* **2016**, *17*, 688. [CrossRef] [PubMed]
4. Hayward, A.; Cornwallis, C.K.; Jern, P. Pan-vertebrate comparative genomics unmasks retrovirus macroevolution. *Proc. Natl. Acad. Sci. USA* **2015**, *112*, 464–469. [CrossRef] [PubMed]
5. Sperber, G.O.; Airola, T.; Jern, P.; Blomberg, J. Automated recognition of retroviral sequences in genomic data—Retrotector. *Nucleic Acids Res.* **2007**, *35*, 4964–4976. [CrossRef] [PubMed]
6. Feschotte, C.; Gilbert, C. Endogenous viruses: Insights into viral evolution and impact on host biology. *Nat. Rev. Genet.* **2012**, *13*, 283–296. [CrossRef] [PubMed]
7. Rebollo, R.; Romanish, M.T.; Mager, D.L. Transposable elements: An abundant and natural source of regulatory sequences for host genes. *Annu. Rev. Genet.* **2012**, *46*, 21–42. [CrossRef] [PubMed]

8. Stoye, J.P. Studies of endogenous retroviruses reveal a continuing evolutionary saga. *Nat. Rev. Microbiol.* **2012**, *10*, 395–406. [CrossRef] [PubMed]
9. Boi, S.; Rosenke, K.; Hansen, E.; Hendrick, D.; Malik, F.; Evans, L.H. Endogenous retroviruses mobilized during friend murine leukemia virus infection. *Virology* **2016**, *499*, 136–143. [CrossRef] [PubMed]
10. Belshaw, R.; Watson, J.; Katzourakis, A.; Howe, A.; Woolven-Allen, J.; Burt, A.; Tristem, M. Rate of recombinational deletion among human endogenous retroviruses. *J. Virol.* **2007**, *81*, 9437–9442. [CrossRef] [PubMed]
11. Copeland, N.G.; Hutchison, K.W.; Jenkins, N.A. Excision of the DBA ecotropic provirus in dilute coat-color revertants of mice occurs by homologous recombination involving the viral LTRs. *Cell* **1983**, *33*, 379–387. [CrossRef]
12. Hughes, J.F.; Coffin, J.M. Evidence for genomic rearrangements mediated by human endogenous retroviruses during primate evolution. *Nat. Genet.* **2001**, *29*, 487–489. [CrossRef] [PubMed]
13. Kamp, C.; Hirschmann, P.; Voss, H.; Huellen, K.; Vogt, P.H. Two long homologous retroviral sequence blocks in proximal Yq11 cause AZFa microdeletions as a result of intrachromosomal recombination events. *Hum. Mol. Genet.* **2000**, *9*, 2563–2572. [CrossRef] [PubMed]
14. Wildschutte, J.H.; Williams, Z.H.; Montesion, M.; Subramanian, R.P.; Kidd, J.M.; Coffin, J.M. Discovery of unfixed endogenous retrovirus insertions in diverse human populations. *Proc. Natl. Acad. Sci. USA* **2016**, *113*, E2326–E2334. [CrossRef] [PubMed]
15. Holloway, J.R.; Williams, Z.H.; Freeman, M.M.; Bulow, U.; Coffin, J.M. Gorillas have been infected with the HERV-K (HML-2) endogenous retrovirus much more recently than humans and chimpanzees. *Proc. Natl. Acad. Sci. USA* **2019**, *116*, 1337–1346. [CrossRef] [PubMed]
16. Dunnington, E.A.; Siegel, P.B. Long-term divergent selection for eight-week body weight in white plymouth rock chickens. *Poult. Sci.* **1996**, *75*, 1168–1179. [CrossRef] [PubMed]
17. Andersson, L. Genetic dissection of phenotypic diversity in farm animals. *Nat. Rev. Genet.* **2001**, *2*, 130–138. [CrossRef] [PubMed]
18. Rubin, C.J.; Zody, M.C.; Eriksson, J.; Meadows, J.R.; Sherwood, E.; Webster, M.T.; Jiang, L.; Ingman, M.; Sharpe, T.; Ka, S.; et al. Whole-genome resequencing reveals loci under selection during chicken domestication. *Nature* **2010**, *464*, 587–591. [CrossRef] [PubMed]
19. Weiss, R.A. The discovery of endogenous retroviruses. *Retrovirology* **2006**, *3*, 67. [CrossRef] [PubMed]
20. Weiss, R.A. On the concept and elucidation of endogenous retroviruses. *Philos. Trans. R. Soc. Lond. B Biol. Sci.* **2013**, *368*, 20120494. [CrossRef] [PubMed]
21. Johansson, A.M.; Pettersson, M.E.; Siegel, P.B.; Carlborg, O. Genome-wide effects of long-term divergent selection. *PLoS Genet.* **2010**, *6*, e1001188. [CrossRef] [PubMed]
22. Jacobsson, L.; Park, H.B.; Wahlberg, P.; Fredriksson, R.; Perez-Enciso, M.; Siegel, P.B.; Andersson, L. Many qtls with minor additive effects are associated with a large difference in growth between two selection lines in chickens. *Genet. Res.* **2005**, *86*, 115–125. [CrossRef] [PubMed]
23. David, M.; Dzamba, M.; Lister, D.; Ilie, L.; Brudno, M. Shrimp2: Sensitive yet practical short read mapping. *Bioinformatics* **2011**, *27*, 1011–1012. [CrossRef] [PubMed]
24. Benachenhou, F.; Jern, P.; Oja, M.; Sperber, G.; Blikstad, V.; Somervuo, P.; Kaski, S.; Blomberg, J. Evolutionary conservation of orthoretroviral long terminal repeats (LTRs) and ab initio detection of single LTRs in genomic data. *PLoS ONE* **2009**, *4*, e5179. [CrossRef] [PubMed]
25. Hayward, A.; Grabherr, M.; Jern, P. Broad-scale phylogenomics provides insights into retrovirus-host evolution. *Proc. Natl. Acad. Sci. USA* **2013**, *110*, 20146–20151. [CrossRef] [PubMed]
26. Rivas-Carrillo, S.D.; Pettersson, M.E.; Rubin, C.J.; Jern, P. Whole-genome comparison of endogenous retrovirus segregation across wild and domestic host species populations. *Proc. Natl. Acad. Sci. USA* **2018**, *115*, 11012–11017. [CrossRef] [PubMed]
27. Price, M.N.; Dehal, P.S.; Arkin, A.P. Fasttree 2—Approximately maximum-likelihood trees for large alignments. *PLoS ONE* **2010**, *5*, e9490. [CrossRef] [PubMed]
28. van de Lagemaat, L.N.; Medstrand, P.; Mager, D.L. Multiple effects govern endogenous retrovirus survival patterns in human gene introns. *Genome Biol.* **2006**, *7*, R86. [CrossRef] [PubMed]
29. Sheng, Z.; Pettersson, M.E.; Honaker, C.F.; Siegel, P.B.; Carlborg, O. Standing genetic variation as a major contributor to adaptation in the virginia chicken lines selection experiment. *Genome Biol.* **2015**, *16*, 219. [CrossRef] [PubMed]

30. Johnson, W.E. Endogenous retroviruses in the genomics era. *Annu. Rev. Virol.* **2015**, *2*, 135–159. [CrossRef] [PubMed]
31. Bolisetty, M.; Blomberg, J.; Benachenhou, F.; Sperber, G.; Beemon, K. Unexpected diversity and expression of avian endogenous retroviruses. *mBio* **2012**, *3*, e00344-12. [CrossRef] [PubMed]
32. Lee, J.; Mun, S.; Kim, D.H.; Cho, C.S.; Oh, D.Y.; Han, K. Chicken (*Gallus gallus*) endogenous retrovirus generates genomic variations in the chicken genome. *Mob. DNA* **2017**, *8*, 2. [CrossRef] [PubMed]

© 2019 by the authors. Licensee MDPI, Basel, Switzerland. This article is an open access article distributed under the terms and conditions of the Creative Commons Attribution (CC BY) license (http://creativecommons.org/licenses/by/4.0/).

Article

The Integrity of piRNA Clusters is Abolished by Insulators in the *Drosophila* Germline

Elizaveta Radion [1], Olesya Sokolova [1], Sergei Ryazansky [1], Pavel A. Komarov [1,2], Yuri Abramov [1] and Alla Kalmykova [1,*]

[1] Institute of Molecular Genetics, Russian Academy of Sciences, 123182 Moscow, Russia; sradion-radion.90@mail.ru (R.E.); sokolova@img.ras.ru (O.S.); ryazansky@img.ras.ru (S.R.); pkom94@gmail.com (P.A.K.); abramov75@rambler.ru (Y.A.)
[2] Present Address: Friedrich Miescher Institute for Biomedical Research, Maulbeerstrasse 66, 4058 Basel, Switzerland
* Correspondence: allakalm@img.ras.ru; Tel.: +7(499)-1960019; Fax: +7(499)-1960221

Received: 29 January 2019; Accepted: 6 March 2019; Published: 11 March 2019

Abstract: Piwi-interacting RNAs (piRNAs) control transposable element (TE) activity in the germline. piRNAs are produced from single-stranded precursors transcribed from distinct genomic loci, enriched by TE fragments and termed piRNA clusters. The specific chromatin organization and transcriptional regulation of *Drosophila* germline-specific piRNA clusters ensure transcription and processing of piRNA precursors. TEs harbour various regulatory elements that could affect piRNA cluster integrity. One of such elements is the suppressor-of-hairy-wing (Su(Hw))-mediated insulator, which is harboured in the retrotransposon *gypsy*. To understand how insulators contribute to piRNA cluster activity, we studied the effects of transgenes containing *gypsy* insulators on local organization of endogenous piRNA clusters. We show that transgene insertions interfere with piRNA precursor transcription, small RNA production and the formation of piRNA cluster-specific chromatin, a hallmark of which is Rhino, the germline homolog of the heterochromatin protein 1 (HP1). The mutations of Su(Hw) restored the integrity of piRNA clusters in transgenic strains. Surprisingly, Su(Hw) depletion enhanced the production of piRNAs by the domesticated telomeric retrotransposon *TART*, indicating that Su(Hw)-dependent elements protect *TART* transcripts from piRNA processing machinery in telomeres. A genome-wide analysis revealed that Su(Hw)-binding sites are depleted in endogenous germline piRNA clusters, suggesting that their functional integrity is under strict evolutionary constraints.

Keywords: drosophila; retrotransposons; transgene; piRNA cluster; insulator; Su(Hw); Rhino; germline; transcription; *HeT-A* and *TART* telomeric retrotransposons

1. Introduction

The Piwi-interacting RNA (piRNA) pathway is an essential mechanism that protects genome integrity by suppressing transposable element (TE) activity in animal gonads [1]. In *Drosophila*, piRNA precursors are derived from distinct genomic regions termed piRNA clusters, which are enriched in TE fragments [2]. The specific chromatin structure of piRNA clusters ensures the recruitment of the noncanonical transcriptional machinery that drives piRNA precursor expression [3–6]. The chromatin of piRNA clusters is enriched in a common heterochromatic histone mark, trimethylated lysine 9 of histone H3 (H3K9me3) and by two chromodomain-containing proteins, heterochromatic protein 1 (HP1) and its germline-specific ortholog Rhino (Rhi) [4,7–9]. The protein Maelstrom represses canonical transcription from TEs and neighbouring gene promoters in dual-strand piRNA clusters [10]. Instead, noncanonical convergent transcription from both genomic strands initiated at multiple random sites facilitates the transcription of piRNA precursors from dual strand piRNA clusters [6,7]. The initiation of such noncanonical transcription within the heterochromatin of piRNA clusters is mediated by the germline-specific TFIIA-L paralog Moonshiner, which forms an alternative RNA Polymerase II preinitiation complex in Rhi-enriched domains [6]. Rhi binding suppresses the splicing of piRNA precursors [11]. In addition, Rhi recruits Cutoff (Cuff) protein [4], which mediates the generation of long read-through transcripts from piRNA clusters by inhibiting termination at poly(A) sites [12]. Finally, the transcription-export (TREX) complex participates in the export of unspliced piRNA precursors from the nucleus to the cytoplasmic piRNA processing machinery [13,14]. The Piwi-dependent establishment of piRNA cluster identity occurs during early embryogenesis, which is crucial for TE repression at later developmental stages [9]. The integrity of piRNA clusters is an important factor in antitransposon defence, since the adaptivity of the piRNA system is based on the ability of alien sequences inserted within piRNA clusters to become their integral part and produce cognate piRNAs [15–18].

Insulators and their binding proteins play an essential role in transcription regulation by limiting inappropriate enhancer-promoter interactions of neighbouring genes or by blocking repressive chromatin spreading [19]. Insulators are found in the regulatory regions of genes and at homeotic gene loci and the boundaries of topologically associating domains, TADs [20,21]. Some *Drosophila* retrotransposons also contain insulators [22–24]. One of the best-characterized TE insulators is located in the regulatory region of the *gypsy* long terminal repeat (LTR) retrotransposon and contains binding sites for the suppressor of hairy wing (Su(Hw)) zinc-finger protein [25]. This DNA-binding protein establishes the multicomponent chromatin complex important for transcriptional regulation and germline development [19,26]. piRNA clusters contain different TEs, including those demonstrating insulator activity; however, the hierarchical relationship between the chromatin of piRNA clusters and insulator complex formation is not clear. We show here that transgenes bearing Su(Hw) recognition sites embedded in endogenous pericentromeric and telomeric piRNA clusters interfere with the local transcription of piRNA precursors, production of small RNAs and formation of specific chromatin structure.

2. Materials and Methods

2.1. Drosophila Transgenic Strains

The transgenic strain *KG10047* carrying the insertion of the P{SUPor-P} element in the *HeT-A* 3′ UTR was described previously [27]. Transgenic strain *KG09351* (Bloomington Drosophila Stock Centre #16481; the strain was terminated) carries a P{SUPor-P} insertion in the *42AB* locus at the position 2R:2,160,357 [-] (according to the dm3 genome assembly). The transgenic strain KG02245 (Bloomington Drosophila Stock Centre #12975) carries a P{SUPor-P} insertion in the *49E* locus. Transheterozygous $su(Hw)^V/su(Hw)^f$ flies were used in the study.

2.2. Small RNA Library Preparation and Analysis

Small RNAs 19–29 nt in size from total ovarian RNA extracts were cloned as previously described [18]. The libraries were barcoded according to Illumina TrueSeq Small RNA sample prep kit instructions and submitted for sequencing using the Illumina HiSeq-2000 sequencing system (San Diego, CA, USA). After clipping the Illumina 3′-adapter sequence, small RNA reads that passed quality control and minimal length filter (>18 nt) were mapped (allowing 0 mismatches) to the Drosophila melanogaster genome (Apr. 2006, BDGP assembly R5/dm3) or transgenes by bowtie2 [28]. The plotting of size distributions, read coverage and nucleotide biases were performed as described previously [18]. To identify piRNAs (24–29 nt reads) or siRNAs (21 nt reads) derived from TEs and piRNA clusters, small RNA reads were mapped to the canonical sequences of transposable elements (http://www.fruitfly.org/p_disrupt/TE.html) or to the piRNA clusters [2] by bowtie2 [28].

Ovarian small RNA-seq data for *KG10047;+/+*, *KG10047;su(Hw)^V/su(Hw)^f*, *KG09351;+/+*; *KG09351;su(Hw)^V/su(Hw)^f* ; *KG02245;+/+* and *KG02245;su(Hw)^V/su(Hw)^f* were deposited at Gene Expression Omnibus (GEO), accession number GSE125173.

2.3. Chromatin Immunoprecipitation (ChIP)

ChIP was performed according to the published procedure [7]. Chromatin was immunoprecipitated with the following antibodies: anti-HP1a (C1A9 Developmental Studies Hybridoma Bank, Iowa Sity, IA, USA), anti-trimethyl-histone H3 Lys9 (Millipore, Burlington, MA, USA), Rhi antiserum [29] and anti-Su(Hw) [30]. Primers used in the study are listed in Table S1. Quantitative PCR was conducted with a LightCycler 96 (Roche, Basel, Switzerland). Obtained values were normalized to input and compared with values at *rp49* gene as a control genomic region. Standard error of mean (SEM) of triplicate PCR measurements for three biological replicates was calculated.

2.4. RT-PCR

RNA was isolated from the ovaries of three-day-old females. cDNA was synthesized using random hexamers or strand-specific primers and SuperScriptII reverse transcriptase (Life Technologies, Carlsbad, CA, USA). cDNA samples were analysed by real-time quantitative PCR using SYTO-13 dye on a LightCycler96 (Roche, Basel, Switzerland). Values were averaged and normalized to the expression level of the ribosomal protein gene *rp49*. The primers used are described in Table S1.

2.5. Motif Finding

To estimate the frequency of the Su(Hw) insulator sites, the corresponding PWM profile (MA0533.1 from JASPAR_2016 database) was searched against the dm6 genome assembly or piRNA cluster regions [2] by using fimo 4.11.1 from the MEME suite [31]. The *p*-value 1×10^{-5} was used as the threshold level.

3. Results and Discussion

3.1. P{SUPor-P} Transgenic Constructs Inserted into piRNA Clusters Do Not Produce piRNAs

To study how insulators affect piRNA cluster integrity, we used P{SUPor-P} transgenic constructs carrying two *gypsy* insulators and located within endogenous piRNA clusters. KG10047 and KG09351 transgenes were inserted into the 3′UTR of telomeric retrotransposon *HeT-A* and the major pericentromeric piRNA cluster in the *42AB* locus, respectively (Figure 1A).

Both integrated loci were previously described as potent piRNA clusters able to adapt new insertions for piRNA production [12,15,16,32]. The euchromatic KG02245 transgene was used as a control. To determine whether the constructs carrying *gypsy* insulators are able to become a part of piRNA clusters and produce piRNAs, we sequenced small RNAs from the ovaries of the transgenic strains. The mapping of small RNAs to P{SUPor-P} revealed a negligible amount of the transgene-derived small RNAs in both cases (Figure 1B). We suggested that Su(Hw) binding could impede piRNA production and performed ovarian small RNA sequencing of transgenic strains bearing *su(Hw)^V/su(Hw)^f* mutations. These mutations cause the loss of Su(Hw) binding to the *gypsy* insulator [33,34] and, as we show in the next section, to P{SUPor-P} transgene (Figure 2B). Su(Hw) mutations result in the production of abundant transgenic small RNAs, most of which are 24–29 nt long and demonstrate 5′ terminal uridine bias (1U bias), which is a characteristic of piRNAs (Figure 1B,C). However, we did not find the sense/antisense piRNA pairs overlapping by 10 nt, which is a signature of the ping-pong piRNA amplification cycle [2,35] (Figure 1D). This result indicates that primary processing plays a major role in transgenic piRNA production. The most likely explanation is that a low abundance of transgenic transcripts prevents efficient ping-pong amplification. The euchromatic transgene KG02245 (control) produces a negligible amount of small RNAs in wild type and Su(Hw) mutant backgrounds (Figure 1).

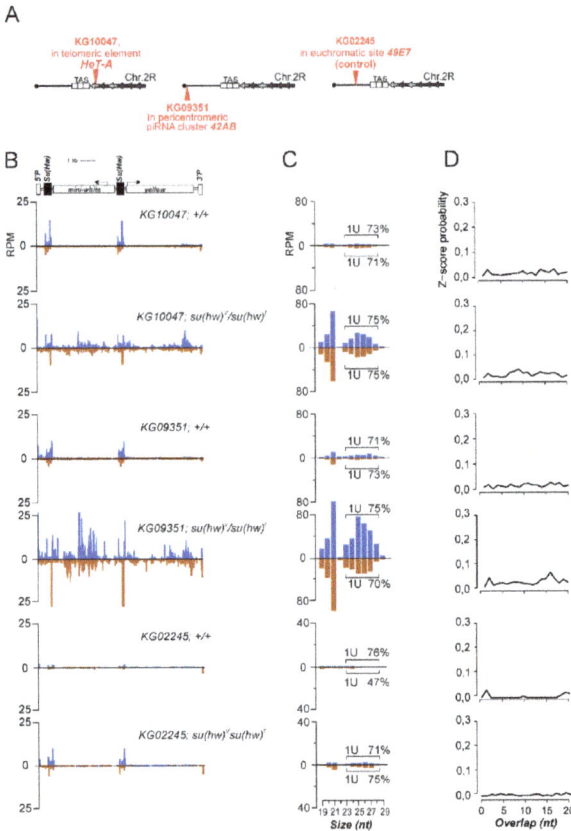

Figure 1. Generation of small RNAs by transgenes containing Su(Hw)-binding sites and located in piRNA clusters. (**A**) Schematic representation of transgenic insertion sites. Insertion sites of transgenes are indicated as triangles situated either up or below the schemes, which correspond to their genomic orientation. (**B**) Scheme of SUPor-P construct is shown above. Normalized numbers of small RNAs (19–29 nt, in reads per million, RPM) mapped to transgenic constructs (blue—sense; brown—antisense; no mismatches allowed) in wild type *Drosophila* ovaries and in $su(Hw)^V/su(Hw)^f$ mutants. (**C**) Length distribution of transgenic small RNAs. Percentage of transgenic reads excluded Su(Hw) sites having 1U is indicated for each strand (only 24–29-nt reads were considered). (**D**) Relative frequencies (Z-score) of 5′ overlap for sense and antisense 24–29-nt transgenic piRNAs excluded Su(Hw) sites (ping-pong signature).

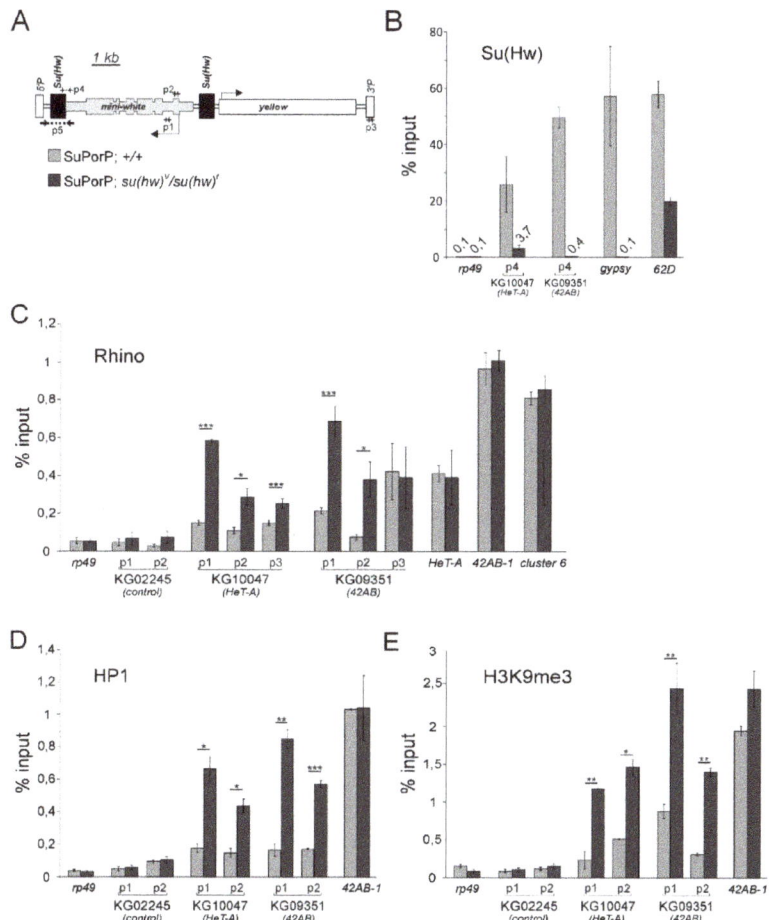

Figure 2. Chromatin components and transcription state of the transgenes containing Su(Hw) sites and located within piRNA clusters. (**A**) Schematic representation of the P{SUPor-P} transgene and the positions of the primers used in ChIP (p1, p2, p3, p4) and RT-PCR (p1, p3, p5) are shown. (**B**) Su(Hw) binds P{SUPor-P} transgenes in Su(Hw) wild type but not in mutant ovaries. The *rp49* is used as a control devoid of Su(Hw) binding sites. As expected, the *gypsy* insulator lost Su(Hw) binding in the $su(Hw)^V/su(Hw)^f$ background but the insulator in the 62D locus retained Su(Hw) association [33,34]. Mean values are indicated only for low levels of Su(Hw) binding. (**C–E**) Rhi (**C**), HP1 (**D**) and H3K9me3 (**E**) occupancies at P{SUPor-P} transgenes in wild type and $su(Hw)^V/su(Hw)^f$ mutants were estimated by ChIP-qPCR using indicated primers (p1, p2, p3). The regions of the endogenous *42AB* and #6 piRNA clusters and telomeric retrotransposon *HeT-A* are enriched by all studied chromatin components and used as positive controls. *rp49* is used as a negative control. Asterisks indicate statistically significant differences in chromatin protein enrichments between wild type and $su(Hw)^V/su(Hw)^f$ mutants (* $p < 0.05$ to 0.01, ** $p < 0.01$ to 0.001, *** $p < 0.001$, unpaired *t*-test). Error bars represent SEM of 3 biological replicate experiments. For HP1 and H3K9me3 binding to KG02245, the error bars represent SD of three technical replicates.

Interestingly, a significant fraction of the small RNAs produced by both P{SUPor-P} transgenes in Su(Hw) mutants are 21-nt endogenous small interfering RNAs (endo-siRNAs) (Figure 1C). Indeed, it has been reported that endogenous and transgenic piRNA clusters also produce significant levels of endo-siRNAs in wild type ovaries [18,32,36,37]. In Su(Hw) mutants, P{SUPor-P} transgenes become part of the endogenous piRNA clusters, producing both pi- and endo-siRNAs.

Previously, it was reported that P{lArB}, pW8-hsp-pA and P{EPgy2} transgenes lacking *gypsy* insulators inserted in *Drosophila* subtelomeric and telomeric piRNA clusters are incorporated in piRNA production and acquire chromatin properties of their surrounding regions [16,18,32]. Similar to transgenes, TE insertions into the germline piRNA clusters result in the production of cognate piRNAs ensuring silencing of mobilized TEs [15]. We show here, that Su(Hw)-mediated *gypsy* insulators prevent piRNA production from P{SUPor-P} transgenic sequences, even if the latter are inserted into endogenous piRNA-producing regions.

3.2. The Su(Hw) Complex Prevents the Assembly of the Chromatin Structure and Read-Through Transcription Typical of piRNA Clusters

To learn more about the mechanism of the Su(Hw)-mediated prevention of piRNA production, we compared the chromatin state of transgenes in the wild type and Su(Hw)-mutant backgrounds. In our experiments, we used transheterozygous flies with two Su(Hw) alleles: the Su(Hw)^V null allele and the Su(Hw)^f, which carry a defective zinc finger 10 [34]. Su(Hw)^f protein demonstrated the loss of binding to *gypsy* insulator and the reduced occupancy of many non-*gypsy* Su(Hw)-binding sites [34,38]. ChIP data obtained using anti-Su(Hw) antibodies demonstrate that P{SUPor-P} transgenes binds Su(Hw) in wild type ovaries (Figure 2B). ChIP performed with $su(Hw)^V/su(Hw)^f$ ovaries, shows a dramatic decrease of Su(Hw) binding to the *gypsy* insulator and P{SUPor-P} transgenes. As it was reported previously, mutant Su(Hw)^f is retained at some insulators including site located in *62D* locus [33,34]. Indeed, ChIP demonstrated that mutant Su(Hw) bound this region, that serves as a positive control (Figure 2B).

The data from ChIP using Rhi, HP1 and H3K9me3 antibodies show that P{SUPor-P} transgenes inserted in both the *42AB* locus and telomeric retroelement *HeT-A* lack these chromatin hallmarks in the presence of Su(Hw) protein (in the wild type genetic background) but acquire them in Su(Hw)-mutant ovaries (Figure 2C–E). At the same time, Rhi binding to 3' P-element region, located ~5 kb apart from the *gypsy* insulators, is not affected by Su(Hw) mutation in *KG09351* and only 1.5-fold increases in *KG10047;su(Hw)$^{V/f}$* ovaries (Figure 2C, p3 primer pair). Chromatin of the telomeric element *HeT-A* and dual-strand piRNA clusters (*42AB* and cluster #6) was not affected by Su(Hw) mutation (Figure 2). Therefore, the *gypsy* insulator complex is established upstream of the Piwi-dependent chromatin formation of piRNA clusters and Su(Hw) binding mostly affects local chromatin conformation.

Our data suggest that Su(Hw) binding to P{SUPor-P} transgenes inserted into piRNA clusters should block transcription of long piRNA precursors. To verify this suggestion, we compared expression of *mini-white* gene of the transgenes inserted into *HeT-A* (*KG10047* strain) and *42AB* (*KG09351* strain) in the wild-type and mutant background using transgene-specific primers (Figure 3A). It should be noted that the *white* promoter shows a very low activity in ovaries. In contrast, we observed that the level of *mini-white* transcripts was significantly increased in Su(Hw) mutants for both insertions (Figure 3A). In addition, Su(Hw) mutations lead to the increased transcript levels of 3'P transgenic regions located 5 kb downstream Su(Hw) binding sites (Figure 3A). Strand-specific RT-PCR analysis of 3'P transcription in *KG09351* strain demonstrated a lowered level of only sense transgenic transcripts suggesting that the effect of *gypsy* insulator on transcription is strand-specific (Figure 3B). We therefore suggest that the Su(Hw)-insulator complex should directly interfere with transcription of piRNA clusters by blocking read-through transcription of piRNA precursors. To test this, we conducted RT-PCR using the primers corresponding to the upstream and downstream regions of the transgenic Su(Hw) site (Figure 2A, p5 primer pair). Accordingly, we revealed transgenic transcripts only in the ovaries of Su(Hw) mutants (Figure 3C). These data indicate that the insulator complex blocks the read-through transcription of transgenic piRNA precursors. Taken together, our data suggest that insertions of the transgenes containing *gypsy* insulators into endogenous piRNA clusters affect local chromatin conformation, causing the disruption of long piRNA precursor transcription and that this effect is mediated by Su(Hw) binding (Figure 3D). However, the function of the *gypsy* insulator also requires Centrosomal Protein 190 kD (CP190) and Modifier of mdg4 (Mod67.2) [39,40], the insulator proteins which were not considered here. Therefore, strictly speaking, we could not unambiguously conclude whether or not the effect of Su(Hw) binding to piRNA clusters demonstrated here was dependent on the entire insulator complex assembly.

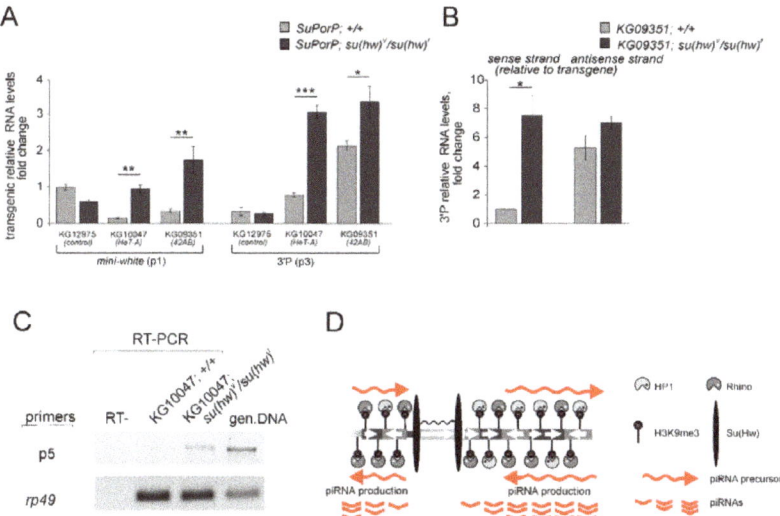

Figure 3. Transcription state of the transgenes containing Su(Hw) sites and located within piRNA clusters. The positions of primers are indicated in Figure 2A. (**A**) RT-qPCR analysis of transgenic *mini-white* and 3′P transcripts in the ovaries of transgenic strains in wild type and Su(Hw) mutant backgrounds. The P1 primer pair used for RT-PCR specifically detects unspliced transgenic *mini-white* transcripts. (**B**) Strand-specific RT-PCR using the P3 primers corresponding to the transgenic 3′P region showed a decreased level of RNA from sense but not from antisense transgenic strand downstream of the Su(Hw) binding sites. Asterisks indicate statistically significant differences in the expression levels between wild type and $su(Hw)^V/su(Hw)^f$ mutants (* $p < 0.05$ to 0.01, ** $p < 0.01$ to 0.001, *** $p < 0.001$, unpaired *t*-test). (**C**) Agarose gel electrophoresis of RT-PCR products shows the presence of read-through transgenic transcripts comprising Su(Hw)-binding sites only in Su(Hw) mutants. Samples without reverse transcriptase were used as RT⁻ controls. PCR on genomic DNA served as a positive control. (**D**) Scheme showing that insertion of the Su(Hw) insulator into the piRNA cluster disrupts local transcription of piRNA precursors, production of small RNAs and formation of specific chromatin structure.

Given the functional integrity of piRNA clusters, we suggested that *gypsy* insulators might impair the functioning of the cluster regions in the close vicinity of transgenic insertion. To verify this suggestion, we estimated piRNA production in *42AB* regions located around the KG09351 insertion.

Due to the presence of highly degenerated TE fragments, piRNA clusters produce abundant piRNAs uniquely mapped to the genome, allowing their mapping to repeat-rich regions [2]. We found that the amount of piRNAs uniquely mapped to the 10-kb region flanking the 3′-end of the transgene was dramatically lower in the presence of the transgene insertion than in the native *42AB* locus (Figure 4A).

However, Su(Hw) mutations restored the level of transgene-flanking piRNAs in the *42AB* locus up to the level observed in the native *42AB* region in *KG10047* strain (Figure 4A). We also show that the production of piRNAs derived from the same region is not affected by Su(Hw) mutations in *KG10047* strain. In addition, the transgenic insertion in *42AB* leads to a reduction in transcript levels downstream of the transgene (Figure 4B, transgene is located in the minus genomic strand). Strand-specific RT-PCR analysis using single-mapped primers [41] demonstrated a dramatic decrease in the level of transcripts from the negative genomic strand downstream the transgene (Figure 4C). This fact is in agreement with previous result, demonstrating the lowered levels of transgenic 3'P sense transcripts (corresponding to the minus genomic strand) in the ovaries of *KG09351* strain (Figure 3B). Thus, the Su(Hw) insulator blocks piRNA precursor transcription at least within a 10-kb neighbouring region. Our observations suggest that transgenes containing Su(Hw)-binding sites disrupt long transcription units within piRNA clusters and that this effect is strand-specific. Surprisingly, transgene insertion resulting in lowered level of only antisense transcripts led to the decreased level of both sense and antisense piRNAs uniquely mapped to transgene flanking region in *42AB* (Figure 4A). This fact can be explained by an impaired efficiency of ping-pong amplification between piRNA precursors derived from this region. Indeed, it was reported that sense and antisense transcripts originated from the heterochromatic piRNA cluster are involved in the reciprocal cleavage in the course of ping-pong piRNA amplification [42]. Nevertheless, the total abundance of *I*-specific piRNAs is not affected by the KG09351 insertion in *42AB* (Figure S1) because numerous active *I*-element copies participate in piRNA production in this strain.

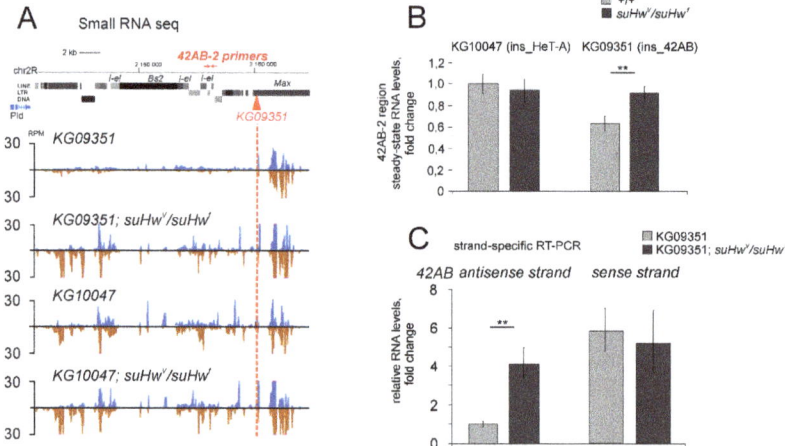

Figure 4. Su(Hw)-binding sites disrupt the integrity of the endogenous piRNA cluster. (**A**) The effect of SUPor-P transgene inserted in the *42AB* locus on piRNA expression. Profile of ovarian small RNA density at the *42AB* region adjacent to the KG09351 transgene in wild type and $su(Hw)^V/su(Hw)^f$ mutant flies and in the *KG10047* strain without a transgene insertion in this region. The position of the transgene in the minus genomic strand is designated by a red rectangle. Genomic coordinates are indicated according to the dm6 genome assembly. (**B**) RT-qPCR analysis of the expression levels of the *42AB* region located 4 kb downstream of the KG09351 transgene. The positions of the primers used for RT-PCR are schematically indicated in (**A**). (**C**) Strand-specific RT-PCR using the primers indicated in (**A**) showed a decreased level of RNA from the negative genomic strand downstream of the transgene insertion. Asterisks indicate statistically significant differences in the expression levels between wild type and $su(Hw)^V/su(Hw)^f$ mutants (* $p < 0.05$ to 0.01, ** $p < 0.01$ to 0.001, *** $p < 0.001$, unpaired *t*-test).

Taken together, these data explain why Su(Hw) binding results in a local decrease in small RNA production not only from the *mini-white* located between the insulator sequences but also from the transgenic and genomic flanking regions.

We could not perform the analysis of flanking piRNAs near the KG10047 transgene inserted in the *HeT-A* 3'UTR at 2R telomere because the unique mapping of small RNAs to poorly assembled and highly repetitive telomeric regions was technically impossible.

To a certain extent, insertion of P{SUPor-P} transgene in *42AB* was helpful for understanding of the transcription regulation of the major uni-strand *flamenco* (*flam*) locus that controls TE expression in ovarian follicular cells [43,44]. In contrast to the germline dual-strand piRNA clusters that generate piRNAs corresponding to both genomic strands, the *flam* locus produces primary piRNAs from single strand precursors [2]. Insulator-harbouring TEs, such as *gypsy*, *ZAM* and *Idefix* [22–24], are exceptionally arranged in antisense orientation relative to *flam* transcription. It is believed that piRNAs complementary to the coding transcripts of these TEs are produced from the single strand precursors transcribed by the *flam* sense strand [2]. Apparently, *gypsy*, *Zam* and *Idefix* insulators do not interfere with the transcription of the *flam* piRNA precursors, likely due to a strand-specific mode of insulator influence on transcription.

3.3. Su(Hw) Restricts piRNA Production from Telomeric TART Retrotransposons

The main structural telomeric element, *HeT-A*, is a non-autonomous retroelement and reverse transcriptase (RT) activity is likely to be provided by *TART* or *TAHRE* telomeric retrotransposons. In *Drosophila* germline, telomeric regions are organized in the piRNA clusters, although telomeric elements are heterogeneous in piRNA production and Rhi binding: *TART* retrotransposons are less susceptible to piRNA production and Rhi deposition than *HeT-A* [32]. This implies that *TART* transcripts may be protected by an unknown mechanism from piRNA biogenesis machinery to provide stable expression of RT essential for telomere elongation in the germline. Here, we present the data suggesting a role for Su(Hw) in this mechanism.

Using small RNAseq data, we studied the genome-wide impact of Su(Hw) depletion on transposon-derived piRNA production. The mapping of small RNAs to a canonical set of TEs does not demonstrate global changes caused by Su(Hw) mutations (Figure 5A, Figure S2).

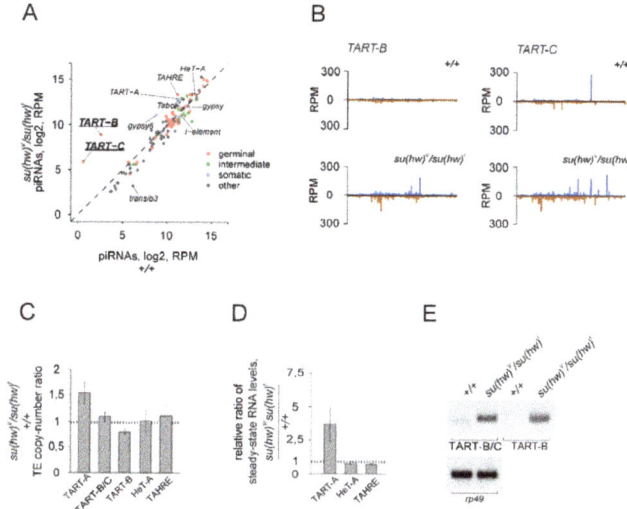

Figure 5. Su(Hw) depletion leads to the increased production of *TART* piRNAs. (**A**) Scatter plots of log2-transformed and RPM-normalized small RNAseq reads in the ovaries of wild type and *su(Hw)* mutant flies mapped to the canonical TE sequences. The colour of the dots indicates the type of TEs according to their capacity for maternal deposition in embryos according to [45]. (**B**) Small RNA mapping to canonical *TART-B* and *TART-C* telomeric retrotransposons. Reads mapped to the sense strand are shown in blue and antisense in brown. Analysis of ovarian small RNA libraries from *KG10047;+/+* and *KG10047;su(Hw)^V/su(Hw)^f* strains (0–3 mismatches allowed) is shown (**A,B**). (**C**) qPCR on the genomic DNA was done to estimate relative copy number of telomeric retrotransposons in *KG10047;su(Hw)^V/su(Hw)^f* compared to *KG10047;+/+* strain. Normalization to the single-copy *rp49* gene was done. (**D**) RT-qPCR analysis of transcript levels of *TART-A*, *HeT-A* and *TAHRE* telomeric elements normalized to *rp49* in the ovaries. Shown are fold changes of steady-state RNA levels in KG10047; *su(Hw)^V/su(Hw)^f* compared to *KG10047;+/+* strain. (**E**) Agarose gel electrophoresis of RT-PCR products demonstrates increased levels of *TART-B* and *TART-C* transcripts in Su(Hw) mutants. *Rp49* is used as a loading control.

However, piRNA production from few TEs was strongly affected (Table S2). In particular, Su(Hw) mutations caused a 100-fold increase in the abundance of piRNAs specific to *TART-B* and *TART-C* subfamilies of telomeric retrotransposons (Figure 5B, Table S2). To exclude the influence of copy number polymorphism, we evaluated the relative copy number of telomeric retrotransposons in $KG10047;+/+$ and $KG10047;su(Hw)^V/su(Hw)^f$ strains. To this end, we performed PCR on genomic DNA and showed that the relative copy numbers of *HeT-A, TAHRE, TART-A, TART-B* and *TART-C* are very similar in both strains (Figure 5C). These data strongly suggest that *TART-B/TART-C* transcripts are protected by Su(Hw)-dependent border elements from the piRNA production. Next, we examined the RNA levels of telomeric retroelements in the ovaries of Su(Hw) mutants by RT-qPCR. We found an increased level of *TART-A* transcripts in the ovaries of Su(Hw) mutants, while *HeT-A* and *TAHRE* expression was not affected (Figure 5D). However, the levels of *TART-B* and *TART-C* transcripts in the wild type ovaries were undetectable by RT-qPCR. Probably, *TART* expression is limited by a time window during oogenesis, resulting in low levels of *TART* RNA in the total ovarian RNA. Then, we performed semiquantitative RT-PCR and observed that *TART-B* and *TART-C* transcripts were barely revealed in the wild type ovaries but readily detected in the ovaries of Su(Hw) mutants (Figure 5E). These data suggest that Su(Hw)-dependent insulators provide appropriate levels of *TART*-encoding transcripts in telomeres. Su(Hw) serves as a transcriptional repressor of coding genes in the ovary [26]. Notably, in wild-type ovaries, *TART* transcripts are less abundant than upon Su(Hw) depletion. Thus, Su(Hw) likely mediates transcriptional repression of *TART* in the germline.

3.4. Su(Hw)-Binding Sites Are Depleted from Dual-Strand piRNA Clusters

What could be happened if the insulator is inserted into the piRNA cluster? To some extent, this situation is simulated by the insertion of P{SUPor-P} transgene into the *42AB* cluster (Figure 4). Interestingly, the region of the KG09351 insertion in *42AB* harbours remnants of ancestral *I*-related retrotransposon, producing abundant piRNAs playing a key role in the control of *I*-element activity [41,46]. The integration of insulator-containing TE nearby *I*-element fragments in *42AB* would strongly decrease the abundance of piRNAs derived from this region and the resistance to *I*-element mobilization in the strain. One may suggest that shaping endogenous piRNA clusters should be under the constraint of adaptive evolution and, therefore, the germline-specific piRNA clusters should be depleted of insulator-binding sites. To examine this idea, we estimated the average density of Su(Hw)-binding sites in the whole genome and in piRNA clusters (see Materials and Methods). We found that the Su(Hw) site density was lower in piRNA clusters (one site per 45 kb) than in the genome (one site per 14 kb, p-value $< 1 \times 10^{-5}$). It is tempting to speculate that the spectrum of regulatory sequences associated with TEs in piRNA clusters is subjected to strict selection to provide functional integrity of piRNA-producing loci.

Indeed, mapping of small RNAseq reads to the annotated piRNA clusters [2] did not reveal global changes in the abundance of pi- and endo-siRNAs caused by Su(Hw) mutations (Figure 6A, Figure S3). Of those affected are many piRNA clusters related to telomeric regions which contain *TART* and *TAHRE* retrotransposons (Table S2). However, motif search analyses with the same parameters as above (p-value $< 1 \times 10^{-5}$) failed to identify the Su(Hw) binding sites in canonical copies of telomeric retrotransposons. Thus, the nature of Su(Hw) binding sites in telomeres is still to be determined.

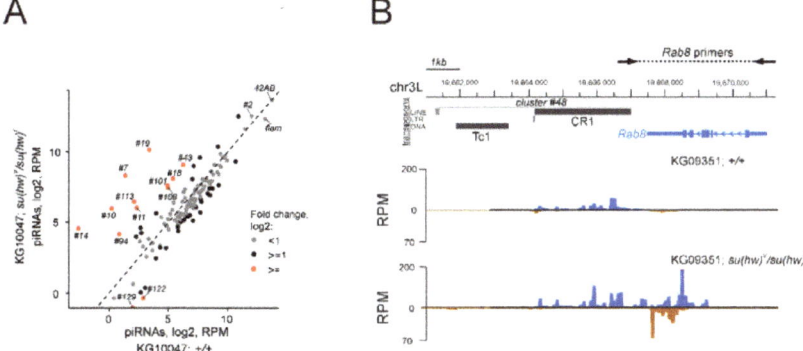

Figure 6. Su(Hw) depletion leads to increased production of piRNAs in distinct genomic sites. (**A**) Scatter plot of log2-transformed and RPM-normalized small RNAseq reads in the ovaries of wild type and $su(Hw)^V/su(Hw)^f$ mutant flies mapped to piRNA clusters (**B**) Profile of the ovarian small RNA density at the *Rab8* gene region located in close proximity to cluster #48, comprising of *Tc1* and *CR1* TEs, in wild type and $su(Hw)^V/su(Hw)^f$ mutant flies. The increased production of small RNAs by both genomic strands in the Su(Hw) mutant background is shown. Small RNAseq data from the ovaries of *KG10047;+/+* and *KG10047;su(Hw)^V/su(Hw)^f* strains were used in the figure. Genomic coordinates are indicated according to the dm6 genome assembly. PCR on genomic DNA using indicated primers did not reveal variations in the length of the *Rab8* gene region.

3.5. The Su(Hw) Complex Protects Coding Genes from Spurious piRNA Production in the Germline

Although the loss of Su(Hw) causes female sterility, its particular role in the female germline development is not well understood [34,38]. Su(Hw) mutations lead to the increased expression of many target genes in the ovary, suggesting that Su(Hw) serves as a transcriptional repressor during oogenesis [26].

The genome-wide analysis of small RNAseq data demonstrated additional functions of Su(Hw) in the germline. We revealed a genome-wide effect of $Su(Hw)^V/Su(Hw)^f$ mutations on the abundance of piRNAs derived from coding genes (Table S3). Most of the affected genes do not produce piRNAs in a wild-type ovary. The piRNAs corresponding to these genes were observed in $Su(Hw)^V/Su(Hw)^f$ ovaries, suggesting that Su(Hw) might prevent the spread of piRNA production from piRNA clusters/TEs to neighbouring genes. Indeed, Su(Hw) depletion causes the appearance of piRNAs antisense to the transcripts of the *Rab8* gene located next to the cluster #48 (Figure 6B). PCR by using genomic DNA did not reveal variations in the length of the gene region producing genic piRNAs, rejecting the possibility of transposon insertion polymorphism. Thus, the insulator most likely blocks transcription of long piRNA precursors, thus protecting coding gene transcripts from entering the piRNA biogenesis machinery in the germline. However, annotated TEs were found in the close vicinity of the affected genes only in a few cases. Surprisingly, we revealed an effect of Su(Hw) mutations on the abundance of the piRNAs derived from dozens of the coding genes located far from annotated TEs or piRNA clusters (Table S3). At least partly, genic piRNA production may be explained by strain-specific transposon insertions [47]. In most cases, the molecular mechanisms responsible for generation of genic piRNAs in the ovaries of Su(Hw) mutants remain unclear. It is tempting to speculate that Su(Hw)-mediated complexes perform a barrier function to protect coding gene transcripts from spurious piRNA production in the germline.

4. Conclusions

Genomic regions containing damaged transposon fragments were for a long time considered as waste dumps. Most of these regions are piRNA-producing loci that play an essential role in antitransposon defence. The complex regulation of piRNA clusters has evolved to provide piRNA production from the entire piRNA cluster. This mechanism ensures an adaptive response to insertions of alien transposons in a piRNA cluster. We show that Su(Hw) binding sites disrupt the integrity of endogenous piRNA clusters, indicating that the assembly of insulator complexes occurs upstream of cluster-specific chromatin formation. Considering that distinct TE families comprise insulator-binding sites and other regulatory sequences, the TE content of the piRNA clusters should be under strict evolutionary constraints. Moreover, Su(Hw)-mediated complexes likely protect telomeric retrotransposon *TART* and coding gene transcripts from spurious piRNA production in the germline.

Supplementary Materials: The following are available online at http://www.mdpi.com/2073-4425/10/3/209/s1, Figure S1: Small RNA mapping to the canonical *I*-element. Figure S2: Genome-wide analysis of Su(Hw) depletion on TE piRNA production (related to Figure 5). Figure S3: Genome-wide analysis of Su(Hw) depletion on small RNA production from piRNA clusters and genes (related to Figure 6). Table S1. Primers used in the study (5′-to-3′). Table S2. Su(Hw) mutations affect piRNA production from some TEs and piRNA clusters. Table S3. Su(Hw) mutations cause accumulation of genic piRNAs.

Author Contributions: A.K. conceived and designed the experiments; E.R., O.S. and P.A.K. performed the experiments; Y.A. performed genetic experiments; S.R. and P.A.K. performed bioinformatic analysis; A.K. wrote the paper.

Funding: This work was supported by the Russian Science Foundation (grant no. 16-14-10167 to A.K.) and the Russian Foundation for Basic Researches (grant no. 18-34-00415 to E.R.).

Acknowledgments: We thank J. Mason for KG10047 strain, M. Savitsky for helpful advice on genetic crosses, A. Golovnin for Su(Hw) mutant strains and anti-Su(Hw) antibodies, N. Akulenko for assistance with small RNA library preparation. We also thank the Bloomington Stock Centre for fly strains and the Developmental Studies Hybridoma bank for antibodies.

Conflicts of Interest: The authors declare no conflict of interest.

References

1. Iwasaki, Y.W.; Siomi, M.C.; Siomi, H. PIWI-Interacting RNA: Its Biogenesis and Functions. *Annu. Rev. Biochem.* **2015**, *84*, 405–433. [CrossRef] [PubMed]
2. Brennecke, J.; Aravin, A.A.; Stark, A.; Dus, M.; Kellis, M.; Sachidanandam, R.; Hannon, G.J. Discrete small RNA-generating loci as master regulators of transposon activity in Drosophila. *Cell* **2007**, *128*, 1089–1103. [CrossRef] [PubMed]
3. Klattenhoff, C.; Xi, H.; Li, C.; Lee, S.; Xu, J.; Khurana, J.S.; Zhang, F.; Schultz, N.; Koppetsch, B.S.; Nowosielska, A.; et al. The Drosophila HP1 homolog Rhino is required for transposon silencing and piRNA production by dual-strand clusters. *Cell* **2009**, *138*, 1137–1149. [CrossRef] [PubMed]
4. Mohn, F.; Sienski, G.; Handler, D.; Brennecke, J. The rhino-deadlock-cutoff complex licenses noncanonical transcription of dual-strand piRNA clusters in Drosophila. *Cell* **2014**, *157*, 1364–1379. [CrossRef] [PubMed]
5. Pane, A.; Jiang, P.; Zhao, D.Y.; Singh, M.; Schupbach, T. The Cutoff protein regulates piRNA cluster expression and piRNA production in the Drosophila germline. *Embo J.* **2011**, *30*, 4601–4615. [CrossRef] [PubMed]
6. Andersen, P.R.; Tirian, L.; Vunjak, M.; Brennecke, J. A heterochromatin-dependent transcription machinery drives piRNA expression. *Nature* **2017**, *549*, 54–59. [CrossRef] [PubMed]
7. Akulenko, N.; Ryazansky, S.; Morgunova, V.; Komarov, P.A.; Olovnikov, I.; Vaury, C.; Jensen, S.; Kalmykova, A. Transcriptional and chromatin changes accompanying de novo formation of transgenic piRNA clusters. *RNA* **2018**, *24*, 574–584. [CrossRef] [PubMed]
8. Rangan, P.; Malone, C.D.; Navarro, C.; Newbold, S.P.; Hayes, P.S.; Sachidanandam, R.; Hannon, G.J.; Lehmann, R. piRNA Production Requires Heterochromatin Formation in Drosophila. *Curr. Biol.* **2011**, *21*, 1373–1379. [CrossRef] [PubMed]
9. Akkouche, A.; Mugat, B.; Barckmann, B.; Varela-Chavez, C.; Li, B.; Raffel, R.; Pelisson, A.; Chambeyron, S. Piwi Is Required during Drosophila Embryogenesis to License Dual-Strand piRNA Clusters for Transposon Repression in Adult Ovaries. *Mol. Cell* **2017**, *66*, 411–419 e414. [CrossRef] [PubMed]
10. Chang, T.H.; Mattei, E.; Gainetdinov, I.; Colpan, C.; Weng, Z.; Zamore, P.D. Maelstrom Represses Canonical Polymerase II Transcription within Bi-directional piRNA Clusters in Drosophila melanogaster. *Mol. Cell* **2018**. [CrossRef] [PubMed]

11. Zhang, Z.; Wang, J.; Schultz, N.; Zhang, F.; Parhad, S.S.; Tu, S.; Vreven, T.; Zamore, P.D.; Weng, Z.; Theurkauf, W.E. The HP1 homolog rhino anchors a nuclear complex that suppresses piRNA precursor splicing. *Cell* **2014**, *157*, 1353–1363. [CrossRef] [PubMed]
12. Chen, Y.C.; Stuwe, E.; Luo, Y.; Ninova, M.; Le Thomas, A.; Rozhavskaya, E.; Li, S.; Vempati, S.; Laver, J.D.; Patel, D.J.; et al. Cutoff Suppresses RNA Polymerase II Termination to Ensure Expression of piRNA Precursors. *Mol. Cell* **2016**, *63*, 97–109. [CrossRef] [PubMed]
13. Hur, J.K.; Luo, Y.; Moon, S.; Ninova, M.; Marinov, G.K.; Chung, Y.D.; Aravin, A.A. Splicing-independent loading of TREX on nascent RNA is required for efficient expression of dual-strand piRNA clusters in Drosophila. *Genes Dev.* **2016**, *30*, 840–855. [CrossRef] [PubMed]
14. Zhang, F.; Wang, J.; Xu, J.; Zhang, Z.; Koppetsch, B.S.; Schultz, N.; Vreven, T.; Meignin, C.; Davis, I.; Zamore, P.D.; et al. UAP56 couples piRNA clusters to the perinuclear transposon silencing machinery. *Cell* **2012**, *151*, 871–884. [CrossRef] [PubMed]
15. Khurana, J.S.; Wang, J.; Xu, J.; Koppetsch, B.S.; Thomson, T.C.; Nowosielska, A.; Li, C.; Zamore, P.D.; Weng, Z.; Theurkauf, W.E. Adaptation to P element transposon invasion in Drosophila melanogaster. *Cell* **2011**, *147*, 1551–1563. [CrossRef] [PubMed]
16. Muerdter, F.; Olovnikov, I.; Molaro, A.; Rozhkov, N.V.; Czech, B.; Gordon, A.; Hannon, G.J.; Aravin, A.A. Production of artificial piRNAs in flies and mice. *RNA* **2012**, *18*, 42–52. [CrossRef] [PubMed]
17. Todeschini, A.L.; Teysset, L.; Delmarre, V.; Ronsseray, S. The epigenetic trans-silencing effect in Drosophila involves maternally-transmitted small RNAs whose production depends on the piRNA pathway and HP1. *PLoS ONE* **2010**, *5*, e11032. [CrossRef] [PubMed]
18. Olovnikov, I.; Ryazansky, S.; Shpiz, S.; Lavrov, S.; Abramov, Y.; Vaury, C.; Jensen, S.; Kalmykova, A. De novo piRNA cluster formation in the Drosophila germ line triggered by transgenes containing a transcribed transposon fragment. *Nucleic Acids Res.* **2013**, *41*, 5757–5768. [CrossRef] [PubMed]
19. Kyrchanova, O.; Georgiev, P. Chromatin insulators and long-distance interactions in Drosophila. *FEBS Lett.* **2014**, *588*, 8–14. [CrossRef] [PubMed]
20. Chetverina, D.; Aoki, T.; Erokhin, M.; Georgiev, P.; Schedl, P. Making connections: Insulators organize eukaryotic chromosomes into independent cis-regulatory networks. *Bioessays* **2014**, *36*, 163–172. [CrossRef] [PubMed]
21. Vogelmann, J.; Valeri, A.; Guillou, E.; Cuvier, O.; Nollmann, M. Roles of chromatin insulator proteins in higher-order chromatin organization and transcription regulation. *Nucleus* **2011**, *2*, 358–369. [CrossRef] [PubMed]
22. Brasset, E.; Hermant, C.; Jensen, S.; Vaury, C. The Idefix enhancer-blocking insulator also harbors barrier activity. *Gene* **2010**, *450*, 25–31. [CrossRef] [PubMed]
23. Minervini, C.F.; Ruggieri, S.; Traversa, M.; D'Aiuto, L.; Marsano, R.M.; Leronni, D.; Centomani, I.; De Giovanni, C.; Viggiano, L. Evidences for insulator activity of the 5′UTR of the Drosophila melanogaster LTR-retrotransposon ZAM. *Mol. Genet. Genom.* **2010**, *283*, 503–509. [CrossRef]
24. Geyer, P.K.; Corces, V.G. DNA position-specific repression of transcription by a Drosophila zinc finger protein. *Genes Dev.* **1992**, *6*, 1865–1873. [CrossRef] [PubMed]
25. Gdula, D.A.; Gerasimova, T.I.; Corces, V.G. Genetic and molecular analysis of the gypsy chromatin insulator of Drosophila. *Proc. Natl. Acad. Sci. USA* **1996**, *93*, 9378–9383. [CrossRef] [PubMed]
26. Soshnev, A.A.; Baxley, R.M.; Manak, J.R.; Tan, K.; Geyer, P.K. The insulator protein Suppressor of Hairy-wing is an essential transcriptional repressor in the Drosophila ovary. *Development* **2013**, *140*, 3613–3623. [CrossRef] [PubMed]
27. Biessmann, H.; Prasad, S.; Semeshin, V.F.; Andreyeva, E.N.; Nguyen, Q.; Walter, M.F.; Mason, J.M. Two distinct domains in Drosophila melanogaster telomeres. *Genetics* **2005**, *171*, 1767–1777. [CrossRef] [PubMed]
28. Langmead, B.; Salzberg, S.L. Fast gapped-read alignment with Bowtie 2. *Nat. Methods* **2012**, *9*, 357–359. [CrossRef] [PubMed]
29. Radion, E.; Ryazansky, S.; Akulenko, N.; Rozovsky, Y.; Kwon, D.; Morgunova, V.; Olovnikov, I.; Kalmykova, A. Telomeric Retrotransposon HeT-A Contains a Bidirectional Promoter that Initiates Divergent Transcription of piRNA Precursors in Drosophila Germline. *J. Mol. Biol.* **2017**, *429*, 3280–3289. [CrossRef] [PubMed]

30. Golovnin, A.; Volkov, I.; Georgiev, P. SUMO conjugation is required for the assembly of Drosophila Su(Hw) and Mod(mdg4) into insulator bodies that facilitate insulator complex formation. *J. Cell Sci.* **2012**, *125*, 2064–2074. [CrossRef] [PubMed]
31. Grant, C.E.; Bailey, T.L.; Noble, W.S. FIMO: Scanning for occurrences of a given motif. *Bioinformatics* **2011**, *27*, 1017–1018. [CrossRef] [PubMed]
32. Radion, E.; Morgunova, V.; Ryazansky, S.; Akulenko, N.; Lavrov, S.; Abramov, Y.; Komarov, P.A.; Glukhov, S.I.; Olovnikov, I.; Kalmykova, A. Key role of piRNAs in telomeric chromatin maintenance and telomere nuclear positioning in Drosophila germline. *Epigenetics Chromatin* **2018**, *11*, 40. [CrossRef] [PubMed]
33. Melnikova, L.; Kostyuchenko, M.; Parshikov, A.; Georgiev, P.; Golovnin, A. Role of Su(Hw) zinc finger 10 and interaction with CP190 and Mod(mdg4) proteins in recruiting the Su(Hw) complex to chromatin sites in Drosophila. *PLoS ONE* **2018**, *13*, e0193497. [CrossRef] [PubMed]
34. Baxley, R.M.; Soshnev, A.A.; Koryakov, D.E.; Zhimulev, I.F.; Geyer, P.K. The role of the Suppressor of Hairy-wing insulator protein in Drosophila oogenesis. *Dev. Biol.* **2011**, *356*, 398–410. [CrossRef] [PubMed]
35. Gunawardane, L.S.; Saito, K.; Nishida, K.M.; Miyoshi, K.; Kawamura, Y.; Nagami, T.; Siomi, H.; Siomi, M.C. A slicer-mediated mechanism for repeat-associated siRNA 5′ end formation in Drosophila. *Science* **2007**, *315*, 1587–1590. [CrossRef] [PubMed]
36. Czech, B.; Malone, C.D.; Zhou, R.; Stark, A.; Schlingeheyde, C.; Dus, M.; Perrimon, N.; Kellis, M.; Wohlschlegel, J.A.; Sachidanandam, R.; et al. An endogenous small interfering RNA pathway in Drosophila. *Nature* **2008**, *453*, 798–802. [CrossRef] [PubMed]
37. De Vanssay, A.; Bouge, A.L.; Boivin, A.; Hermant, C.; Teysset, L.; Delmarre, V.; Antoniewski, C.; Ronsseray, S. Paramutation in Drosophila linked to emergence of a piRNA-producing locus. *Nature* **2012**, *490*, 112–115. [CrossRef] [PubMed]
38. Soshnev, A.A.; He, B.; Baxley, R.M.; Jiang, N.; Hart, C.M.; Tan, K.; Geyer, P.K. Genome-wide studies of the multi-zinc finger Drosophila Suppressor of Hairy-wing protein in the ovary. *Nucleic Acids Res.* **2012**, *40*, 5415–5431. [CrossRef] [PubMed]
39. Georgiev, P.; Kozycina, M. Interaction between mutations in the suppressor of Hairy wing and modifier of mdg4 genes of Drosophila melanogaster affecting the phenotype of gypsy-induced mutations. *Genetics* **1996**, *142*, 425–436. [PubMed]
40. Pai, C.Y.; Lei, E.P.; Ghosh, D.; Corces, V.G. The centrosomal protein CP190 is a component of the gypsy chromatin insulator. *Mol. Cell* **2004**, *16*, 737–748. [CrossRef] [PubMed]
41. Ryazansky, S.; Radion, E.; Mironova, A.; Akulenko, N.; Abramov, Y.; Morgunova, V.; Kordyukova, M.Y.; Olovnikov, I.; Kalmykova, A. Natural variation of piRNA expression affects immunity to transposable elements. *PLoS Genet.* **2017**, *13*, e1006731. [CrossRef] [PubMed]
42. Grentzinger, T.; Armenise, C.; Brun, C.; Mugat, B.; Serrano, V.; Pelisson, A.; Chambeyron, S. piRNA-mediated transgenerational inheritance of an acquired trait. *Genome Res.* **2012**, *22*, 1877–1888. [CrossRef] [PubMed]
43. Desset, S.; Meignin, C.; Dastugue, B.; Vaury, C. COM, a heterochromatic locus governing the control of independent endogenous retroviruses from Drosophila melanogaster. *Genetics* **2003**, *164*, 501–509. [PubMed]
44. Sarot, E.; Payen-Groschene, G.; Bucheton, A.; Pelisson, A. Evidence for a piwi-dependent RNA silencing of the gypsy endogenous retrovirus by the Drosophila melanogaster flamenco gene. *Genetics* **2004**, *166*, 1313–1321. [CrossRef] [PubMed]
45. Malone, C.D.; Brennecke, J.; Dus, M.; Stark, A.; McCombie, W.R.; Sachidanandam, R.; Hannon, G.J. Specialized piRNA pathways act in germline and somatic tissues of the Drosophila ovary. *Cell* **2009**, *137*, 522–535. [CrossRef] [PubMed]
46. Brennecke, J.; Malone, C.D.; Aravin, A.A.; Sachidanandam, R.; Stark, A.; Hannon, G.J. An epigenetic role for maternally inherited piRNAs in transposon silencing. *Science* **2008**, *322*, 1387–1392. [CrossRef] [PubMed]
47. Shpiz, S.; Ryazansky, S.; Olovnikov, I.; Abramov, Y.; Kalmykova, A. Euchromatic transposon insertions trigger production of novel Pi- and endo-siRNAs at the target sites in the drosophila germline. *PLoS Genet.* **2014**, *10*, e1004138. [CrossRef] [PubMed]

© 2019 by the authors. Licensee MDPI, Basel, Switzerland. This article is an open access article distributed under the terms and conditions of the Creative Commons Attribution (CC BY) license (http://creativecommons.org/licenses/by/4.0/).

Review

Integrative rDNAomics—Importance of the Oldest Repetitive Fraction of the Eukaryote Genome

Radka Symonová

Faculty of Science, Department of Biology, University of Hradec Králové, 500 03 Hradec Králové, Czech Republic; radka.symonova@gmail.com

Received: 3 February 2019; Accepted: 25 April 2019; Published: 7 May 2019

Abstract: Nuclear ribosomal RNA (rRNA) genes represent the oldest repetitive fraction universal to all eukaryotic genomes. Their deeply anchored universality and omnipresence during eukaryotic evolution reflects in multiple roles and functions reaching far beyond ribosomal synthesis. Merely the copy number of non-transcribed rRNA genes is involved in mechanisms governing e.g., maintenance of genome integrity and control of cellular aging. Their copy number can vary in response to environmental cues, in cellular stress sensing, in development of cancer and other diseases. While reaching hundreds of copies in humans, there are records of up to 20,000 copies in fish and frogs and even 400,000 copies in ciliates forming thus a literal subgenome or an rDNAome within the genome. From the compositional and evolutionary dynamics viewpoint, the precursor 45S rDNA represents universally GC-enriched, highly recombining and homogenized regions. Hence, it is not accidental that both rDNA sequence and the corresponding rRNA secondary structure belong to established phylogenetic markers broadly used to infer phylogeny on multiple taxonomical levels including species delimitation. However, these multiple roles of rDNAs have been treated and discussed as being separate and independent from each other. Here, I aim to address nuclear rDNAs in an integrative approach to better assess the complexity of rDNA importance in the evolutionary context.

Keywords: nuclear rDNA; rRNA; GC-content; secondary structure; nucleolus

---

1. The Eukaryotic rDNAome

RNA is essential for information flow from DNA to protein being the dominant macromolecule in protein synthesis [1]. Of the major RNA types, mRNA, tRNA, rRNA, and numerous short non-coding snRNAs, our focus here is on the nuclear rRNA encoded by ribosomal DNA (rDNA), i.e., by rRNA genes. In eukaryotic cells, up to 80% of RNA synthesis belongs to rRNA transcription indispensable to the preservation of ribosome biogenesis and protein synthesis [2]. There are about 1.5–3 million ribosomes per eukaryotic cell [3]. Hence, ribosome biogenesis consumes a tremendous amount of cellular energy and rRNA synthesis is tightly linked to cell growth and proliferation, and as such, it is responsive to general metabolism and environmental challenges [4]. In Eukaryotes, rRNA genes consist of several distinct multigene families tandemly arrayed as repeats composed of tens to hundreds or even thousands of copies. Beside two mitochondrial rRNAs, i.e., the 12S and 16S rRNA in eukaryotes, there are two fractions of nuclear rDNAs—a large, nucleolus-forming 45/47S rDNA unit and a substantially smaller extra-nucleolar 5S rDNA (Figure 1). Both the 45S and 5S rDNAs are organized into clusters of repeats often enabling their cytogenetic visualization on chromosomes [5]. The 5S rDNA can also (co)exist scattered separately within the genome e.g., in the spotted gar as shown in Figure 2a and in other organisms in S1–S4. In budding yeast, the rDNA has a very peculiar organization—the 5S rDNA unit is present in the intergenic spacers (IGSs) of the 45S rDNA and thus, alternating with 45S rDNA units [6]. The coding rDNA sequence is highly conserved among eukaryotes, while the IGSs (Figure 1) that separate the proper units of the 45S rDNA cluster can differ in length and sequence.

In budding yeast, where rDNAs are particularly well-described, IGSs contain three unique elements that are common: an origin of replication, a replication fork blocking site and a promoter that directs the synthesis of noncoding transcripts [7]. In mammals, IGSs contain regulatory regions called UCE (upstream control element), CP (core promoter) and T (termination of transcription site) [8]. Only a fraction of the numerous rDNA copies is transcribed into rRNA. The non-transcribed rDNA copies are extremely important for integrity of the entire genome [7]. In yeast, strains with artificially reduced rDNA copy numbers became sensitive to DNA damage by chemicals and ultraviolet light. This sensitivity further increased as the number of rDNA repeats decreased [7]. In rats, mice, and clawed frog *Xenopus*, the IGS contains one or more RNA polymerase I (Pol I) promoters with high homology to the core region of the main rDNA promoter [9]. Transcripts originating from spacer promoters are co-directional with pre-rRNA synthesis and enhance transcription from the main rDNA promoter, possibly by releasing Pol I [10,11]. Intergenic spacers rRNA have a crucial function in rDNA silencing. In mice, intergenic transcripts originating from a promoter located approximately 2 kb upstream from the pre-rRNA start site are processed into a heterogeneous population of 150–250 nucleotide RNAs, dubbed promoter RNA (pRNA) as their sequence matches the rDNA promoter [6,8,9,11].

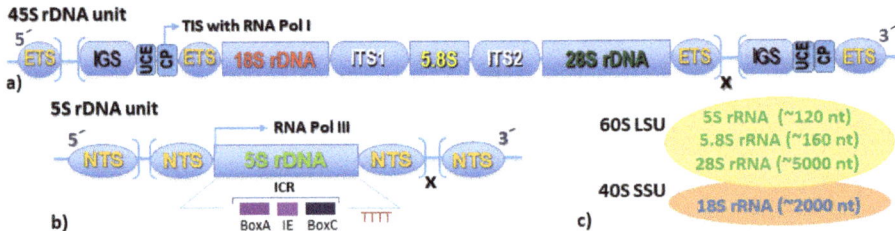

Figure 1. Brief guide to eukaryotic rDNAome—the genomic organization of the rDNA loci. (**a**) Structural organization of the 45S rDNA gene cluster (or rRNA transcription unit); the repeating or single clusters of rDNA can be found scattered throughout genome, they form the precursor pre-rRNA since ribonucleases remove spacers and release separate rRNA molecules in nucleolus—the site of ribosome biogenesis to polysome ribosome formation; (**b**) Structural organization of the 5S rDNA unit (the 5S rDNA can be also dispersed in the genome in many species); (**c**) 80S eukaryotic ribosome composed of the large subunit (LSU) and the small subunit (SSU) with outlined rRNAs. CP—core promoter, ETS—external transcribed spacer, ICR—internal control region, IE—internal element, IGS—intergenic spacer, ITS1, ITS2—internal transcribed spacer 1 and 2, RNA Pol I and III—RNA polymerase I and III, LSU—large (ribosomal) subunit, nt—nucleotides, NTS—non-transcribed spacer, SSU—small subunit, TIS—transcription initiation site, TTTT—polyT transcription termination site, UCE—upstream control element.

2. The Multifaceted Nucleolus

Multiple copies of rRNA gene clusters form nucleolar organizer regions (NORs), the NORs, around which nucleoli are built in the interphase nucleus. The nucleolar 45S rDNAs are transcribed by RNA polymerase I into rRNAs, further processed and assembled with ribosomal proteins into ribosomes [12]. Nucleoli form at the end of mitosis and persist until the onset of the next mitosis. Active nucleoli, where the pre-rRNA transcription takes place, can be visualized in nuclei by silver impregnation, the argyrophilic Ag-NOR staining [13]. From the ultrastructural viewpoint, avian and mammalian nucleoli contain three components (fibrillar centers, dense fibrillar component, and granular component) and differ from all other eukaryotes that possess bipartite nucleoli (i.e., a network of fibrillary strands embedded within granules) [14]. Interestingly, both types of nucleolar arrangement occur among living reptiles: a bicompartmentalized nucleolus in turtles and a tricompartmentalized nucleolus in lizards, crocodiles, and snakes [15]. From the functional viewpoint, the nucleolus was long regarded as a mere ribosome-producing factory. However, during recent decades numerous and crucial non-ribosomal roles were described for the nucleolus [4]. Now, there is a still growing body of evidence that the nucleolus is central to cellular processes as varied as stress response, cell cycle regulation, RNA modification, cell metabolism, and genome stability and integrity [7]. All organisms sense and respond to stressing conditions by downregulating the transcription of rDNA to rRNA and ribosome biogenesis as these processes are extremely energy-consuming [4].

3. The Nucleolus Forming 45S rDNA

The 45S rDNA transcription unit forms a precursor pre-rRNA consisting of 18S, 5.8S, and 28S rRNAs separated by two internal transcribed spacers (ITS1, ITS2) that are removed during the rRNAs maturation process. The entire unit is further delimited by external transcribed spacers (ETS). Intergenic spacers (IGS; Figure 1a) separate each such unit with both of its sides bearing important regulatory elements [16]. This nomenclature applies to the Animal Kingdom. In plants, there is a 25S rDNA gene (instead of the 28S rDNA) within the large nucleolar rDNA multigene family 35S rDNA (instead of the 45S rDNA [17]). In unicellular organisms, where budding yeasts are the most important model system, the 35S rDNA consists of 25S, 5.8S, and 18S together with the 5S rDNA localized into the intergenic spacer within the 35S rDNA [18]. By addition of some 50–60 ribosomal proteins, the 25S/28S, 5.8S, and 5S rRNAs are fashioned into the large ribosomal subunit, 60S LSU. The 18S rRNA associates with 30–40 ribosomal proteins to form the small ribosomal subunit, 40S SSU (Figure 1c). Molecular cytogenetic localization of the rDNAs 28S rDNA fraction of the 45S rDNA unit on chromosomes is shown on Figure 2b.

The importance of nucleolus-forming rDNA and its proper functioning can be seen in the phenomenon of nucleolar dominance. Nucleolar dominance (or a nucleolus under-development due to expression of rDNA from just one parent) is a dramatic disruption in the formation of nucleoli and epigenetically controlled silencing of 45S rDNA in one of the progenitors in an interspecies hybrid. It is characteristic of some plant and animal distant hybrids and represents an example of a non-mammalian maternal imprinting of 45S rDNA [19,20]. Among animals it has been so far evidenced in details in intra-generic hybrids of *Xenopus* [20,21], in inter-generic hybrids of cyprinid fish [22,23] and in two lines of mouse-human somatic hybrids, where the human ribosomal genes were repressed, and only mouse ribosomal genes were expressed [24]. In the species *Drosophila melanogaster*, a special example of allelic inactivation resembling nucleolar dominance exists [25]. *D. melanogaster* carries its rDNA array on the X and on the Y chromosome [26], but the entire X chromosome rDNA array is normally silenced in *D. melanogaster* males, while the Y chromosome rDNA array is dominant and expressed [25].

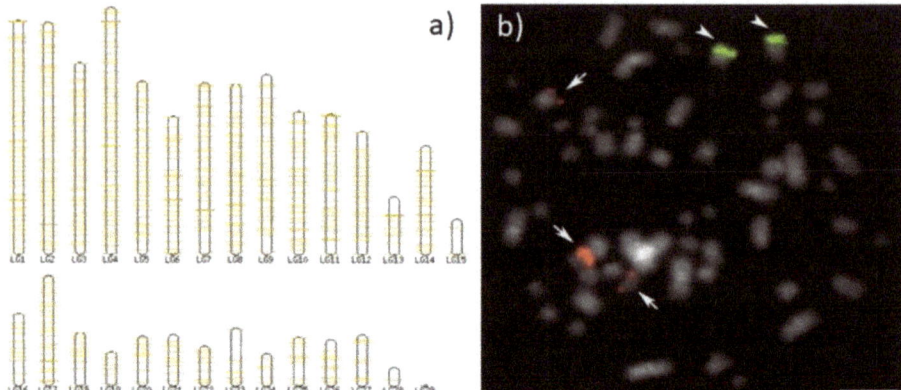

Figure 2. Comparison of two approaches of localization of rDNA on linkage groups and chromosomes in an ancient non-teleost ray-finned fish, spotted gar *(Lepisosteus oculatus)*. (**a**) An in silico approach of visualization of the genomic position of rDNA loci utilizing the Ensembl genome browser tool BioMart to map 5S rDNA on linkage groups (LGs); (**b**) molecular cytogenetic localization of 5S (green, arrowheads) and 28S rDNA (red, arrows) on chromosomes by means of *fluorescence* in situ hybridization (FISH). Bar equals 5 μm (From [27], online Supplementary Material). This comparison shows the sensitivity of the in silico approach. The method enables detection of a single 5S rDNA molecule. It is possible to visualize dispersed molecules across the genome and their pseudogenes in this case. The FISH approach is limited only to huge clusters of accumulated rDNAs and has been utilized for decades particularly in cytotaxonomy in fishes, where most other cytogenetic markers work poorly. Both approaches have their own importance and justification and limits of their mutual interconnection at the current level of genomic data quality—chromosome pairs have not yet been assigned to their corresponding LGs in still too many of the sequenced species. More examples of 5S rDNA localization across LGs are available in Supplementary Materials Figures S1–S4.

4. The Extra-Nucleolar 5S rDNA

The 5S rDNA is much shorter and far less complex within its tandem array structure in comparison with the 45S rDNA. The 5S rDNA consists of a highly conserved sequence of about 120 bp coding for the 5S rRNA and including following functional elements: Box A, IE, Box C [28]. This transcribed sequence is separated at both of its ends from other transcriptional units by a highly variable non-transcribed spacer (NTS; Figure 1b). The participation of 5S rRNA on the ribosome structure is shown on Figure 1c. The 5S rRNA enhances protein synthesis by stabilizing the ribosomal structure and the peptidyl transferase activity, and potentially transmits and coordinates functional centres of the ribosome [29,30]. Two ways for visualization of 5S rDNA sites in silico on linkage groups (LGs) utilizing genomic data and employing methods of molecular cytogenetics on chromosomes are shown in Figure 2. More examples of in silico visualization of 5S rDNA on LGs are in the Supplementary Materials Figures S1–S4.

Two tissue and developmentally specific types of 5S rDNA exist in lower vertebrates, namely the somatic and the oocytes-specific ones. In fish, the oocyte type is lost during development completely (e.g., [31]). Whereas in a frog, it is lost largely [32]. The oocyte repeat comprises a 120 bp oocyte-type 5S rRNA gene placed within the few hundred bp long native AT-rich flanks, whereas the somatic repeat, i.e., a similar 120 bp somatic-type 5S rRNA gene is placed within native GC-rich flanks [33]. Instability of the oocyte 5S rRNA gene transcription complex contributes to the inactivation of the oocyte 5S rRNA gene during embryogenesis [34].

Moreover, in bony and cartilaginous fish, two types of co-occurring 5S rDNA can be distinguished. These probably paralogous type I and II were described in bony fish [35,36] and in elasmobranchs [28]. These types differ by the length and the sequence of their NTS region whereby the longer version is designated as type II [28]. Three functional variants of 5S rRNA genes exist in all life stages of common sea urchin *Paracentrotus lividus* [37]. Systematic study of 5S rDNA sequence diversity in 97 metazoan species [38] describe several paralogous 5S rDNA sequences in 58 of the examined organisms and a flexible genome organization of 5S rDNA in animals. This study also describes three different types of termination signals and variable distances between the coding regions and the typical termination signal. Importantly, a consensus sequence and secondary structure of metazoan 5S rRNA is presented in this study [38], which can be very useful in more detailed future studies of both 5S rDNA and rRNA.

5. Copy Number Really Matters

Gene duplication is an important and frequent evolutionary process [39] and the resulting copy number variation (CNV) is the most frequent type of genetic variation per base pair in the population [40]. Although alteration of gene copy number or gene dosage has deleterious effects for a significant fraction of the genome, changes in dosage are well tolerated in many genes (reviewed by [41]). CNV of rDNA is highly studied in rDNAomics since it provides a mechanism for cellular homeostasis and for rapid and above all reversible adaptation [42–44]. Due to the tandem repetitive structure of rDNA, the repeat number can be easily reduced by homologous recombination among the repeats. However, there is a finely tuned ´gene amplification system´ compensating for these losses and another highly sophisticated system controlling the ´proper´ rDNA copy number [7]. Moreover, these systems are capable of linking external nutrients availability with rDNA copy number [45] that proves the role of rDNA in the cellular energy metabolism as described for nucleolus above. This illustrates how crucial the right copy number of rDNA is for each cell. Moreover, as uncovered at least in humans, the CNV of rDNA represents a novel and cryptic source of hypervariable genomic diversity with far-reaching global regulatory consequences [42]. However, we have accumulated only limited understanding of the immense importance of these seemingly passive and simple phenomena tightly linked with regulation of nuclear as well as mitochondrial genes expression [42] and probably with many more essential cellular mechanisms. There is an inconsistency in the quantification of rDNA copy number even in the human genome. One important study states that the rDNA copy number varies among healthy humans as a result of natural genetic diversity between 14–410 copies of the 45S rDNA unit per genome [41]. A more recent study reported that the number of rDNA repeats varies from 250 to 670 copies per diploid genome [46]. Therefore, data provided by Gibbons et al. [41,45] should be treated with caution as the low limit number of 14 copies has not been otherwise found in mammals. The genomes of higher eukaryotes harbour hundreds and thousands of copies and only prokaryotic genomes can carry fewer copies of ribosomal genes. Moreover, given that there are five pairs of clusters of ribosomal genes, located on five pairs of human acrocentric chromosomes (13, 14, 15, 21, 22), it is logically impossible that ten clusters could totally count 14 copies, i.e., just 1.4 copies per cluster. These obviously underestimated values can be explained as artifacts caused by a poor suitability of PCR-based techniques for the quantification of GC-rich moderate repeats used by Gibbons et al., 2014, 2015 [42,47] compared to more suitable nonradioactive quantitative hybridization (NQH) used by Chestkov et al., 2018 [46]. The reason is that rDNA is a specific region often forming non-canonical hairpin and loop structures and prone to oxidation in vivo and after extraction from cells [46]. Phenotypic effects of rDNA copy number were recently summarized by [48]. CNV of rDNA (loss as well as amplification) is linked to tumorigenesis [49,50].

A substantial intra-species CNV of rDNA was, among others, evidenced in a freshwater microcrustacean *Daphnia* [51]. rDNA copy number does change among tissues and during ontogenesis in multicellular organisms [31,52] and it is age-dependent in single-cell yeasts [7]. Substantial differences in rDNA CN and number of their sites on chromosomes have been repeatedly recorded in vertebrates and invertebrates on the inter-population and inter-species level of comparison [5]. Such genomic differences might also contribute to genome diversifications, reproductive barriers formation, speciation events and finally to an increased biodiversity, e.g., [53,54]. Numerous examples from fish cytogenetics show that the variation in rDNA repeats prove to be highly informative as it is subject to a more relaxed regulation than in higher vertebrates [55]. Here, traditional cytogenetics meet the currently booming genomics to mutual usage and benefit from each other. The Animal rDNA database currently contains 539 records on fish rDNAs, namely 5S rDNA in 417 species and 45S rDNA in 479 species [5]. However, a detailed analysis of rDNA sequence organization and variation and CNV on the molecular level exists only for a handful of fish species including both 5S and 45S rDNA of zebrafish [31,51], 5S rDNA and only partial 45S rDNA of pike [56], only 5S rDNA of tilapia [57], molecular organization of the 5S rDNA type II of elasmobranchs (i.e., sharks, rays, and skates [28]) and cichlids [58]. In *Drosophila* germline stem cells, rDNA copy number decreases during aging and this age-dependent decrease in rDNA copy number is transgenerationally heritable. However, young animals are capable of recovering the normal rDNA copy number [59]. The copy number obviously plays a functional role: in *Xenopus*: the somatic 5S rDNA has about 400 copies, while the oocyte 5S rDNA has about 20,000 copies [32]. Locati et al. [31] detected about 9000 5S rRNA genes in the zebrafish genome assembly GRCz10 [31] and Symonova et al. detected about 20,000 copies of 5S rRNA genes in the Northern pike *Esox lucius* and its congener *E. cisalpinus* [56]. However, the record holders are currently protists, namely ciliates: Oligotrichia and Peritrichia [60] and representatives of the ciliate group Spirotrichea - *Oxytricha nova* with about 200,000 rDNA copies [61] and *Stylonychia lemnae* with estimated 400,000 copies of rDNA [62]. The already mentioned single-cell ciliate protozoan *Tetrahymena* amplifies its rDNA 9000-fold during development of the somatic macronucleus [63]. Whereas the copy number of 45S and 5S rDNA units is tightly coupled in mouse and human [47], such a control is apparently missing in fish [52,56,64]. This fact together with the aforementioned difference in nucleolar organization between higher and lower vertebrates and also other genomic traits (e.g., genomic GC heterogeneity) indicate that another major evolutionary transition *sensu* [65] occurred in evolution from anamniotes towards amniotes. This huge copy number variation might be linked to (or might have resulted in) the heterogeneity in rRNA genes and their variants that had been considered a peculiarity of some plants. Only recently, this heterogeneity was proved also in animal ribosomal genes, including human and mouse, where variant rRNA alleles exhibit tissue-specific expression and ribosomes bearing variant rRNA alleles are present in the actively translating ribosome pool [66].

One special topic of the rDNA CNV is based on molecular cytogenetic localization of both 5S and 45S rDNAs using FISH. FISH with rDNAs represents one of the most important chromosomal markers particularly in non-model organisms and especially in cold-blooded vertebrates, where methods like R-banding do not yield any usable and reproducible pattern. For these reasons, a heavy body of literature on molecular cytogenetics of rDNA has accumulated (for plants [17], for animals [5]). Since rDNA was omitted from many genome sequencing projects due to issues with its assembling [50,67], any precise quantification of rDNA copies is mostly still impossible. On the other hand, the still increasing availability of long-read sequencing can overcome assembling issues and provides opportunity to link the numerous results from molecular cytogenetics with genomics as was successfully demonstrated in fish cytogenomics [54,56].

A very special chapter of the rDNAomics book deals with rDNA of eukaryotic microorganisms [67]. In their 2010 review, Torres-Machorro et al. present available information on both rDNA fractions from about hundred microbial eukaryotes and show an unexpected diversity in their genomic organization [68]. Later, Drouin and Tsang [69] focus their review of 5S rDNA in protists on adaptive potential of its organization. Microbial eukaryotic rDNAs may be coded alone, in tandem repeats, linked to each other or linked to other genes. They exist in the chromosome or extrachromosomally in linear or circular units and rDNA coding regions may contain introns, sequence insertions, protein-coding genes, or additional spacers [68]. The atypical structures of rDNA have been considered as exceptions. However, it is rather likely that these organisms have preserved variations in the organization of these versatile genes that may be considered as living records of evolution [68]. A huge step in establishing the functional significance of rDNA in evolution and in ecology of organisms has been performed in protists [60].

6. Overview of Important Facts about rDNA

The most import facts about rDNA can be summarized as follows: rDNA is ubiquitous and universal across Prokaryotes, Archaea, and Eukaryotes [70]. It has a high degree of functional and sequence conservation of rDNA genes [71]. At the same time, rDNA belongs to the most copy number-hypervariable genomic segments [42] and the tandemly repeated rDNA arrays are among the most evolutionary dynamic loci of eukaryotic genomes in terms of copy number. Due to its heavy transcription, repetitive structure, and programmed replication fork pauses, the rDNA is one of the most unstable regions in the genome [7,18]. Their high genomic copy number relative to other genes appears to be much larger than required, however, unlike protein-coding genes, rDNA cannot undergo additional rounds of amplification via translation when organisms require more rRNA transcripts [72]. Copy number of 45S units is balanced with that of the 5S rDNA in mouse and human [47] but not in fish, summarized by [5]. These multiple copies of rDNA evolve in a highly coordinated manner, through unequal crossing over and/or gene conversion, two mechanisms related to homologous recombination [73]. The rRNA gene repeats use a unique gene amplification system to restore the copy number after this has been reduced due to recombination [7]. The RFB (replication fork barrier) coordinates replication and recombination, and through the latter, mediates a possible increase in the number of rDNA repeats. rDNA loci are dynamic genetic elements, their copy number changes dynamically and transgenerationally yet is maintained through a recovery mechanism in the germline (for *Drosophila* see [59]). In plants, extensive variation can exist in both rDNA copy number and rRNA expression. Among maize inbred lines, thousands of genes co-regulate with rRNA expression, including genes participating in ribosome biogenesis and other functionally relevant pathways [74]. Not only the rDNA copy number [45] but also the rRNA expression variation is a valuable source of functional diversity that affects gene expression variation and field-based phenotypic changes [74]. The intra-genomic homogenization of rDNA mostly occurs through 'concerted evolution' [75]. rDNA also shows high rates of meiotic recombination [75,76] and rDNA sites are hotspots for genome rearrangements [77]. Copy number of rDNA arrays modulates genome-wide expression of hundreds to thousands of genes and subtle changes in rDNA copy number between individuals may contribute to biologically relevant phenotypic variation also in humans [78]. rDNA contributes to global chromatin regulation and thus to a balance between heterochromatin and euchromatin in the nucleus [79]. The enormous variation in the number of rDNA copies per eukaryotic genome correlates with genome size [80] and the copy number of the 45S rDNA fraction was shown to negatively correlate with mtDNA abundance [42]. Hence, rDNA copy number variation, CNV ("rDNA dosage") is a major determinant of naturally occurring genome-wide gene expression variation in humans [42]. Ribosomal RNAs (rRNAs) account for up to 80% of all RNAs in eukaryotic cells [50]. Growth-activated rRNA synthesis may be mediated by the up-regulation of individual rDNA units, in addition to the activation of silent gene copies see e.g., Banditt et al. [81]. In mammals, 5S and 45S rDNA arrays are non-homologous, physically unlinked, transcribed by different RNA Polymerases and encode functionally interdependent RNA components of the ribosome [47]. Clusters of the 45S rDNA unit give origin to the nucleolus, the nuclear organelle that is the site of pre-45S rDNA transcription and ribosome biogenesis, see e.g., [8,82]. The rDNA contact map shows that 5S and 45S arrays each have thousands of contacts in the folded genome, with rDNA-associated regions and genes dispersed across all chromosomes [83,84]. Due to its highly repetitive nature, rDNA has been excluded from most mammalian genome-wide studies because of challenges associated with its analysis, and thus remains understudied. There is an unusual and universal GC richness of the 45S rDNA fraction in cold- as well as warm-blooded vertebrates (more details below) [72,84]. An extensive range of epigenetic modifications regulating rRNA genes transcription [67,85–87] results in only a mere subset of the multiple copies being transcribed however with far reaching implications for the entire genome (elucidation of the epigenetics of rDNA in sufficient detail would require a lot of research). On top of it, rDNA loci serve as a specialized niche for mobile elements [88].

rDNA units (so called 'rDNA-like signal') can be found scattered throughout the genome in humans [89]. These units can be described as follows: 1) highly degraded, but near full length, rDNA units, including both 45S and Intergenic Spacer (IGS), can be found at multiple sites in the human genome on chromosomes without rDNA arrays; 2) these rDNA sequences have a propensity for being centromere proximal; and 3) sequence at all human functional rDNA array ends is divergent from canonical rDNA to the point that it is pseudogenic. For this in fish, see Figure 2a and in other chordates see the Supplementary Materials Figures S1–S4.

rDNA represents a cryptic source of hypervariable genomic diversity with global regulatory consequences (ribosomal quantitative trait loci (eQTL)) in humans. The variation provides a mechanism for cellular homeostasis and for rapid and reversible adaptation [42,47].

7. GC Content of rDNA

The 45S rDNA gene clusters form the GC-richest genomic fraction particularly in Eukaryotes [90] with humans having 60%–80% GC in different parts of the rDNA [91], whereas the median genomic GC is 40.9% (NCBI, human genome assembly). This GC-richness is ascribed to the recombination rate based process known as GC-biased gene conversion [73]. On the other hand, some studies link the extremely high GC levels in rDNAs to particular requirements for stem-and-loop systems in rRNA that have an effect on the composition of the corresponding genes to thermal adaptation [92]. This line of explanations belongs to the Thermodynamic Stability Hypothesis attempting to account for the overall AT/GC heterogeneity in birds and mammals and the AT/GC homogeneity in the remaining vertebrates and the other Eukaryotes [90]. However, although Wang et al., 2006 showed support for such a thermal adaptation in Bacteria and Archaea, they did not find any for warm-blooded birds and mammals with only a slightly higher GC content of 18S (55.7%) versus cold-blooded fishes and amphibians with approximately 53.5% of GC. Their partitioning of the GC content across the 18S rRNA sequences into stem and loop regions demonstrated [93] that the differences are not concentrated in the paired stem regions as expected by Bernardi [90]. Interesting and relevant aspects of rDNA GC content exist in so-called expansion segments (ES) in 28S and 18S rRNA molecules [94,95]. Expansion of the 28S rRNA shows a clear phylogenetic increase, with a dramatic rise in mammals and especially in hominids. Here, a GC- or AU-biased expansion of rRNAs has developed in both plants and metazoans, with the GC-bias largely being preferred in extremely GC-rich ES of vertebrate 28S rRNA. This compositional bias towards GC is linked to potential roles of GC-rich rRNA during protein synthesis [96,97] and could contribute to the discussion whether the genomic GC content is driven by neutral versus selective processes. An interesting explanation of the universal GC richness of 45S rDNA comes from the GC biology—these multicopy genes should all be in a DNA region with a homogenous GC composition to allow concerted evolution and to prevent divergence through generations [98].

8. Phylogeny, Species Delimitation, and Secondary Structure of rRNAs—The Way How to Determine in Silico Whether Two Lineages Can Successfully Cross

The ITS2 sequence already belongs to the most popular and well established phylogenetic and DNA barcoding markers [99]. The rDNA sequence and its corresponding rRNA secondary structure is one of the few universal features of life without any known case of horizontal transfer and above all, identifying the organism to a unique species, making it uniquely suited to assess phylogenetic relationships [100,101]. The secondary structure of the ITS regions is well known for a wide variety of eukaryotes and have been used to aid in the alignment of these sequences for phylogenetic comparisons [101]. The RNA sequence of the ITS2 possesses another special trait so far not fully examined, namely compensatory base changes (CBCs, Figure 3). CBCs are mutations occurring simultaneously on both sides of a nucleotide pair in the ITS2 secondary structure with retention of the paired nucleotide bond, whereas hemi-CBC is a mutation of a single nucleotide of the pair still retaining the bond [102]. CBC analyses have been primarily performed in fungi and plants [92–94]. This is the reason why the majority of literature references, including methods descriptions, are on plants (e.g., estimating structure-based phylogenetic trees from ITS2 data by [103–105]). CBCs analyses have been already successfully used to verify taxonomy of closely related species and to distinguish morphologically indistinct species in insects [102]. That shows the huge potential of mining for CBCs in ITS2 rRNA secondary structures also in the Animal Kingdom.

CBC		hemi-CBC	
G-C	G-C	G-C	G-C
U-A	U-A	U-A	U-A
A-U	A-U	A-U	A-U
C-G →	A-U	C-G →	A-G
G-C	G-C	G-C	G-C
G A	G A	G A	G A
A G	A G	A G	A G
A U	A U	A U	A U
U-A	U-A	U-A	U-A
G-C	G-C	G-C	G-C
A-U	A-U	A-U	A-U

Figure 3. Visualization of compensatory base changes (CBCs) on a hypothetical internal transcribed spacer 2 (ITS2) secondary structure. The left panel depicts a conserved helix segment in which a CBC occurs, where both nucleotides of the pair underwent a mutation that resulted in retaining the paired nucleotide bond. The right panel shows a hemi-CBC, where only one nucleotide of the pair, i.e., C to A, underwent a mutation while the pair retained the nucleotide bond. Redrawn according to [106].

However, any detailed and particularly systematic survey of rRNA secondary structure in the Animal Kingdom is still in its infancy although it would be highly desirable in numerous areas of biology. Moreover, analyses of rRNA secondary structure could represent another intersection between rDNAomics based on molecular cytogenetics and on genomics since molecular cytogenetic studies frequently provide DNA sequences of rDNA fragments used in FISH experiments and further DNA sequences, in the meanwhile, became available in NCBI GenBank or could be retrieved from whole-genome datasets. Ideally, such integrative studies could contain cytogenetic results accompanied by details on rDNA/rRNA sequence and rRNA secondary structure as shown in Figure 4 to better explore any potential sequence polymorphism.

Figure 4. Text and graphic representation of prediction of rRNA secondary structure. (**a**) the Xfasta or "dot-bracket notation" way of representation of the ITS2 rRNA secondary structure of channel catfish, (**b**) visualization of the corresponding secondary structure of the sequence in (**a**) with one longest helix and four short helices, (**c**) stickleback, prediction of 5S rRNA secondary structure, Ensemble.

9. Concluding Remarks

The field of rDNAomics is extremely rapidly evolving further and has far-reaching implications for numerous areas of current biology and medicine. Medical aspects represent another crucial chapter of rDNAomics that already exceeds the scope of this review. On the other hand, being aware of this fact might help scientists from the area of fundamental research working on non-model organisms to provide justification of their work. There are numerous diseases associated with rDNA dysfunction, particularly cancer [39,40,61,66]. Ribosomopathies are diseases caused by abnormalities in the structure or function of ribosomal component proteins or rRNA genes, or other genes whose products are involved in ribosome biogenesis [107]. Not only sequence, but also copy number of rDNAs is of particular importance in cancer—human cancer genomes show a loss of copies, accompanied by global copy number co-variation [50]. Even more relevant is the fact that rDNA repeat instability coincides with predisposition to cancer, premature aging and neurological impairment in ataxia-telangiectasia and Bloom syndrome (Warmerdam and Wolthuis, 2018). Additionally, it was shown that cancers undergo coupled 5S rDNA array expansion and 45S rDNA loss that is accompanied by increased proliferation rate and nucleolar activity. Somatic changes in rDNA copy number can exceed 10-fold the naturally occurring copy number variation across individuals [49]. Malfunction of nucleoli can be the cause of several human conditions called nucleolopathies [108]. The nucleolus is being investigated as a target for cancer chemotherapy [109,110]. Moreover, rDNA copy number may be a simple and useful indicator of whether a cancer will be sensitive to DNA damaging treatments [50]. Hence, it is desirable to understand rDNA organization, function, and its impact on the entire nucleolus and other genes' regulation (as briefly outlined here) in a broader evolutionary context.

Supplementary Materials: The following are available online at http://www.mdpi.com/2073-4425/10/5/345/s1, Figure S1: *Tetraodon nigroviridis*, karyogram with labeled 5S rDNA loci (red arrowheads). Figure S2: Zebrafish (*Danio rerio*), complete karyogram with labeled 5S rDNA loci (pink lines). Figure S3: A tunicate sea squirt (*Ciona intestinalis*), C ≈ 0.2, the thirty-five 5S rDNA site detected on four chromosomes could be assigned to a linkage group. Figure S4: Human genome illustrates here the best assembled vertebrate genome, C ≈ 3.5. Genomic localization of annotated 5S rDNA showing 5S rDNA scattered throughout fish genomes visualized on karyograms of species assembled to the chromosome level (5S rDNA sequences were filtered using the BioMart tool and the Ensemble Genes Database version 92 from Ensembl.org version 92). For comparison, two model mammalian genomes (Hsa, Mmu) and one avian genome are shown.

Funding: This study was supported by IRP PřF UHK 1903/2018.

Acknowledgments: I would like to thank Mike W. Howell for his insightful comments to this manuscript.

Conflicts of Interest: The author declares no conflict of interest.

References

1. Bernhardt, H.S.; Tate, W.P. A Ribosome Without RNA. *Front. Ecol. Evol.* **2015**, *3*, 1–6. [CrossRef]
2. Boisvert, F.-M.; van Koningsbruggen, S.; Navascués, J.; Lamond, A.I. The multifunctional nucleolus. *Nat. Rev. Mol. Cell Biol.* **2007**, *8*, 574–585. [CrossRef] [PubMed]
3. Porokhovnik, L.; Gerton, J.L. Ribosomal DNA-connecting ribosome biogenesis and chromosome biology. *Chromosome Res. Int. J. Mol. Supramol. Evol. Asp. Chromosome Biol.* **2019**. [CrossRef] [PubMed]
4. Grummt, I. The nucleolus—guardian of cellular homeostasis and genome integrity. *Chromosoma* **2013**, *122*, 487–497. [CrossRef] [PubMed]
5. Sochorová, J.; Garcia, S.; Gálvez, F.; Symonová, R.; Kovařík, A. Evolutionary trends in animal ribosomal DNA loci: introduction to a new online database. *Chromosoma* **2018**, *127*, 141–150. [CrossRef]
6. James, S.A.; O'Kelly, M.J.T.; Carter, D.M.; Davey, R.P.; van Oudenaarden, A.; Roberts, I.N. Repetitive sequence variation and dynamics in the ribosomal DNA array of Saccharomyces cerevisiae as revealed by whole-genome resequencing. *Genome Res.* **2009**, *19*, 626–635. [CrossRef] [PubMed]
7. Kobayashi, T. Ribosomal RNA gene repeats, their stability and cellular senescence. *Proc. Jpn. Acad. Ser. B* **2014**, *90*, 119–129. [CrossRef]
8. Russell, J.; Zomerdijk, J.C.B.M. RNA-polymerase-I-directed rDNA transcription, life and works. *Trends Biochem. Sci.* **2005**, *30*, 87–96. [CrossRef] [PubMed]
9. Santoro, R.; Schmitz, K.-M.; Sandoval, J.; Grummt, I. Intergenic transcripts originating from a subclass of ribosomal DNA repeats silence ribosomal RNA genes in trans. *EMBO Rep.* **2010**, *11*, 52–58. [CrossRef] [PubMed]

10. Grimaldi, G.; Di Nocera, P.P. Multiple repeated units in Drosophila melanogaster ribosomal DNA spacer stimulate rRNA precursor transcription. *Proc. Natl. Acad. Sci. USA* **1988**, *85*, 5502–5506. [CrossRef]
11. Mayer, C.; Neubert, M.; Grummt, I. The structure of NoRC-associated RNA is crucial for targeting the chromatin remodelling complex NoRC to the nucleolus. *EMBO Rep.* **2008**, *9*, 774–780. [CrossRef]
12. Németh, A.; Längst, G. Genome organization in and around the nucleolus. *Trends Genet. TIG* **2011**, *27*, 149–156. [CrossRef] [PubMed]
13. Howell, W.M.; Black, D.A. Controlled silver-staining of nucleolus organizer regions with a protective colloidal developer: A 1-step method. *Experientia* **1980**, *36*, 1014–1015. [CrossRef] [PubMed]
14. Thiry, M.; Lafontaine, D.L.J. Birth of a nucleolus: The evolution of nucleolar compartments. *Trends Cell Biol.* **2005**, *15*, 194–199. [CrossRef]
15. Thiry, M.; Lamaye, F.; Lafontaine, D.L.J. The nucleolus: When 2 became 3. *Nucleus* **2011**, *2*, 289–293. [CrossRef]
16. Henras, A.K.; Plisson-Chastang, C.; O'Donohue, M.-F.; Chakraborty, A.; Gleizes, P.-E. An overview of pre-ribosomal RNA processing in eukaryotes. *Wiley Interdiscip. Rev. RNA* **2015**, *6*, 225–242. [CrossRef] [PubMed]
17. Garcia, S.; Garnatje, T.; Kovařík, A. Plant rDNA database: Ribosomal DNA loci information goes online. *Chromosoma* **2012**, *121*, 389–394. [CrossRef] [PubMed]
18. Kobayashi, T.; Sasaki, M. Ribosomal DNA stability is supported by many 'buffer genes'—introduction to the Yeast rDNA Stability Database. *FEMS Yeast Res.* **2017**, *17*. [CrossRef]
19. Pikaard, C.S. Nucleolar dominance: Uniparental gene silencing on a multi-megabase scale in genetic hybrids. *Plant Mol. Biol.* **2000**, *43*, 163–177. [CrossRef] [PubMed]
20. Michalak, K.; Maciak, S.; Kim, Y.B.; Santopietro, G.; Oh, J.H.; Kang, L.; Garner, H.R.; Michalak, P. Nucleolar dominance and maternal control of 45S rDNA expression. *Proc. R. Soc. B Biol. Sci.* **2015**, *282*, 20152201. [CrossRef]
21. Maciak, S.; Michalak, K.; Kale, S.D.; Michalak, P. Nucleolar Dominance and Repression of 45S Ribosomal RNA Genes in Hybrids between *Xenopus borealis* and *X. muelleri* (2n = 36). *Cytogenet. Genome Res.* **2016**, *149*, 290–296. [CrossRef] [PubMed]
22. Xiao, J.; Hu, F.; Luo, K.; Li, W.; Liu, S. Unique nucleolar dominance patterns in distant hybrid lineage derived from Megalobrama Amblycephala × Culter Alburnus. *BMC Genet.* **2016**, *17*. [CrossRef]
23. Cao, L.; Qin, Q.; Xiao, Q.; Yin, H.; Wen, J.; Liu, Q.; Huang, X.; Huo, Y.; Tao, M.; Zhang, C.; et al. Nucleolar Dominance in a Tetraploidy Hybrid Lineage Derived From Carassius auratus red var. () × Megalobrama amblycephala (). *Front. Genet.* **2018**, *9*. [CrossRef] [PubMed]
24. Onishi, T.; Berglund, C.; Reeder, R.H. On the mechanism of nucleolar dominance in mouse-human somatic cell hybrids. *Proc. Natl. Acad. Sci. USA* **1984**, *81*, 484–487. [CrossRef] [PubMed]
25. Greil, F.; Ahmad, K. Nucleolar Dominance of the Y Chromosome in *Drosophila melanogaster*. *Genetics* **2012**, *191*, 1119–1128. [CrossRef] [PubMed]
26. Tautz, D.; Hancock, J.M.; Webb, D.A.; Tautz, C.; Dover, G.A. Complete sequences of the rRNA genes of Drosophila melanogaster. *Mol. Biol. Evol.* **1988**, *5*, 366–376. [PubMed]
27. Symonová, R.; Majtánová, Z.; Arias-Rodriguez, L.; Mořkovský, L.; Kořínková, T.; Cavin, L.; Pokorná, M.J.; Doležálková, M.; Flajšhans, M.; Normandeau, E.; et al. Genome Compositional Organization in Gars Shows More Similarities to Mammals than to Other Ray-Finned Fish: CYTOGENOMICS OF GARS. *J. Exp. Zoolog. B Mol. Dev. Evol.* **2017**, *328*, 607–619. [CrossRef]
28. Castro, S.I.; Hleap, J.S.; Cárdenas, H.; Blouin, C. Molecular organization of the 5S rDNA gene type II in elasmobranchs. *RNA Biol.* **2016**, *13*, 391–399. [CrossRef]
29. Szymanski, M.; Barciszewska, M.Z.; Erdmann, V.A.; Barciszewski, J. 5S Ribosomal RNA Database. *Nucleic Acids Res.* **2002**, *30*, 176–178. [CrossRef] [PubMed]
30. Dinman, J.D. 5S rRNA: Structure and Function from Head to Toe. *Int. J. Biomed. Sci. IJBS* **2005**, *1*, 2–7.
31. Locati, M.D.; Pagano, J.F.B.; Ensink, W.A.; van Olst, M.; van Leeuwen, S.; Nehrdich, U.; Zhu, K.; Spaink, H.P.; Girard, G.; Rauwerda, H.; et al. Linking maternal and somatic 5S rRNA types with different sequence-specific non-LTR retrotransposons. *RNA* **2017**, *23*, 446–456. [CrossRef]
32. Peterson, R.C.; Doering, J.L.; Brown, D.D. Characterization of two xenopus somatic 5S DNAs and one minor oocyte-specific 5S DNA. *Cell* **1980**, *20*, 131–141. [CrossRef]

33. Tomaszewski, R.; Jerzmanowski, A. The AT-rich flanks of the oocyte-type 5S RNA gene of Xenopus laevis act as a strong local signal for histone H1-mediated chromatin reorganization in vitro. *Nucleic Acids Res.* **1997**, *25*, 458–466. [CrossRef] [PubMed]
34. Wolffe, A.P.; Brown, D.D. Developmental regulation of two 5S ribosomal RNA genes. *Science* **1988**, *241*, 1626–1632. [CrossRef]
35. Martins, C.; Galetti, P.M. Two 5S rDNA arrays in neotropical fish species: is it a general rule for fishes? *Genetica* **2001**, *111*, 439–446. [CrossRef] [PubMed]
36. *Fish Cytogenetics*; Pisano, E. (Ed.) Science Publishers: Enfield, NH, USA, 2007; ISBN 978-1-57808-330-5.
37. Dimarco, E.; Cascone, E.; Bellavia, D.; Caradonna, F. Functional variants of 5S rRNA in the ribosomes of common sea urchin Paracentrotus lividus. *Gene* **2012**, *508*, 21–25. [CrossRef] [PubMed]
38. Vierna, J.; Wehner, S.; Höner zu Siederdissen, C.; Martínez-Lage, A.; Marz, M. Systematic analysis and evolution of 5S ribosomal DNA in metazoans. *Heredity* **2013**, *111*, 410–421. [CrossRef] [PubMed]
39. Conrad, B.; Antonarakis, S.E. Gene Duplication: A Drive for Phenotypic Diversity and Cause of Human Disease. *Annu. Rev. Genomics Hum. Genet.* **2007**, *8*, 17–35. [CrossRef]
40. The Wellcome Trust Case Control Consortium; Conrad, D.F.; Pinto, D.; Redon, R.; Feuk, L.; Gokcumen, O.; Zhang, Y.; Aerts, J.; Andrews, T.D.; Barnes, C.; et al. Origins and functional impact of copy number variation in the human genome. *Nature* **2010**, *464*, 704–712. [CrossRef]
41. Rice, A.M.; McLysaght, A. Dosage sensitivity is a major determinant of human copy number variant pathogenicity. *Nat. Commun.* **2017**, *8*, 14366. [CrossRef] [PubMed]
42. Gibbons, J.G.; Branco, A.T.; Yu, S.; Lemos, B. Ribosomal DNA copy number is coupled with gene expression variation and mitochondrial abundance in humans. *Nat. Commun.* **2014**, *5*. [CrossRef]
43. Long, E.O.; Dawid, I.B. Repeated genes in eukaryotes. *Annu. Rev. Biochem.* **1980**, *49*, 727–764. [CrossRef] [PubMed]
44. Oakes, M.; Siddiqi, I.; Vu, L.; Aris, J.; Nomura, M. Transcription factor UAF, expansion and contraction of ribosomal DNA (rDNA) repeats, and RNA polymerase switch in transcription of yeast rDNA. *Mol. Cell. Biol.* **1999**, *19*, 8559–8569. [CrossRef]
45. Jack, C.V.; Cruz, C.; Hull, R.M.; Keller, M.A.; Ralser, M.; Houseley, J. Regulation of ribosomal DNA amplification by the TOR pathway. *Proc. Natl. Acad. Sci.* **2015**, *112*, 9674–9679. [CrossRef] [PubMed]
46. Chestkov, I.V.; Jestkova, E.M.; Ershova, E.S.; Golimbet, V.E.; Lezheiko, T.V.; Kolesina, N.Y.; Porokhovnik, L.N.; Lyapunova, N.A.; Izhevskaya, V.L.; Kutsev, S.I.; et al. Abundance of ribosomal RNA gene copies in the genomes of schizophrenia patients. *Schizophr. Res.* **2018**, *197*, 305–314. [CrossRef] [PubMed]
47. Gibbons, J.G.; Branco, A.T.; Godinho, S.A.; Yu, S.; Lemos, B. Concerted copy number variation balances ribosomal DNA dosage in human and mouse genomes. *Proc. Natl. Acad. Sci. USA* **2015**, *112*, 2485–2490. [CrossRef]
48. Porokhovnik, L.N.; Lyapunova, N.A. Dosage effects of human ribosomal genes (rDNA) in health and disease. *Chromosome Res.* **2019**, *27*, 5–17. [CrossRef]
49. Wang, M.; Lemos, B. Ribosomal DNA copy number amplification and loss in human cancers is linked to tumor genetic context, nucleolus activity, and proliferation. *PLoS Genet.* **2017**, *13*, e1006994. [CrossRef]
50. Xu, B.; Li, H.; Perry, J.M.; Singh, V.P.; Unruh, J.; Yu, Z.; Zakari, M.; McDowell, W.; Li, L.; Gerton, J.L. Ribosomal DNA copy number loss and sequence variation in cancer. *PLoS Genet.* **2017**, *13*, e1006771. [CrossRef] [PubMed]
51. Eagle, S.H.; Crease, T.J. Copy number variation of ribosomal DNA and Pokey transposons in natural populations of Daphnia. *Mob. DNA* **2012**, *3*, 4. [CrossRef] [PubMed]
52. Locati, M.D.; Pagano, J.F.B.; Girard, G.; Ensink, W.A.; van Olst, M.; van Leeuwen, S.; Nehrdich, U.; Spaink, H.P.; Rauwerda, H.; Jonker, M.J.; et al. Expression of distinct maternal and somatic 5.8S, 18S, and 28S rRNA types during zebrafish development. *RNA* **2017**, *23*, 1188–1199. [CrossRef] [PubMed]
53. Dion-Cote, A.-M.; Symonova, R.; Rab, P.; Bernatchez, L. Reproductive isolation in a nascent species pair is associated with aneuploidy in hybrid offspring. *Proc. R. Soc. B Biol. Sci.* **2015**, *282*, 20142862. [CrossRef] [PubMed]
54. Dion-Côté, A.-M.; Symonová, R.; Lamaze, F.C.; Pelikánová, Š.; Ráb, P.; Bernatchez, L. Standing chromosomal variation in Lake Whitefish species pairs: The role of historical contingency and relevance for speciation. *Mol. Ecol.* **2017**, *26*, 178–192. [CrossRef]

55. Symonová, R.; Howell, W. Vertebrate Genome Evolution in the Light of Fish Cytogenomics and rDNAomics. *Genes* **2018**, *9*, 96. [CrossRef]
56. Symonová, R.; Ocalewicz, K.; Kirtiklis, L.; Delmastro, G.B.; Pelikánová, Š.; Garcia, S.; Kovařík, A. Higher-order organisation of extremely amplified, potentially functional and massively methylated 5S rDNA in European pikes (*Esox* sp.). *BMC Genomics* **2017**, *18*. [CrossRef]
57. Martins, C.; Wasko, A.P.; Oliveira, C.; Porto-Foresti, F.; Parise-Maltempi, P.P.; Wright, J.M.; Foresti, F. Dynamics of 5S rDNA in the tilapia *(Oreochromis niloticus)* genome: Repeat units, inverted sequences, pseudogenes and chromosome loci. *Cytogenet. Genome Res.* **2002**, *98*, 78–85. [CrossRef] [PubMed]
58. Nakajima, R.T.; Cabral-de-Mello, D.C.; Valente, G.T.; Venere, P.C.; Martins, C. Evolutionary dynamics of rRNA gene clusters in cichlid fish. *BMC Evol. Biol.* **2012**, *12*, 198. [CrossRef]
59. Lu, K.L.; Nelson, J.O.; Watase, G.J.; Warsinger-Pepe, N.; Yamashita, Y.M. Transgenerational dynamics of rDNA copy number in Drosophila male germline stem cells. *eLife* **2018**, *7*. [CrossRef] [PubMed]
60. Gong, J.; Dong, J.; Liu, X.; Massana, R. Extremely High Copy Numbers and Polymorphisms of the rDNA Operon Estimated from Single Cell Analysis of Oligotrich and Peritrich Ciliates. *Protist* **2013**, *164*, 369–379. [CrossRef]
61. Prescott, D.M. The DNA of ciliated protozoa. *Microbiol. Rev.* **1994**, *58*, 233–267. [CrossRef] [PubMed]
62. Heyse, G.; Jönsson, F.; Chang, W.-J.; Lipps, H.J. RNA-dependent control of gene amplification. *Proc. Natl. Acad. Sci. USA* **2010**, *107*, 22134–22139. [CrossRef]
63. Pan, W.-C.; Orias, E.; Flacks, M.; Blackburn, E.H. Allele-specific, selective amplification of a ribosomal RNA gene in tetrahymena thermophila. *Cell* **1982**, *28*, 595–604. [CrossRef]
64. Symonová, R.; Majtánová, Z.; Sember, A.; Staaks, G.B.; Bohlen, J.; Freyhof, J.; Rábová, M.; Ráb, P. Genome differentiation in a species pair of coregonine fishes: An extremely rapid speciation driven by stress-activated retrotransposons mediating extensive ribosomal DNA multiplications. *BMC Evol. Biol.* **2013**, *13*, 42. [CrossRef]
65. Szathmáry, E.; Smith, J.M. The major evolutionary transitions. *Nature* **1995**, *374*, 227–232. [CrossRef] [PubMed]
66. Parks, M.M.; Kurylo, C.M.; Dass, R.A.; Bojmar, L.; Lyden, D.; Vincent, C.T.; Blanchard, S.C. Variant ribosomal RNA alleles are conserved and exhibit tissue-specific expression. *Sci. Adv.* **2018**, *4*, eaao0665. [CrossRef]
67. Bughio, F.; Maggert, K.A. The peculiar genetics of the ribosomal DNA blurs the boundaries of transgenerational epigenetic inheritance. *Chromosome Res. Int. J. Mol. Supramol. Evol. Asp. Chromosome Biol.* **2018**. [CrossRef]
68. Torres-Machorro, A.L.; Hernández, R.; Cevallos, A.M.; López-Villaseñor, I. Ribosomal RNA genes in eukaryotic microorganisms: Witnesses of phylogeny? *FEMS Microbiol. Rev.* **2010**, *34*, 59–86. [CrossRef]
69. Drouin, G.; Tsang, C. 5S rRNA Gene Arrangements in Protists: A Case of Nonadaptive Evolution. *J. Mol. Evol.* **2012**, *74*, 342–351. [CrossRef]
70. Mallatt, J.; Chittenden, K.D. The GC content of LSU rRNA evolves across topological and functional regions of the ribosome in all three domains of life. *Mol. Phylogenet. Evol.* **2014**, *72*, 17–30. [CrossRef] [PubMed]
71. Agrawal, S.; Ganley, A.R.D. The conservation landscape of the human ribosomal RNA gene repeats. *PLOS ONE* **2018**, *13*, e0207531. [CrossRef] [PubMed]
72. Kobayashi, T. A new role of the rDNA and nucleolus in the nucleus—rDNA instability maintains genome integrity. *BioEssays* **2008**, *30*, 267–272. [CrossRef]
73. Escobar, J.S.; Glémin, S.; Galtier, N. GC-Biased Gene Conversion Impacts Ribosomal DNA Evolution in Vertebrates, Angiosperms, and Other Eukaryotes. *Mol. Biol. Evol.* **2011**, *28*, 2561–2575. [CrossRef]
74. Li, B.; Kremling, K.A.G.; Wu, P.; Bukowski, R.; Romay, M.C.; Xie, E.; Buckler, E.S.; Chen, M. Coregulation of ribosomal RNA with hundreds of genes contributes to phenotypic variation. *Genome Res.* **2018**, *28*, 1555–1565. [CrossRef]
75. Ganley, A.R.D.; Kobayashi, T. Highly efficient concerted evolution in the ribosomal DNA repeats: Total rDNA repeat variation revealed by whole-genome shotgun sequence data. *Genome Res.* **2007**, *17*, 184–191. [CrossRef]
76. Stults, D.M.; Killen, M.W.; Pierce, H.H.; Pierce, A.J. Genomic architecture and inheritance of human ribosomal RNA gene clusters. *Genome Res.* **2007**, *18*, 13–18. [CrossRef] [PubMed]

77. Stults, D.M.; Killen, M.W.; Williamson, E.P.; Hourigan, J.S.; Vargas, H.D.; Arnold, S.M.; Moscow, J.A.; Pierce, A.J. Human rRNA Gene Clusters Are Recombinational Hotspots in Cancer. *Cancer Res.* **2009**, *69*, 9096–9104. [CrossRef]
78. Paredes, S.; Branco, A.T.; Hartl, D.L.; Maggert, K.A.; Lemos, B. Ribosomal DNA deletions modulate genome-wide gene expression: "rDNA-sensitive" genes and natural variation. *PLoS Genet.* **2011**, *7*, e1001376. [CrossRef]
79. Paredes, S.; Maggert, K.A. Ribosomal DNA contributes to global chromatin regulation. *Proc. Natl. Acad. Sci. USA* **2009**, *106*, 17829–17834. [CrossRef]
80. Prokopowich, C.D.; Gregory, T.R.; Crease, T.J. The correlation between rDNA copy number and genome size in eukaryotes. *Genome* **2003**, *46*, 48–50. [CrossRef] [PubMed]
81. Banditt, M.; Koller, T.; Sogo, J.M. Transcriptional activity and chromatin structure of enhancer-deleted rRNA genes in Saccharomyces cerevisiae. *Mol. Cell. Biol.* **1999**, *19*, 4953–4960. [CrossRef] [PubMed]
82. Pederson, T. The Nucleolus. *Cold Spring Harb. Perspect. Biol.* **2011**, *3*, a000638. [CrossRef]
83. Yu, S.; Lemos, B. A Portrait of Ribosomal DNA Contacts with Hi-C Reveals 5S and 45S rDNA Anchoring Points in the Folded Human Genome. *Genome Biol. Evol.* **2016**, *8*, 3545–3558. [CrossRef] [PubMed]
84. Yu, S.; Lemos, B. The long-range interaction map of ribosomal DNA arrays. *PLoS Genet.* **2018**, *14*, e1007258. [CrossRef] [PubMed]
85. Bierhoff, H.; Postepska-Igielska, A.; Grummt, I. Noisy silence: Non-coding RNA and heterochromatin formation at repetitive elements. *Epigenetics* **2014**, *9*, 53–61. [CrossRef] [PubMed]
86. McStay, B.; Grummt, I. The Epigenetics of rRNA Genes: From Molecular to Chromosome Biology. *Annu. Rev. Cell Dev. Biol.* **2008**, *24*, 131–157. [CrossRef]
87. Schöfer, C.; Weipoltshammer, K. Nucleolus and chromatin. *Histochem. Cell Biol.* **2018**, *150*, 209–225. [CrossRef] [PubMed]
88. Eickbush, T.H.; Eickbush, D.G. Finely Orchestrated Movements: Evolution of the Ribosomal RNA Genes. *Genetics* **2007**, *175*, 477–485. [CrossRef] [PubMed]
89. Robicheau, B.M.; Susko, E.; Harrigan, A.M.; Snyder, M. Ribosomal RNA Genes Contribute to the Formation of Pseudogenes and Junk DNA in the Human Genome. *Genome Biol. Evol.* **2017**, *9*, 380–397. [CrossRef]
90. Bernardi, G. *Structural and Evolutionary Genomics: Natural Selection in Genome Evolution*; Elsevier: Amsterdam, The Netherlands, 2005; ISBN 978-0-08-046187-8.
91. Galtier, N.; Piganeau, G.; Mouchiroud, D.; Duret, L. GC-content evolution in mammalian genomes: The biased gene conversion hypothesis. *Genetics* **2001**, *159*, 907–911.
92. Varriale, A.; Torelli, G.; Bernardi, G. Compositional properties and thermal adaptation of 18S rRNA in vertebrates. *RNA* **2008**, *14*, 1492–1500. [CrossRef]
93. Wang, H.-C.; Xia, X.; Hickey, D. Thermal adaptation of the small subunit ribosomal RNA gene: A comparative study. *J. Mol. Evol.* **2006**, *63*, 120–126. [CrossRef]
94. Parker, M.S.; Balasubramaniam, A.; Sallee, F.R.; Parker, S.L. The Expansion Segments of 28S Ribosomal RNA Extensively Match Human Messenger RNAs. *Front. Genet.* **2018**, *9*. [CrossRef]
95. Parker, M.S.; Sallee, F.R.; Park, E.A.; Parker, S.L. Homoiterons and expansion in ribosomal RNAs. *FEBS Open Bio* **2015**, *5*, 864–876. [CrossRef]
96. Demeshkina, N.; Repkova, M.; Ven'Yaminova, A.; Graifer, D.; Karpova, G. Nucleotides of 18S rRNA surrounding mRNA codons at the human ribosomal A, P, and E sites: A crosslinking study with mRNA analogs carrying an aryl azide group at either the uracil or the guanine residue. *RNA* **2000**, *6*, 1727–1736. [CrossRef]
97. Barendt, P.A.; Shah, N.A.; Barendt, G.A.; Kothari, P.A.; Sarkar, C.A. Evidence for Context-Dependent Complementarity of Non-Shine-Dalgarno Ribosome Binding Sites to *Escherichia coli* rRNA. *ACS Chem. Biol.* **2013**, *8*, 958–966. [CrossRef]
98. Forsdyke, D.R. *Evolutionary Bioinformatics*, 3rd ed.; Springer: Cham, Switzerland, 2016; ISBN 978-3-319-28755-3.
99. Coleman, A.W. ITS2 is a double-edged tool for eukaryote evolutionary comparisons. *Trends Genet.* **2003**, *19*, 370–375. [CrossRef]
100. Coleman, A.W. Pan-eukaryote ITS2 homologies revealed by RNA secondary structure. *Nucleic Acids Res.* **2007**, *35*, 3322–3329. [CrossRef]

101. Coleman, A.W. Nuclear rRNA transcript processing versus internal transcribed spacer secondary structure. *Trends Genet.* **2015**, *31*, 157–163. [CrossRef]
102. Ruhl, M.W.; Wolf, M.; Jenkins, T.M. Compensatory base changes illuminate morphologically difficult taxonomy. *Mol. Phylogenet. Evol.* **2010**, *54*, 664–669. [CrossRef] [PubMed]
103. Muller, T.; Philippi, N.; Dandekar, T.; Schultz, J.; Wolf, M. Distinguishing species. *RNA* **2007**, *13*, 1469–1472. [CrossRef] [PubMed]
104. Song, J.; Shi, L.; Li, D.; Sun, Y.; Niu, Y.; Chen, Z.; Luo, H.; Pang, X.; Sun, Z.; Liu, C.; et al. Extensive Pyrosequencing Reveals Frequent Intra-Genomic Variations of Internal Transcribed Spacer Regions of Nuclear Ribosomal DNA. *PLoS ONE* **2012**, *7*, e43971. [CrossRef] [PubMed]
105. Schultz, J.; Wolf, M. ITS2 sequence–structure analysis in phylogenetics: A how-to manual for molecular systematics. *Mol. Phylogenet. Evol.* **2009**, *52*, 520–523. [CrossRef] [PubMed]
106. Ruhl, W.M. Compensatory Base Changes Illuminate Morphologically Difficult Taxonomy. Master Thesis, University of Georgia in Athens, Athens, Greece, 2009.
107. Nakhoul, H.; Ke, J.; Zhou, X.; Liao, W.; Zeng, S.X.; Lu, H. Ribosomopathies: mechanisms of disease. *Clin. Med. Insights Blood Disord.* **2014**, *7*, 7–16. [CrossRef] [PubMed]
108. Hetman, M. Role of the nucleolus in human diseases. *Biochim. Biophys. Acta BBA Mol. Basis Dis.* **2014**, *1842*, 757. [CrossRef] [PubMed]
109. Quin, J.E.; Devlin, J.R.; Cameron, D.; Hannan, K.M.; Pearson, R.B.; Hannan, R.D. Targeting the nucleolus for cancer intervention. *Biochim. Biophys. Acta BBA Mol. Basis Dis.* **2014**, *1842*, 802–816. [CrossRef] [PubMed]
110. Woods, S.J.; Hannan, K.M.; Pearson, R.B.; Hannan, R.D. The nucleolus as a fundamental regulator of the p53 response and a new target for cancer therapy. *Biochim. Biophys. Acta BBA Gene Regul. Mech.* **2015**, *1849*, 821–829. [CrossRef] [PubMed]

© 2019 by the author. Licensee MDPI, Basel, Switzerland. This article is an open access article distributed under the terms and conditions of the Creative Commons Attribution (CC BY) license (http://creativecommons.org/licenses/by/4.0/).

Article

Diversification of Transposable Elements in Arthropods and Its Impact on Genome Evolution

Changcheng Wu and Jian Lu *

State Key Laboratory of Protein and Plant Gene Research, Center for Bioinformatics, School of Life Sciences, Peking University, Beijing 100871, China; ccwu@pku.edu.cn
* Correspondence: LUJ@pku.edu.cn

Received: 18 February 2019; Accepted: 26 April 2019; Published: 6 May 2019

Abstract: Transposable elements (TEs) are ubiquitous in arthropods. However, analyses of large-scale and long-term coevolution between TEs and host genomes remain scarce in arthropods. Here, we choose 14 representative *Arthropoda* species from eight orders spanning more than 500 million years of evolution. By developing an unbiased TE annotation pipeline, we obtained 87 to 2266 TE reference sequences in a species, which is a considerable improvement compared to the reference TEs previously annotated in Repbase. We find that TE loads are diversified among species and were previously underestimated. The highly species- and time-specific expansions and contractions, and intraspecific sequence diversification are the leading driver of long terminal repeat (LTR) dynamics in Lepidoptera. Terminal inverted repeats (TIRs) proliferated substantially in five species with large genomes. A phylogenetic comparison reveals that the loads of multiple TE subfamilies are positively correlated with genome sizes. We also identified a few horizontally transferred TE candidates across nine species. In addition, we set up the Arthropod Transposable Elements database (ArTEdb) to provide TE references and annotations. Collectively, our results provide high-quality TE references and uncover that TE loads and expansion histories vary greatly among arthropods, which implies that TEs are an important driving force shaping the evolution of genomes through gain and loss.

Keywords: transposable elements; evolution; arthropods; genome size; horizontal transfer; database

1. Introduction

Transposable elements (TEs) are DNA sequences that can jump in host genomes [1]. TEs are widespread in eukaryotic organisms and occupy more than 45% of the human genome [2]. Previous studies showed that TEs mainly adopt two mechanisms in replication: "copy and paste" and "cut and paste" [3]. The first class of TEs are mainly retrotransposable elements that require RNA intermediates, and the second class of TEs are mainly DNA transposons (terminal inverted repeats or TIRs). As TE translocation might cause genomic instabilities or waste energy of the host organisms, TEs used to be regarded as "junk DNA" [4].

The genomes of many arthropods have been sequenced in the past decades, which suggests the contents of TEs are highly variable in *Arthropoda* [5–18]. For instance, *Locusta migratoria* has a huge genome, which is larger than 6.5 Gb [17], while *Tetranychus urticae* has a much smaller genome which is less than 0.1 Gb [9]. As the number of genes does not differ significantly between these two species [9,17], the 65-fold genome size variation might be mainly due to the rapid evolution of TEs. Consistently, a very recent study surveyed TEs in 62 insects and 11 non-insect outgroup species and found TE contents vary considerably in insect genomes, suggesting the variation in genome size is shaped by the expansion and contraction of TEs in arthropods [19]. TE gain and loss is one of the major drivers of genome size changes, as previously shown in mammals [20], avians [21] and *Drosophila* [22].

Nevertheless, several unaddressed gaps remain in our understanding of the evolutionary dynamics of TEs and their impact on the evolution of arthropod genomes. First, most of the identified TE sequences in arthropods are based on the reference sequences in the Repbase database, and many species-specific TEs that are under-represented in Repbase are not well recovered. Second, many TE annotation programs such as RepeatModeler does not consider the global structure of a TE, which might identify a partial but not the full-length sequence of a TE. Third, the expansion and contraction dynamics for most TE families or subfamilies in arthropods are still not well understood. Forth, it remains unclear how frequently horizontal transfer of TEs (HTTs) occur in arthropods. Arthropods have tremendously diversified phenotypes and abundant genomic resources. Answers to the above questions might help understand the roles of TEs during the diversification of arthropods.

We explored the evolutionary dynamics of TEs in fourteen representative *Arthropoda* species spanning eight orders. We first built a high-quality TE reference library for each species by combining sequence homology searching and *de novo* TE identification. Then, we explored TE expansion and contraction in these arthropod species based on the phylogenetic tree. We found that frequent gains and losses, sequence diversification, and HTTs jointly contributed to TE load diversity in arthropods. Finally, we report the database ArTEdb (http://db.cbi.pku.edu.cn/arte), which incorporates the sequences and annotations of TEs identified in this study. The resources provided by this study will benefit future TE studies in arthropods.

2. Materials and Methods

2.1. Transposable Elements Reference Construction

The genome sequences were downloaded from the NCBI, FlyBase, and SilkBase databases (Table S1). The published TEs were downloaded from Repbase (v23.02) [23]. LTRs have several structural features, including target site repeats, long terminal repeats, primer binding sites (PBSs), polypurine tract (PPT) and multiple open reading frames (ORFs). The ORFs in long terminal repeats (LTRs) encode functional domains such as reverse transcriptase (RT), integrase (IN), and RNase H (RH). Reverse transcription of LTR requires tRNA primer that pairs with the PBS. Therefore, the domain profiles and tRNAs will help to identify and classify LTRs better. LTR domain profiles were downloaded from GypsyDB (www.gydb.org) [24]. The tRNAs were annotated using tRNAScan-SE (-G) [25]. Only high-quality tRNAs (score > 40) with clear anticodons were kept and used in the LTR annotation. TE reference libraries were built using both homology-based and *de novo* methods.

2.2. Identification and Annotation of Transposable Elements

Two *de novo* tools were used to identify full-length LTR candidates initially. LTR_Finder uses tRNAs and Pfam domain profiles (-w 2 -l 100 -L 1000 -D 12000 -d 2000) [26], and LTRharvest (-seed 80 -minlenltr 100 -maxlenltr 1000 -mindistltr 2000 -maxdistltr 12000 -overlaps no -similar 80 -mintsd4 -maxtsd 20 -longoutput) is one module of GenomicTools [27]. The tRNAs and Pfam domains profiles are used for identifying PBSs and enzyme domains respectively. LTRdigest (-pptlen 10 30 -pbsoffset 0 3) [28] was applied to refine the identifications using both tRNAs and LTR domain profiles downloaded from GypsyDB. Only LTR candidates with at least one of the five LTR domains (GAG, AP, INT, RT, and RH) were kept. All identified LTR candidates were combined and clustered with the UCLUST (id = 0.9) algorithm [29]. For TEs in each cluster, CLUSTALW2 [30] was applied to perform multiple alignments, and the cons (EMBOSS) tool was used to build consensus sequences. Singletons (only one TE in a cluster) having at least four of the five LTR domains were kept. Both singletons and consensus sequences were masked by RepeatMasker [31] with the built-in libraries of corresponding species. Sequences that overlapped with RepeatMasker libraries (more than 80% of the queries were masked) were collapsed.

Besides the LTR specific annotation programs, we also employed RepeatModeler (www.repeatmasker.org/RepeatModeler) pipeline with the default parameters to identify TEs in each species.

Consensus sequences aligned to known protein-coding genes of *Drosophila melanogaster* were removed. Moreover, the remaining sequences that overlapped with previous annotations of Repbase or LTRs were collapsed. All annotated TE sequences by RepeatModeler were combined, and their classes and subfamilies were further determined based on both sequence similarity to known TEs (Repbase) or TE-specific domains (Pfam_v27 and GypsyDB) by PASTEClassifier [32] and *de novo* classifications by TEclass [33].

For each species, the TEs identified by the LTR specific tools and RepeatModeler were combined. USEARCH [29] was used to obtain the nonredundant TE libraries for each species. All the TE references for each species can be downloaded from the ArTEdb database (http://db.cbi.pku.edu.cn/arte). For each TE reference annotated in this study, we denote it with the TE class followed by the first three letters of the genus and the first three letters of the species name. For example, the homologous sequence of *Gypsy-1_DSim* in *Drosophila melanogaster* is *Gypsy-1_DroMel* in the ArTEdb database.

2.3. Transposable Element Loads and Expansion Analyses

Genomes were masked by RepeatMasker using TE libraries defined in this study. TE loads in each species were calculated using the script ONE_CODE_TO_FIND_THEM_ALL.PL [34]. The Kimura 2-Parameter divergence of TEs was calculated using the RepeatMasker built-in tool calcDivergenceFromAlign.pl, and the distributions of divergence were plotted using ECharts (www.echarts.baidu.com).

2.4. Reconstructing the Phylogenetic Tree

BUSCO [35] was adopted with the insect core genes to annotate single-copy orthologous genes in the fourteen species. Only single-copy genes with intact ORFs were kept. Orthologous protein multiple alignments generated by T-COFFEE [36] were then transformed into codon alignments using RevTrans [37]. A preliminary phylogeny tree for the selected species was firstly reconstructed by MEGA [38] based on the concatenated protein alignments of orthologous genes. Sites with more than 50% gaps were removed from the concatenated alignment, and the phylogeny was built using Maximum-likelihood algorithm with the JTT matrix. The topological position of *L. migratoria* was manually curated based on a previous study [39]. All the codon alignments were concatenated, and CODEML from PAML [40] was used to calculate the *dN* with the free model (runmode = 0; model = 1). The *dN* values of the concatenated sequences were then set as the branch length of the phylogenetic tree. The tree is provided in the Supplementary File S1.

2.5. Time-Calibrated Phylogeny

The non-parametric tool r8s [41] was used to transform the branch lengths into millions of years. Fossil evidence shows that the divergence time between *Bombus terrestris* and *Apis mellifera* is 23 to 28.4 million of years ago (Ma) [39], and the age of the root node was set as 550 to 580 Ma [39,42].

2.6. Fitting Multiple Phylogenetic Comparative Models

Both the TE loads and genomes sizes (gaps excluded) were transformed in natural log(Ln)grams. Their phylogenetic signals were estimated using the phylosig (method = "lambda") function from phytools [43]. The adequacy of four standard phylogenetic comparative models were tested using the fitContinuous function of Geiger [44] in R. These models are Brownian motion (BM) [45], Ornstein-Uhlenbeck (OU) [46], Early-burst (EB; also named as Accelerating-Decelerating (ACDC)) [47], and white noise (non-phylogenetic and normal distribution). The AICc (corrected Akaike information criterion for small sample size) values of these models were evaluated, and the results suggest that the BM model was the most suitable model. Therefore, the BM model was used in the next phylogenetically independent contrasts and ancestral state reconstruction of TE loads.

2.7. Phylogenetically Independent Contrasts

The phylogenetically independent contrasts (PIC) of TE loads and host genome sizes (gaps excluded) were calculated using the pic function of ape [48] in R. The TE loads of all subfamilies were added to the total TE loads. For the four main classes (TIR, LTR, LINE, and SINE), the TE loads of all subfamilies belong to them were added together. Subfamilies that appeared in more than seven species were preserved for the additional subfamily-level PIC analyses. Both the TE loads and genome sizes were in natural log(Ln)grams. The correlation between TE loads and host genome sizes were calculated using Pearson's product-moment correlation test in R. The P values of these subfamilies were corrected using the Holm-Bonferroni correction [49].

2.8. Ancestral State Reconstruction of Transposable Element Loads

The TE loads that are in natural log(Ln)grams were fitted to the Brownian motion model. The TE load states of the ancestor nodes were inferred by the phytools [43] using the maximum-likelihood analysis method [50]. The fastAnc function from phytools was used to reconstruct the ancestral state as previously reported [51]. Similar to the method used in a previous study [51], the TE load change ratio was defined as the TE load of offspring node relative to its ancestral node. The phylogenetic tree with TE load change ratios was plotted using ReproPhylo [52].

2.9. Identifying Shifts of Transposable Element Loads Change Rates in the Phylogeny

BAMM (Bayesian analysis of macroevolutionary mixtures) [53–55] was used to identify shifts of TE loads evolutionary rates in the time-calibrated phylogeny. The betaInitPrior and betaShiftPrior parameters were estimated by the setBAMMpriors function of BAMMtools v2.1.6 [56], and 1,000,000 generations of MCMC (Markov chain Monte Carlo) sampling were conducted. The outputs of BAMM were then post-processed by BAMMtools, and the evolutionary rates of TE loads and the best shift configuration were extracted and highlighted in the phylorate plots for four TE classes and three subfamilies (TIR/Mariner, LTR/Gypsy, and LINE/Jockey).

2.10. Transposable Element Protein Annotation

Proteins of TEs were annotated with the homology-based method. TE proteins extracted from Repbase were aligned to annotated TEs using the tblastn program [57]. For each TE, the aligned protein with the smallest E-value was selected. All preserved query-target pairs were then realigned using exonerate (–model protein2genome:bestfit) to annotated proteins [58].

2.11. The Phylogenetic Analyses of LTRs in LEPIDOPTERA

The *pol*-encoded proteins of LTRs were first aligned through T-COFFEE [36]. For the alignments of *Gypsy* and *Pao*, aligned sites covered by less than 70% of all TE sequences were removed. FastTree was applied to build the phylogenetic trees for *Copia*, *Gypsy*, and *Pao* [59]. The phylogenetic trees were plotted using FigTree (https://github.com/rambaut/figtree).

2.12. Horizontal Transposable Element Transfer

Due to the large evolutionary distances among the fourteen-selected species, we adopted the genome-wide amino acid distance instead of dS (synonymous substitutions per synonymous site) as the cutoff for identifying HTT. Protein sequences of single-copy orthologous genes that were annotated using BUSCO with core genes of Insecta lineage were aligned using CLUSTALW2 [30] between each pair of species. The amino acid distances were calculated using PAML (aaRatefile = jones.dat) and sorted in ascending order. Since the number of orthologous genes between every two arthropods might be larger than 5000 [60], and our BUSCO analyses might have captured the most conserved ones, therefore, we set the genome-wide cutoff as the 100th minimum amino acid distance between two species, which represents the top ~2% of the total orthologous genes (Table S3). TE proteins were aligned between each pair of fourteen-selected species, and only reciprocal best hits were kept. The amino acid distances of the aligned TEs were calculated as described above. TE pairs with lower amino acid distances than genome-wide cutoffs were selected as HTT candidates.

2.13. Analysis of Transposable Elements Distribution in Arthropods

Genomes of 126 extra insect species were downloaded from the InsectBase (http://insect-genome.com). For each pair of HTT candidates, we aligned them to genomes of all 140 arthropods using the fasta36 (-E 1e-5) program of FASTA [61]. All hits that are longer than 80 bp and have more than 80% similarity were kept.

2.14. The ArTEdb Database

The ArTEdb was written in HTML and PHP and hosted in Apache. All TE information was organized using MySQL in the background. Both gene and TE annotations were embedded in JBrowse. Alignment|Blast uses NCBI BLAST (v2.7.1+) [57] in the background with TE references from 14 arthropods as databases. The alignment results are post processed by xmlBLASTparser (www.github.com/AshokHub/xmlBLASTparser) for visualization. Alignment|RepeatMasker runs RepeatMasker in the background.

3. Results

3.1. Construction of Transposable Element References

Although Repbase [23] provides TE annotations for many arthropods, high-quality TE references are available for only a small subset of the 14 species investigated in this study (Table 1). To obtain high-quality and unbiased TE references, we systematically annotated TEs from genomes of all fourteen species using the pipeline described in Figure 1a. We obtained 87 to 2266 TE reference sequences in a species, which is a considerable improvement compared to the reference TEs previously annotated in Repbase (Table 1). Notably, much higher numbers of reference TEs were identified in the species that have large genomes, for instance, *Acanthoscurria geniculata* and *Locusta migratoria*, than the species that have small genomes (Table 1). Eight of the fourteen species had few TEs (≤10) annotated in Repbase, while hundreds of extra reference TEs were identified in this study (Table 1).

Table 1. Genome sizes and the number of transposable elements references of fourteen arthropods.

Class	Order	Species	Genome Sizes [†]	TE Reference Sequences		
				RB/RM/LTR	RTE/DTE/UnC	Total
Insecta	Diptera	*Drosophila melanogaster*	144 Mb	147/79/8	159/47/28	234
	Lepidoptera	*Spodoptera frugiperda*	358 Mb	1/475/12	263/190/35	488
		Bombyx mori [‡]	460 Mb	92/546/183	552/232/37	821
		Papilio xuthus	244 Mb	41/319/60	248/151/21	420
		Melitaea cinxia	390 Mb	0/763/17	433/299/48	780
	Hymenoptera	*Bombus terrestris*	249 Mb	6/520/12	241/267/30	538
		Apis cerana	228 Mb	0/86/1	28/49/10	87
		Apis mellifera	250 Mb	6/136/1	35/98/10	143
	Hemiptera	*Acyrthosiphon pisum*	542 Mb	331/752/75	326/796/36	1158
		Nilaparvata lugens	1.14 Gb	0/1,136/230	872/419/75	1366
	Orthoptera	*Locusta migratoria*	5.6 Gb	1028/1182/56	1144/1018/104	2266
Arachnida	Trombidiformes	*Tetranychus urticae*	90 Mb	10/122/73	105/86/14	205
	Araneae	*Acanthoscurria geniculata*	7.2 Gb	0/1857/167	967/982/75	2024
	Scorpiones	*Mesobuthus martensii*	925 Mb	39/1260/101	476/843/81	1400

[†] Sizes of current versions of genomes used in this study. [‡] Recently assembled genome based on PacBio single-molecule sequencing datasets. RB, RM, and LTR symbolize Repbase, RepeatModeler, and LTR *de novo*, respectively. RTE, DTE, and UnC symbolize retrotransposable elements, TIRs, and unclassified TEs, respectively.

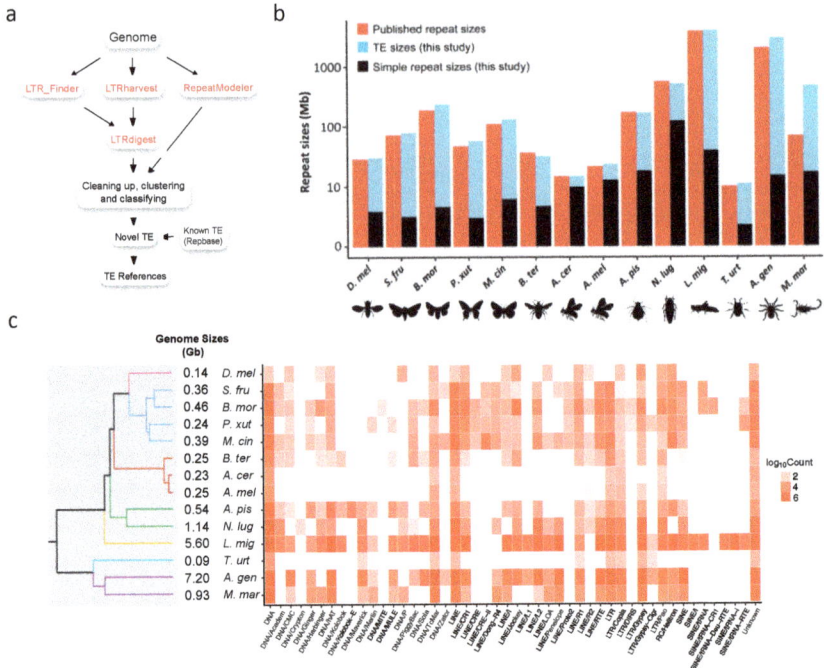

Figure 1. Annotating TEs in arthropods. (**a**) The TE reference construction workflow (see details in Materials and Methods). (**b**) The repeat sizes of fourteen arthropods. Published repeat sizes were adapted from the original genome sequencing studies (Table S1). The repeat sizes (TE and simple repeat) of this study were evaluated using the TE libraries defined in this study. The published repeat size of *B. mor* was adapted from [11]. *D. mel*, *Drosophila melanogaster*; *B. mor*, *Bombyx mori*; *M. cin*, *Melitaea cinxia*; *P. xut*, *Papilio xuthus*; *S. fru*, *Spodoptera frugiperda*; *A. cer*, *Apis cerana*; *A. mel*, *Apis mellifera*; *B. ter*, *Bombus terrestris*; *N. lug*, *Nilaparvata lugens*; *A. pis*, *Acyrthosiphon pisum*; *L. mig*, *Locusta migratoria*; *A. gen*, *Acanthoscurria geniculata*; *M. mar*, *Mesobuthus martensii*; *T. urt*, *Tetranychus urticae*. (**c**) The landscape of TE loads in arthropods.

The new high-quality TE references inspired us to ask whether TE contents were underestimated in previous studies. To answer this question, we evaluated the repeat sizes and TE contents in the selected species with the new TE references (Figure 1b; Table S1). The results show that repeat sizes were previously underestimated in 11 of the 14 selected species, especially those with few TE references annotated in Repbase. For instance, no TE reference had been reported in *A. geniculata* (the original genome sequencing study reported that approximately 60% of the genome excluding N/X runs was composed of TEs, but TE references were not available [16]), and the initial repeat masking by RepeatMasker with the built-in *Arthropoda* library showed that the TE content was only 4.5%. However, 57.17% of the *A. geniculata* genome was masked using the new TE references, which is very close to the number obtained in the original study. In addition, we masked its full genome and identified extra 950 Mb TEs. Another striking species is *Mesobuthus martensii*. Because of the limited number of annotated TEs in Repbase, the TE size of this species was previously underestimated to be 35 Mb (3.1%) [7]. Here, we identified 1400 TE references in *M. martensii* and revealed that approximately half (51.03%; 455 Mb TEs and 17 Mb simple repeats) of its genome consisted of repeats (Figure 1b).

Even for *D. melanogaster*, whose TEs are well annotated in Repbase, our annotation results still identified another 87 TE subfamilies (8 LTRs and 79 non-LTRs). For example, although *Gypsy-1_DSim* is a well known LTR in *Drosophila simulans*, no homologs of this TE have been reported in *D. melanogaster*.

Here we found that the *Gypsy-1_DSim* TE subfamily has 117 copies in the Y chromosome and another 252 copies (at least six full-length ones) in the other five chromosomes (2L/2R/3L/3R/X) of *D. melanogaster* (these TEs were named as *Gypsy-1_DroMel* in our ArTEdb database). Interestingly, the full-length *Gypsy-1_DroMel* was successfully identified by LTR_Finder and LTRharvest, while RepeatModeler only identified a partial sequence of this TE in *D. melanogaster* (Figure S1), which suggests that the approach we employ by combining multiple annotation tools is more powerful than using RepeatModeler alone.

In summary, we annotated TEs for fourteen *Arthropoda* species with the unbiased pipeline and identified many more novel TEs. These high-quality TE libraries might contribute to a better understanding of the evolution and genome contributions of TEs in arthropods.

3.2. Transposable Element Loads Vary Greatly in Arthropods

The huge difference in genome sizes among fourteen arthropods raises the question of how much TEs contribute to their hosts. To answer this question, we masked the genomes of arthropods using RepeatMasker with the new TE references and calculated the TE loads of each species. Hereafter, the TE load is defined as the copy number of TEs in the genome. The loads of multiple TE subfamilies from four main TE classes (TIR, LINE, LTR, and SINE) were summed. Although three (TIR, LINE, and LTR) of the classes existed in all fourteen species, the TE loads varied greatly among species (Figure 1c). The diversity of TEs in arthropods was also reported in a recent study [19]. Next, we will focus on several remarkable insights to show how TEs contribute to genome evolution.

3.2.1. Transposable Element Loss is Prevalent in *Hymenoptera*

The three *Hymenoptera* species, namely, *Apis mellifera*, *Apis cerana*, and *Bombus terrestris*, have the closest evolutionary relationship among the fourteen species. The TE load analysis reveals that these three species have fewer TEs than other species except for *T. urticae* (Figure 1b–c). Especially in the first two species, the repeat sequences account for less than 10% of the host genome, and the real TE content is even lower than 5%. Although we observed several LTRs and LINEs, none of them had coding potential. Only a few *Mariner* (DNA/TcMar) TEs have intact ORFs, consisting with the previous study [8]. Comparing TE loads in *A. mellifera* and *A. cerana* with those in the remaining species, we observed that the TE loss occurred rapidly in *A. mellifera* and *A. cerana*. These two evolutionarily close species had a slight difference in their TE loads, and both the subfamily numbers and the TE loads were significantly reduced (Figure 1c). However, in *B. terrestris*, most TE subfamilies could still be observed, and repeats occupied approximately 13.5% (simple repeat < 2%) of its genome. These results reveal that the extent of TE loss is highly diverse among *Hymenoptera* and somewhat weak in *B. terrestris*.

3.2.2. Recent LTR Expansion in *T. urticae*

In addition to these two *Hymenoptera* species (*A. mellifera* and *A. cerana*), *T. urticae*, which has the smallest genome (90 Mb) among the fourteen species, also has fewer TE subfamilies and loads than the remaining species. In *T. urticae*, TIR is the most abundant class (5.8%), followed by the LTR class (2.9%), while the remaining two classes (LINE and SINE) are quite scarce (<0.5%). A total of 12.4% of the genome of this species consists of TEs (with 2% simple repeats), and the distribution of the sequence divergence presents two distinct peaks (Figure 2a), which is very different from the other species (Figure 2b–d). The first peak in *T. urticae* is from 0% to 1%, and the second one is from 20% to 24%. LTRs are the major component of the first peak, which implies that there is a recent expansion of LTRs in *T. urticae*. For instance, the *Gypsy-6_TetUrt* has at least eight full-length copies. In contrast, the second peak mainly consists of TIRs, which suggests that most TIRs expanded historically and might have diverged thereafter.

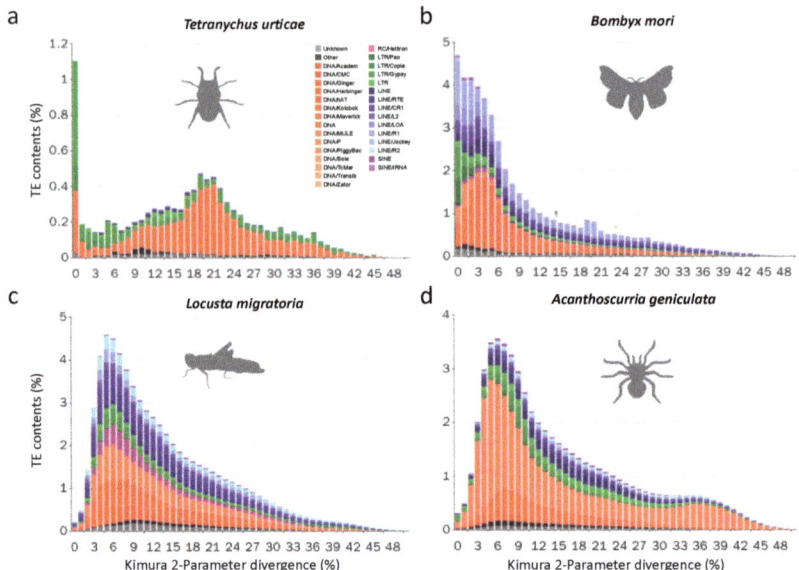

Figure 2. Sequence divergence distribution of TEs. (**a–d**) Distribution of sequence divergence of multiple TE subfamilies in four species. The *y*-axis shows the percentage of the host genomes that is annotated as TEs (TE contents). The *x*-axis shows Kimura 2-Parameter sequence divergence between individual TE copies and consensus references. Unknown, unclassified TEs; Other, Retro and Retroposon.

3.2.3. Lepidopterans Have Diversified Transposable Elements Subfamilies and Large Transposable Elements Loads

For all four *Lepidoptera* species (*Bombyx mori*, *Melitaea cinxia*, *Papilio xuthus*, and *Spodoptera frugiperda*), there are hundreds of TE subfamilies in their genomes. Among these species, *B. mori* has the most abundant TE subfamilies and the largest TE loads (Figure 1c). More than half (51.26%; Table S1) of its genome is occupied by TEs, which is consistent with a previous report [11]. Compared with the three-remaining species, *B. mori* also has the largest genome (Table 1). In addition, the TE divergence distributions of this species show distinct patterns. In *B. mori*, LTRs, TIRs, and LINEs all fall into the 1–10% range, especially the LTRs, almost all of which are located in the 1–5% range (Figure 2b). These results suggest that most LTRs are active and have expanded recently in *B. mori*. However, a distinct distribution is observed in *S. frugiperda*, the evolutionarily closest species to *B. mori* among the 14 selected species. The distribution resembles a classical normal distribution with a peak near 10% (Figure S2A). A similar distribution is observed in *M. cinxia* (Figure S2B). These results suggest that the expansion of TEs are mostly ancient in both *S. frugiperda* and *M. cinxia*. Notably, *P. xuthus* shows a bimodal distribution of TE divergence: one peak is caused by the recent expansion of TEs (the divergence peak is ~4%), and another peak is caused by the historic expansion of rolling circle (RC) TEs (the divergence peak is ~13%, Figure S2C). Altogether, our results suggest that the TE expansion histories in *Lepidoptera* are diverse.

3.2.4. The Evolutionary History of LTRs in *Lepidoptera*

Bombyx mori has the most abundant LTRs among the four *Lepidoptera* species, and the sequence divergence distribution reveals that most LTRs are accumulated due to recent expansion (Figure 2b). However, fewer LTRs are found in the three-remaining species (*M. cinxia*, *P. xuthus*, and *S. frugiperda*; Figure S2A–C). This finding promotes us to ask when, where and which LTRs were gained or lost in *Lepidoptera*. To further explore the details of the evolutionary history of LTRs, we selected intact LTRs and constructed their phylogenies. We annotated proteins of all LTRs and built phylogenetic trees for all three main LTR subfamilies (*BEL/Pao*, *Copia*, and *Gypsy*).

Unlike the TEs in *D. melanogaster* that had been well studied and classified into detailed subfamilies [62,63], the family information of LTRs in *Lepidoptera* are still lacking. Therefore, we arbitrarily divided the *Copia* into three groups (G1–3) according to the phylogeny (Figure 3). Among the four species, *S. frugiperda* has the smallest number of *Copia* (only one member in G2), while *B. mori* has the largest number of *Copia*. Within each group, we frequently detected expansion of *Copia* in *B. mori*. For example, we observed a pair of TEs (*Copia-6_BomMor* and *Copia-20_BomMor* in G3) with extremely low amino acid substitution level in *B. mori*, suggesting that these two could be recently duplicated. Although each of the four species has *Copia* in their genomes, none of the three groups has *Copia* detected in all the four species. A parsimonious explanation is that some *Copia* copies might be lost in a species-specific manner, although we cannot exclude the possibility that the phylogeny is solely caused by TE duplications followed by sequence diversifications.

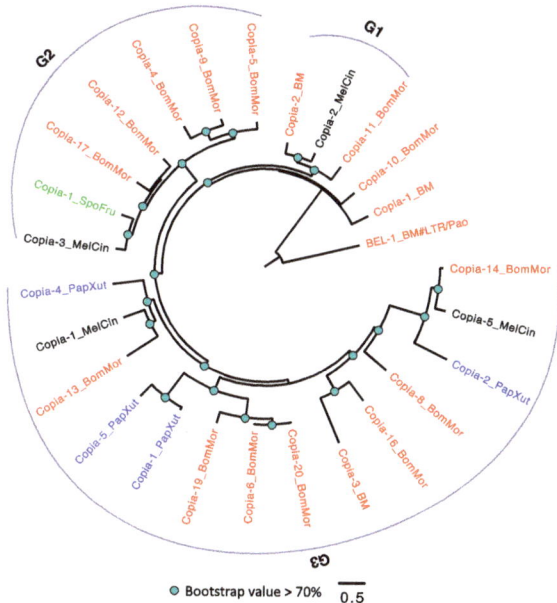

Figure 3. The phylogeny of *Copia* in *Lepidoptera*. The *Copia* were arbitrarily divided into three groups (G1–3). TEs with BM (annotated in Repbase) or BomMor (annotated in this study) suffixes are from *B. mori* (red name); TEs with PapXut, MelCin, and SpoFru suffixes are from *P. xuthus* (blue name), *M. cinxia* (black name), and *S. frugiperda* (green name). Only nodes with bootstrap value not lower than 70% were indicated. BEL-1_BM is the outgroup.

In the *BEL/Pao* subfamily phylogeny, there are four large groups (Figure 4). Similar to *Copia*, *B. mori* also has the largest number of TEs, followed by *P. xuthus*. We frequently observed intraspecific diversifications of *BEL/Pao* (with extremely small amino acid substitution rates) in *B. mori*. It is also possible that the disparity of *BEL/Pao* contents in the four species are caused by the loss of *BEL/Pao* in certain species, although further studies are required to verify this pattern.

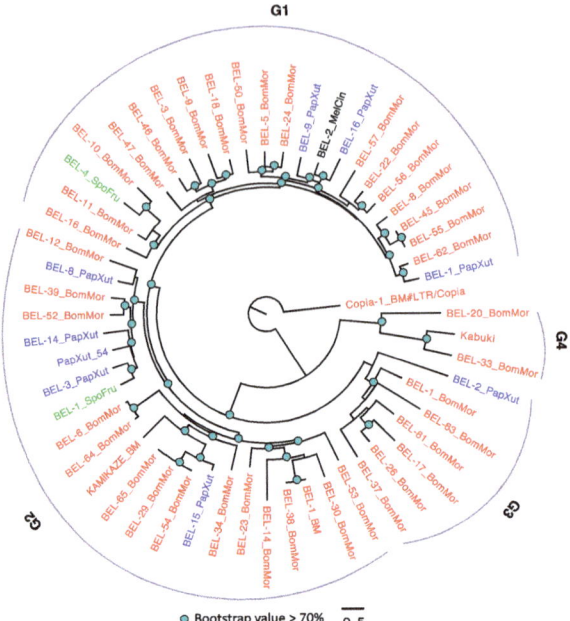

Figure 4. The phylogeny of *BEL/Pao* in *Lepidoptera*. The *BEL/Pao* were arbitrarily divided into four groups (G1–4). TEs with BM (annotated in Repbase) or BomMor (annotated in this study) suffixes are from *B. mori* (red name); TEs with PapXut, MelCin and SpoFru are from *P. xuthus* (blue name), *M. cinxia* (black name) and *S. frugiperda* (green name). *Kabuki* is from *B. mori* and has been annotated in *BmTEdb* [64]. Only nodes with bootstrap value not lower than 70% were indicated. *Copia-1_BM* is the outgroup.

The comparison of the TE contents and loads of *Lepidoptera* revealed that *Gypsy* was most abundant among the three LTR families (Figure 1c, Figure 2b and Figure S2A–C). This family might have the most substantial effect during LTR evolution in *Lepidoptera*. The *Gypsy* phylogeny (Figure 5) reveals three major conclusions. First, *B. mori* has the most abundant *Gypsy* TEs. Second, there are extensive diversifications of *Gypsy* in *B. mori* and *P. xuthus*. Third, we observed one horizontal transfer event between *B. mori* and *M. cinxia* (in G3). The amino acid distance between *Gypsy-6_MelCin* and *Gypsy-32_BomMor* is small (0.052). Considering the long divergence time between *B. mori* and *M. cinxia* (the amino acid distance cutoff between these two species is 0.065; Table S3), the small amino acid distance mostly conflicts with the evolutionary history, confirming the HTTs might be *bona fide*.

Altogether, our results suggest that LTRs evolve expansions and contractions, or intraspecific sequence diversification. All these processes combined to form the current LTR patterns in *Lepidoptera*.

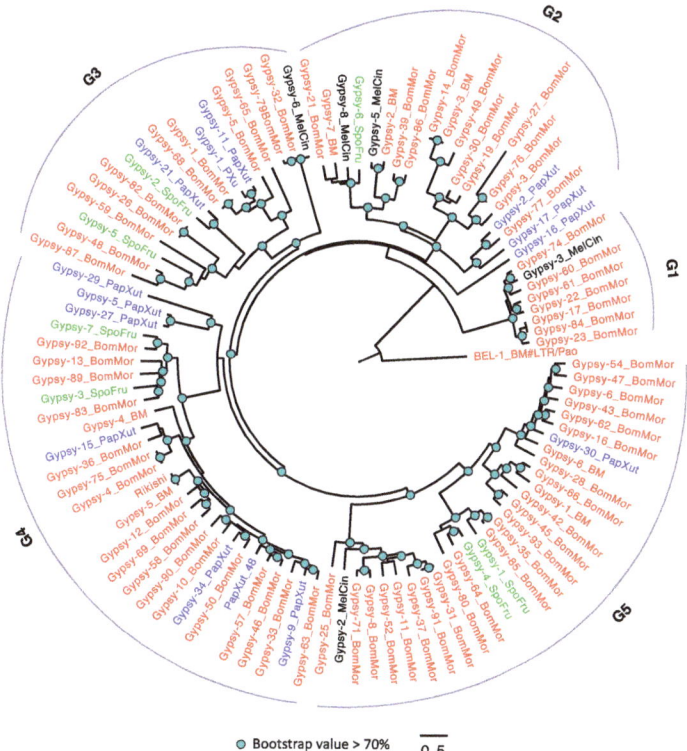

Figure 5. The phylogeny of *Gypsy* in *Lepidoptera*. The *Gypsy* were arbitrarily divided into five groups (G1–5). TEs with BM (annotated in Repbase) or BomMor (annotated in this study) suffixes are from *B. mori* (red name); TEs with PapXut, MelCin and SpoFru are from *P. xuthus* (blue name), *M. cinxia* (black name) and *S. frugiperda* (green name). Rikishi is from *B. mori* and has been annotated in *BmTEdb* [64]. Only nodes with bootstrap value not lower than 70% were indicated. BEL-1_BM is the outgroup.

3.2.5. Non-LTR Transposable Elements Contribute More to Arthropods with Larger Genomes

Given that TEs exist in and occupy large sections of *Arthropod* genomes, we ask how the genome sizes dynamically change due to TE difference. Interestingly, we found the content of TIR (%) in one arthropod genome is positively correlated (Pearson's product-moment correlation: 0.899, $p < 0.0001$) with the genome size (Table S4). In addition, the LINE superfamily shows high abundance in *L. migratoria* (Figure 1c), which may have resulted from the recent rapid expansion of this family (Figure 2c). When the divergence distribution is considered, in *L. migratoria* and *A. geniculata*, the peaks are all close to 6% (Figure 2c,d), suggesting that most of the identified TEs have accumulated recently during their evolutionary history. This pattern suggests that most TEs in their genomes were recently generated through rapid TE expansions.

Altogether, our results show that TE loads vary greatly among arthropods and reveal that the gain and loss of non-LTR TEs are much more prevalent in arthropods than previously thought. The very recent study reported similar conclusions [19], while the underlying causes of diversified TE and lineage-specific activity were not mentioned. Here, we analyzed the TE dynamics of species from the same orders (*Hymenoptera* and *Lepidoptera*) or with extremely large genome sizes. The results reveal that rapid extinction, intraspecific diversification, and HTT are the internal driving forces of the diversity of TEs in arthropods.

3.3. The Expansion and Contraction of Transposable Elements in Arthropods

RNAi is a critical mechanism to inhibit TEs transposition and hence reduces TE loads in metazoans [65,66]. Besides RNAi, the difference in life history, mating system, GC contents have been suggested to account for the difference in TE loads across species [51]. However, the study of long-term TE evolution in 42 nematode species suggest that genetic drift rather than life history or RNAi mainly determined the evolution of TEs [51]. In addition, the authors tested load changes of four main TE classes and found several "expansion hotspots" in the most dynamic LTR TEs [51].

Given the disparity of TE loads across the 14 arthropod species we studied, here we explore the expansion and contraction of TEs in arthropods. We first used phylogenetic comparative analysis to explain the TE diversity over long-time scales. The phylogeny of the fourteen-selected species was inferred using single-copy orthologous genes and time-calibrated using r8s [41]. We evaluated the phylogenetic signals of traits (TE loads and genome sizes transformed in natural log(Ln)grams) using the λ model with phytools [43,50]. The λ is from 0 to 1, and greater values imply stronger phylogenetic signals (trait is highly related to the phylogeny and not random). Most of the kept traits (24/27) have λ values larger than 0.5 (Table S2), indicating that they are highly associated with the phylogeny. To determine the most appropriate comparative model, we fitted four standard phylogenetic comparative models (BM, OU, EB, and white) for the above traits using Geiger. We used both AICc and the weighted AICc (AW) to assess the fitness of the four models. The results showed that the BM model was most suitable (Table S2). Therefore, TE loads of ancestor nodes were inferred using the maximum likelihood method [67] under the BM model.

We found TE expansion (the black branches) and contraction (the red branches) hotspots in all four main TE families (Figure 6a). Although the expansion hotspots (with large change ratios) are slightly different among the four TE classes, most of them are enriched in *L. migratoria*, *A. geniculata*, *N. lugens* and *B. mori*, which is consistent with their TE divergence distributions (Figure 2b–d). Moreover, the genomes of all four species are larger than those of closely related species. Besides, the TE contraction also broadly exists in all four classes, and the most significant species are *T. urticae*, *D. melanogaster*, and the three bees.

The above results suggest that the changes of TE loads were highly variable during arthropod evolution. Therefore, we applied BAMM to analyze the dynamical change of evolutionary rates for TE loads under the phylogeny. In Figure 6b (so-called phylorate plots), although no shift of TE loads evolutionary rates is in all the four classes, the evolutionary rates are broadly variable among both TE classes and clades. Among the four classes, SINE has the highest rates, while TIR has the lowest ones, which helps to explain the largest TE loads variants of SINE (Figure 1c) in arthropods. The rates in *Lepidoptera* clades are broadly lower than the remaining ones, consisting with their low TE loads changes (Figure 6a) relative to ancestor nodes. Besides, three branches (the root branch, *T. urticae*, and *Hymenoptera* clades) have higher rates than the other branches in TIR, LTR, and LINE classes (the root branch and *T. urticae* clade are also higher in SINE). In addition, evolutionary rate dynamics of three representative subfamilies (Figure S3; *Jockey*, *Mariner*, and *Gypsy*) were also evaluated, and the results also supported that evolutionary rates were high in the *Hymenoptera* clade (shift events were identified in *Jockey* and *Gypsy*). These results reveal that the TE loads had rapidly changed at the early radiation of insects and arachnids, and the TE loads change rates could be branch-specific instead of underlying constraints.

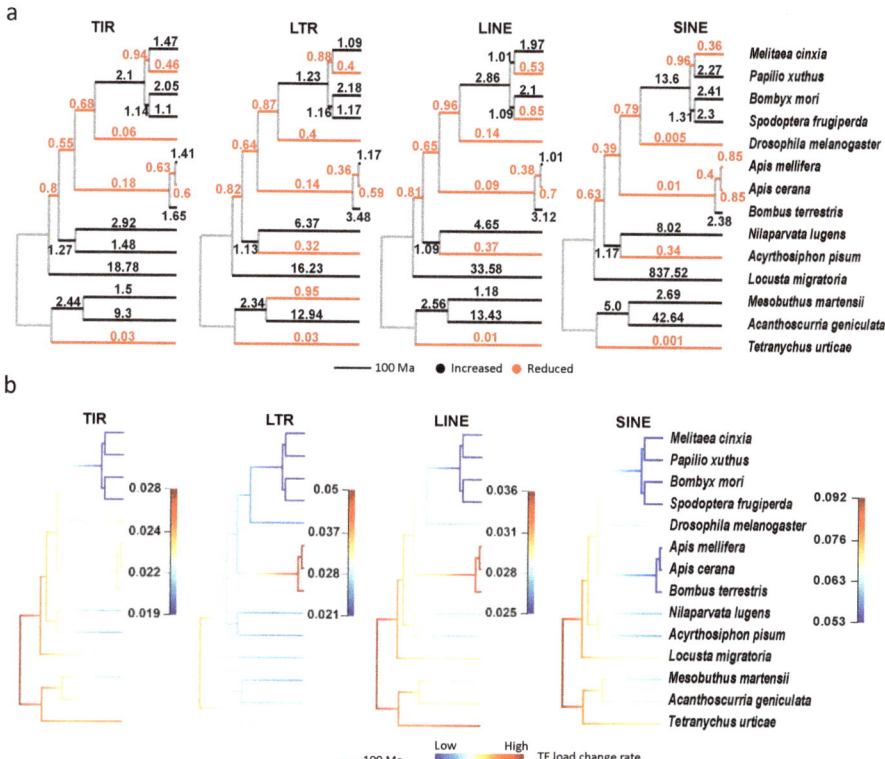

Figure 6. The expansion and contraction of TEs in the 14 arthropod species. (**a**) The ratio of TE load change (the offspring relative to the ancestral node) for each branch. (**b**) The dynamic evolutionary rates of TE loads in the phylogeny. Branch colors are scaled by evolutionary rates of TE loads, and rate increases from the cold color (blue) to warm color (red). TIR, terminal inverted repeat; LTR, long terminal repeat; LINE, long interspersed nuclear element; SINE, short interspersed nuclear element.

3.4. Transposable Element Loads are Significantly Correlated with Genome Sizes

Transposable element contents have been reported positively correlated with the genome sizes in eukaryotes [20,68]. TE contents and the number of subfamilies were found to be correlated with host genome sizes in arthropods [19], indicating that the genomes with larger sizes also have greater TE contents or more TE subfamilies. However, TE contents and the subfamily numbers might be a little more variable than TE loads due to the influences of the large variation of host genome sizes in arthropods. Here, we evaluated correlations between TE loads and TE contents and genome sizes. The results showed that both of these two are positively correlated with genomes sizes and the TE loads (Pearson's product-moment correlation: 0.92; $p = 2.76 \times 10^{-6}$; Figure S4A) have a greater correlation coefficient than TE contents (0.71; $p = 4.92 \times 10^{-3}$; Figure S4B).

The quantitative traits in species from a branching phylogeny are not statistically non-independent due to common ancestry, and the PICs should be conducted [69]. We evaluated the correlations between loads of TEs from multiple subfamilies and host genome sizes. As shown in Table 2, our results show that many TE subfamilies are positively correlated with host genome sizes. Especially, the most abundant *Mariner* (TcMar) subfamily is significantly correlated with the corrected $p < 0.1$. In addition, we also evaluated the correlation between the total TEs of four main families and genome sizes. All the four classes are positively correlated with host genome sizes, for which the corrected $p < 0.1$ (Table S5). After combining all TEs in each species, we observed that total TE loads were also significantly correlated with genome sizes (Pearson's product-moment correlation: 0.83; $p = 4.74 \times 10^{-4}$; Table S5). This result is consistent with the PIC analysis in a recent study [19]. Unlike the TE contents used in the previous study, we used TE loads which might reflect TE abundance in host genomes a little better. Although only 14 species used in this study, the TE loads showed a much stronger correlation than TE contents. Our results reveal that TE expansion is one of the most critical forces driving changes in the sizes of host genomes, which is also consistent with previous reports in flies [22], birds and mammals [21].

Table 2. TE loads are significantly correlated with genome sizes.

TE Subfamilies	Correlation Coefficient	P	Corrected P
LINE/L2	0.868	1.20×10^{-4}	3.01×10^{-3}
SINE/Unclassified	0.784	1.51×10^{-3}	3.18×10^{-2}
LINE/Unclassified	0.760	2.55×10^{-3}	4.85×10^{-2}
DNA/Unclassified	0.756	2.79×10^{-3}	5.02×10^{-2}
DNA/TcMar	0.728	4.75×10^{-3}	8.07×10^{-2}
LTR/Copia	0.724	5.14×10^{-3}	8.22×10^{-2}
DNA/hAT	0.654	1.53×10^{-2}	2.13×10^{-1}
LINE/CR1	0.644	1.75×10^{-2}	2.13×10^{-1}
LINE/RTE	0.652	1.57×10^{-2}	2.13×10^{-1}
LTR/Unclassified	0.654	1.52×10^{-2}	2.13×10^{-1}
LTR/Pao	0.644	1.76×10^{-2}	2.13×10^{-1}
DNA/Ginger	0.559	4.72×10^{-2}	4.25×10^{-1}
LINE/I	0.511	7.45×10^{-2}	5.96×10^{-1}
DNA/CMC	0.489	8.97×10^{-2}	6.28×10^{-1}
DNA/Academ	0.419	1.54×10^{-1}	9.23×10^{-1}
LINE/Dong-R4	0.403	1.73×10^{-1}	9.23×10^{-1}
DNA/Harbinger	0.358	2.30×10^{-1}	9.23×10^{-1}
DNA/PiggyBac	0.371	2.12×10^{-1}	9.23×10^{-1}
LINE/Jockey	0.394	1.83×10^{-1}	9.23×10^{-1}
LINE/R1	0.351	2.40×10^{-1}	9.23×10^{-1}
LTR/Gypsy	0.382	1.98×10^{-1}	9.23×10^{-1}

Unclassified: subfamily information is unavailable.

3.5. Horizontal Transposable Element Transfer in Arthropods

TEs can be transferred by vertical inheritance or horizontal transfer. *P-element* is the first reported and the most famous HTT in *Drosophila* [70] and is the genetic basis of P-M hybrid dysgenesis in *D. melanogaster* [71]. HTT is associated with many phenotypic changes in plants [72]. A recent study identified thousands of HTT events in Insecta [73], and these transferred TEs might be important in driving genome evolution [74]. Here, we identified millions of TE copies from thousands of subfamilies in fourteen arthropods. We asked whether HTTs had been involved in TE evolution, especially in species with large genomes. To solve this problem, we proposed a strategy with which to identify HTTs using genome-wide amino acid distances. Amino acid distance has been widely used in phylogenetic studies, and a smaller distance implies a higher sequence similarity. We expected the amino acid substitution rate of TE (evolving mostly neutrally or being counter selected, thus being permissive to *dN* substitutions) to be higher than orthologous genes (more conserved with a higher rate of negative selection). Therefore, the vertically inherited TEs will have larger amino acid distances than orthologous genes. On the contrary, TEs with lower amino acid distances could be HTTs.

Considering the variable quality of gene annotations in the selected species, we used the BUSCO tool with the built-in Insecta core genes library and annotated genes in all selected species. Using the single-copy homologous genes, we inferred the genome-wide amino acid distance cutoffs (Table S3; see details in Materials and Methods) for each pair of fourteen species. Annotated TE proteins were aligned between each pair of species. The amino acid distances of aligned TE pairs were calculated and compared with the genome-wide cutoff, which was defined as the 100th minimum amino acid distance between two species. According to the recent study in 76 arthropods, the number of genes that present in more than 75% Metazoans are all larger than 5000 [60], which implies that the number of orthologous genes between most arthropods might be larger than 5000. In this study, we set the genome-wide cutoff as the 100th minimum amino acid distance between two species, which equals to at most 2% of the total orthologous genes.

We obtained eight candidate HTTs among nine species (Figure 7a,b and Figure S5; Table S6), and their protein alignments are in Supplementary File S2. All of them are best reciprocal hits between corresponding species. Two of them are from the *Gypsy* subfamily. Six of them are TIRs from the *hAT-Tip100* (n = 1) and *Mariner* (n = 5) subfamilies. The Class II (TIR) TEs are the most frequent HTTs, consisting with previous reports in arthropods [73,75]. The Class II TEs tend to be shorter than Class I TEs, besides their transpositions have weak host dependence [76]. These might help to explain why the Class II (especially the *Mariner* subfamily) have the most abundant HTTs.

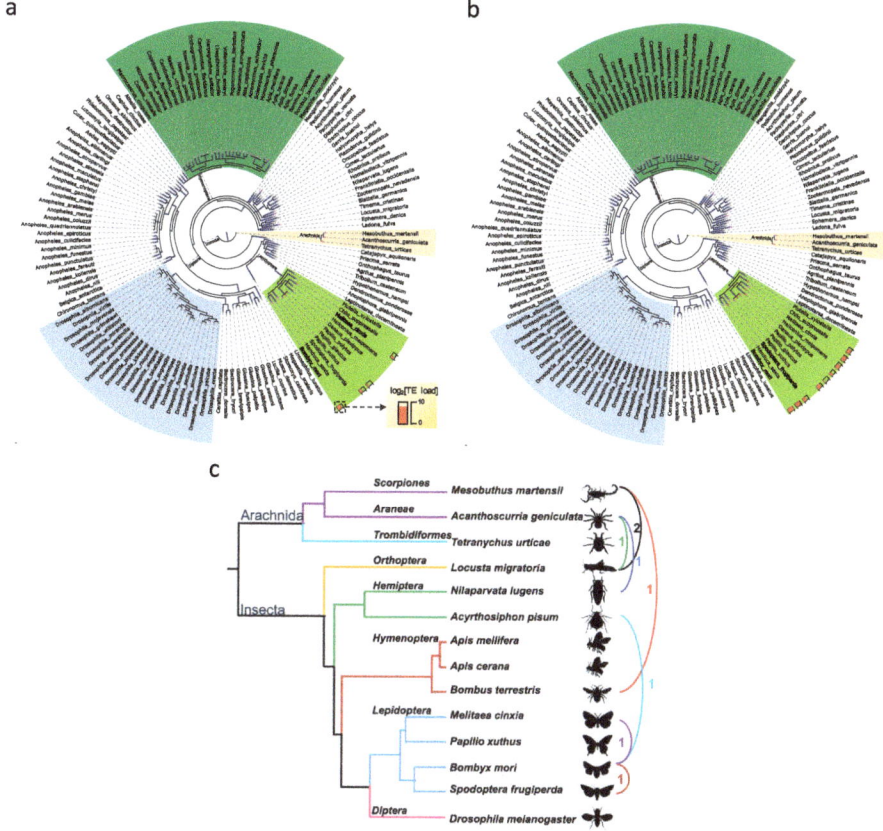

Figure 7. The HTTs in arthropods. (**a**) The HTT of *Gypsy* between *M. cinxia* and *B. mori*. (**b**) The HTT of *Mariner* between *B. mori* and *S. frugiperda*. The species names of hosts of HTT are in bold. (**c**) Eight HTTs in the fourteen arthropods.

In Figure S6A–C, two HTT examples are depicted. These two HTTs all have amino acid distances (the blue line) lower than the genome-wide cutoffs (the gray line). The horizontal transfer event between *Gypsy-6_MelCin* and *Gypsy-32_BomMor* is confirmed because the amino acid distance between them is 0.054, which is lower than the cutoff (0.065) between *B. mori* and *M. cinxia*, which is consistent with the phylogenetic analysis of *Gypsy* in *Lepidoptera* (Figure 5). Besides, these two TEs only appear in four of 140 arthropods (Figure 7a), which implies that the high sequence similarity between them are resulted by horizontal transfer instead of vertical inheritance. Another example is the *Mariner* HTT between *B. mori* and *S. frugiperda* (Figure 7b). The amino distance between TEs is 0.034 which is lower than the genome-wide cutoff 0.047 (Figure S6C; Table S3).

The exact number of HTTs between these species might be underestimated (i) because several TEs were not annotated due to the methodological limitations and low-quality genome assemblies and (ii) because of the strict cutoffs used in HTT identification. Besides, six HTTs were identified in five species (*A. pisum, A. geniculata, L. migratoria, M. martensii,* and *N. lugens*) with genomes larger than 500 Mb (Figure 7c). Considering the large genomes and the TE loads in these species and other arthropods, our and previous results [73,75] both support that HTTs played important roles during their genome expansions and might be another important driving force of genome evolution.

3.6. The ArTEdb

Database technology has been widely used in biological data sharing, especially in the field of TE research (recently reviewed [77]). Several databases provide TE annotations or repeat-masking results [23,24]. Here, to make new TE references and annotations easy to use and contribute to TE studies in arthropods, we set up the ArTEdb. It consists of four main categories: (1) TE landscapes in 14 arthropods, (2) TE references, annotations and downloading of premasked genomes, (3) TE querying based on keywords and sequence similarity, and (4) an online repeat-masking service.

We summarized TE annotations and generated an overview of TE contents for each species. The TE contents were organized by both subfamilies and sequence divergence between each TE copy and TE reference. We provide two methods for querying TEs. People can look up TEs by name, subfamily, and class. Alternatively, one can also use a sequence to search the TE database directly with BLAST. The genome browser helps visualize genomic features more intuitively. We integrated the JBrowse genomic feature visualization tool into the ArTEdb. Using this tool, people can focus on genes or other genomic loci rather than masking repeats by themselves. In addition, we provide an online repeat-masking service, which is very useful for people who want to scan TEs from just a small number of sequences quickly. Finally, all TE annotations and premasked genomes can be downloaded from the ArTEdb directly. We hope that the ArTEdb will benefit future TE studies in arthropods.

4. Discussion

In this study, we chose 14 representative arthropods spanning eight orders to study the coevolution between TEs and host genomes. Using our customized TE annotation pipeline, we both generated high-quality TE references and estimated the TE profiles in these species.

4.1. A New Database of High-Quality Transposable Element References for Arthropods

Characterizing TEs in an unbiased approach is an important task for the non-model organisms. Although Repbase is extensively used for TE identification and annotation [23], the TE reference sequences might not be complete for many non-model organisms. Therefore, many TE annotation tools have recently been developed for unbiased TE characterization [26–28,31,77]. In this study, we built an unbiased TE annotation pipeline (Figure 1a) by combining different TE identification and classification tools. We also present our reference sequences in the ArTEdb database, which provides useful resources for TE annotations in other arthropod genomes.

A very recent study [19] have extensively characterized TEs in the arthropod genomes with the RepeatModeler pipeline. RepeatModeler aims to identify sequences with high-copy numbers in genomes, which is suitable for identifying most types of TEs. However, RepeatModeler does not take into account the structural characteristics of TEs, and sometimes it does not perform as well as the family-specific tools. For example, the novel TE *Gypsy-1_DroMel* was identified by both RepeatModeler and two LTR-specific tools (LTR_Finder and LTRharvest). The sequence alignment indicates that

RepeatModeler only identified a partial sequence of *Gypsy-1_DroMel*, while the combined approach we used (combining RepeatModeler, LTR_Finder, and LTRharvest) successfully identified the full-length *Gypsy-1_DroMel* (Figure S1). Besides TE identification, our ArTEdb database also significantly improved the annotation of TE family information. As RepeatModeler only uses sequence similarity to known TEs in TE classification, most TEs identified by RepeatModeler remain unclassified [19]. In this study, we also used PASTEClassifier [32] and TEclass [33] to annotate the identified TEs. PASTEClassifier takes full use of known TEs (Repbase) and domain annotations (Pfam and Gypsydb), and TEclass utilizes the supporting vector machine based on oligomer frequencies of repeats. These comprehensive classification methods make almost all the TEs be classified at least in class-level. Therefore, our study provides TE profiles with better annotation information.

4.2. Why Does the Transposable Element Loads Can Be So Different Across Arthropod Species?

Previous studies show that TEs gain and loss are important for genome sizes variation in vertebrates [21] and *Drosophila* [22]. Both a very recent study [19] and our results detected frequent gains and losses of TEs in the arthropod genomes. Moreover, our results suggest that multiple evolutionary forces can cause TE profiles to be very different even between closely related species, as shown in the four *Lepidoptera* species we analyzed (Figure 3,Figure 4,Figure 5). Unlike the protein-coding genes which are in general under selective constraints, the sequences of TEs are usually less constrained and evolve rapidly [78]. Therefore, besides the copy number variation caused by TE expansion and contraction, sequence diversification between homologous TE can also lead to TE diversification in the arthropods. Moreover, HTT is not uncommon between arthropods as shown in our results and previous studies [73,75]. Therefore, our results suggest that the TE profiles in the arthropod genomes are shaped by expansion and contraction of TEs, TE sequence diversification, and HTTs.

Then why can the TE loads be so different across arthropod species? According to the classic population genetics framework of TE biology, the content of TEs in a species is shaped by its rapid replication and the selective constraints because TEs are in general deleterious and selected against in most species. Since the selective strength of TEs in a species is determined by the effective population size (Ne) of that species [79], it is possible that the difference in Ne across arthropod species might be important for the variation in TE loads. In addition, DNA methylation [80] and RNA interference pathway [81,82] also suppress TE activities in arthropods, and the suppressive effects of both mechanisms might vary in arthropods. Therefore, the vast difference in TE loads among arthropods can be caused by the difference in natural selection or in the epigenetic regulatory mechanisms, although at this moment we cannot exclude the possibility that the TE difference in the studied arthropod species is mainly shaped by genetic drift as previously shown in nematode [51].

4.3. The Contribution of HTTs to the Transposable Element Repertoire in Arthropod Genomes

Previous studies have reported numerous HTTs in arthropods [70,73,75] and plants [83] (details are reviewed in Ref. [72,84]). In arthropods, most HTTs occurred for the *Mariner* (TcMar) subfamily [73,75], which are generally shorter (1–2 Kb) than LTRs (mostly longer than 5Kb) and might be easily transferred by vectors [85]. Accordingly, among the eight HTTs identified in this study, five of them are caused by the *Mariner* subfamily. Notably, the species with larger genomes have greater TE loads and more HTT events, suggesting that HTT may contribute to the TE expansions. In eukaryotes, TEs are repressed either by suppressing TE transcription or by piRNA-mediated cleavage of TE transcripts [86,87]. As the host organisms take time to develop piRNAs to repress a newly horizontally transferred TE, that TE might replicate rapidly and contribute significantly to the TE repertoire until abundant piRNAs are developed to repress that TE [82].

TEs recently horizontally transferred between two species will have higher sequences similarity than the protein-coding genes, which cause *dS* values to be smaller for the TEs than that for the protein-coding genes between the two species [73,88]. In this study, we used the amino acid distance instead of the *dS* to identify HTTs in arthropods, because the synonymous substitutions in the protein-coding genes between the studied species are usually saturated (*dS* values usually > 1). Since the number of orthologous genes between every two arthropods might be larger than 5000 [60], and our BUSCO analyses might have captured the most conserved ones, therefore, we set the genome-wide cutoff as the 100th minimum amino acid distance between two species, which represents the top ~2% of the total orthologous genes (Table S3). Thus, the HTTs we identified in this study do not necessarily recently occur, but might have occurred anciently.

4.4. Adaptive Transposable Element Insertions in Arthropods

In recent decades, numerous studies have demonstrated that TEs can benefit their hosts in multiple ways. For instance, TEs could be domesticated as promotors [89–91] or enhancers [92–94] to regulate gene expression. Moreover, a few TEs could be co-opted into novel protein-coding genes in the host genomes [95–100]. These findings suggest that TEs are important for providing raw materials of the regulatory elements and proteomes for the hosts. In addition, many studies have shown that TE insertions might increase the fitness of hosts. For example, a *P* element insertion in the promoter of *Hsp70A* significantly increases the fecundity of heat-shocked flies [91], and a *Doc1420* insertion in *CHKov1* significantly increases the pesticide resistance of hosts by disrupting the original gene structure [101]. The carbonaria form of the peppered moth (*Biston betularia*) was reported to be caused by a TE insertion [102] that upregulates a cortex transcript involved in early wing disc development. All these studies indicate that the beneficial effects of TEs are pervasive in eukaryotes. Although millions of TEs had been annotated in our study, the beneficial TE insertions remain unknown. Thus, further studies are required to identify the role of TEs in the adaptation of arthropods.

Supplementary Materials: The following are available online at http://www.mdpi.com/2073-4425/10/5/338/s1, Figure S1. The comparison of one TE identified by LTR-specific tools and RepeatModeler. Figure S2. Sequence divergence distribution of TEs in three *Lepidoptera* species. Figure S3. Rate shifts of TE loads in the phylogeny. Figure S4. The correlation between TEs and host genomes sizes. Figure S5. The copy numbers of six HTTs in 140 arthropods. Figure S6. Two HTTs in arthropods. Table S1. Basic information about the fourteen species. Table S2. The weighted small sample size corrected AIC (AW) and λ of the traits. Table S3. Cutoffs of amino acid distances. Table S4. The content of TIR in the 14 arthropod genomes. Table S5. TE loads are significantly correlated with genome sizes. Table S6. Eight horizontally transferred TEs. File S1. The phylogenetic tree of fourteen species. File S2. Protein alignment of eight HTTs.

Author Contributions: J.L. conceived the study. C.W. conducted the analysis. J.L. and C.W. wrote the manuscript.

Funding: This work was supported by grants from the National Natural Science Foundation of China (Nos. 91731301, 31771411, 31571333, and 91431101), Ministry of Science and Technology of the People's Republic of China (2016YFA0500800) and Peking-Tsinghua Center for Life Science to J.L.

Acknowledgments: The analysis was performed on the Computing Platform of the Center for Life Science at Peking University.

Conflicts of Interest: The authors declare no conflict of interest.

References

1. Finnegan, D.J. Eukaryotic transposable elements and genome evolution. *Trends Genet.* **1989**, *5*, 103–107. [CrossRef]
2. Lander, E.S.; Linton, L.M.; Birren, B.; Nusbaum, C.; Zody, M.C.; Baldwin, J.; Devon, K.; Dewar, K.; Doyle, M.; FitzHugh, W.; et al. Initial sequencing and analysis of the human genome. *Nature* **2001**, *409*, 860–921. [PubMed]
3. Wicker, T.; Sabot, F.; Hua-Van, A.; Bennetzen, J.L.; Capy, P.; Chalhoub, B.; Flavell, A.; Leroy, P.; Morgante, M.; Panaud, O.; et al. A unified classification system for eukaryotic transposable elements. *Nat. Rev. Genet.* **2007**, *8*, 973–982. [CrossRef] [PubMed]
4. Orgel, L.E.; Crick, F. Selfish DNA: The ultimate parasite. *Nature* **1980**, *284*, 604–607. [CrossRef] [PubMed]
5. Adams, M.D. The Genome Sequence of Drosophila melanogaster. *Science* **2000**, *287*, 2185–2195. [CrossRef] [PubMed]

6. Ahola, V.; Lehtonen, R.J.; Somervuo, P.; Salmela, L.; Koskinen, P.; Rastas, P.; Välimäki, N.; Paulin, L.; Kvist, J.; Wahlberg, N.; et al. The Glanville fritillary genome retains an ancient karyotype and reveals selective chromosomal fusions in Lepidoptera. *Nat. Commun.* **2014**, *5*, 4737. [PubMed]
7. Cao, Z.; Yu, Y.; Wu, Y.; Hao, P.; Di, Z.; He, Y.; Chen, Z.; Yang, W.; Shen, Z.; He, X.; et al. The genome of Mesobuthus martensii reveals a unique adaptation model of arthropods. *Nat. Commun.* **2013**, *4*, 2602. [CrossRef]
8. Elsik, C.G.; Worley, K.C.; Bennett, A.K.; Beye, M.; Camara, F.; Childers, C.P.; De Graaf, D.C.; Debyser, G.; Deng, J.; Devreese, B.; et al. Finding the missing honey bee genes: Lessons learned from a genome upgrade. *BMC Genom.* **2014**, *15*, 86. [CrossRef]
9. Grbić, M.; Van Leeuwen, T.; Clark, R.M.; Rombauts, S.; Rouzé, P.; Grbić, V.; Osborne, E.J.; Dermauw, W.; Ngoc, P.C.T.; Ortego, F.; et al. The genome of Tetranychus urticae reveals herbivorous pest adaptations. *Nature* **2011**, *479*, 487–492. [CrossRef] [PubMed]
10. International Aphid Genomics Consortium. Genome sequence of the pea aphid acyrthosiphon pisum. *PLoS Biol.* **2010**, *8*, e1000313.
11. International Aphid Genomics Consortium. The genome of a lepidopteran model insect, the silkworm bombyx mori. *Insect Biochem. Mol. Biol.* **2008**, *38*, 1036–1045. [CrossRef] [PubMed]
12. Kakumani, P.K.; Malhotra, P.; Mukherjee, S.K.; Bhatnagar, R.K. A draft genome assembly of the army worm, Spodoptera frugiperda. *Genomics* **2014**, *104*, 134–143. [CrossRef]
13. Li, X.; Fan, D.; Zhang, W.; Liu, G.; Zhang, L.; Zhao, L.; Fang, X.; Chen, L.; Dong, Y.; Chen, Y.; et al. Outbred genome sequencing and CRISPR/Cas9 gene editing in butterflies. *Nat. Commun.* **2015**, *6*, 8212. [CrossRef]
14. Park, D.; Jung, J.W.; Choi, B.-S.; Jayakodi, M.; Lee, J.; Lim, J.; Yu, Y.; Choi, Y.-S.; Lee, M.-L.; Park, Y.; et al. Uncovering the novel characteristics of Asian honey bee, Apis cerana, by whole genome sequencing. *BMC Genom.* **2015**, *16*, 1. [CrossRef] [PubMed]
15. Sadd, B.M.; Barribeau, S.M.; Bloch, G.; De Graaf, D.C.; Dearden, P.; Elsik, C.G.; Gadau, J.; Grimmelikhuijzen, C.J.; Hasselmann, M.; Lozier, J.D.; et al. The genomes of two key bumblebee species with primitive eusocial organization. *Genome Biol.* **2015**, *16*, 76. [CrossRef] [PubMed]
16. Sanggaard, K.W.; Bechsgaard, J.S.; Fang, X.; Duan, J.; Dyrlund, T.F.; Gupta, V.; Jiang, X.; Cheng, L.; Fan, D.; Feng, Y.; et al. Spider genomes provide insight into composition and evolution of venom and silk. *Nat. Commun.* **2014**, *5*, 3765. [CrossRef]
17. Wang, X.; Fang, X.; Yang, P.; Jiang, X.; Jiang, F.; Zhao, D.; Li, B.; Cui, F.; Wei, J.; Ma, C.; et al. The locust genome provides insight into swarm formation and long-distance flight. *Nat. Commun.* **2014**, *5*, 2957. [CrossRef]
18. Xue, J.; Zhou, X.; Zhang, C.X.; Yu, L.L.; Fan, H.W.; Wang, Z.; Xu, H.J.; Xi, Y.; Zhu, Z.R.; Zhou, W.W.; et al. Genomes of the rice pest brown planthopper and its endosymbionts reveal complex complementary contributions for host adaptation. *Genome Biol.* **2014**, *15*, 521. [CrossRef] [PubMed]
19. Petersen, M.; Armisén, D.; Gibbs, R.A.; Hering, L.; Khila, A.; Mayer, G.; Richards, S.; Niehuis, O.; Misof, B. Diversity and evolution of the transposable element repertoire in arthropods with particular reference to insects. *BMC Evol. Biol.* **2019**, *19*, 11. [CrossRef]
20. Canapa, A.; Barucca, M.; Biscotti, M.A.; Forconi, M.; Olmo, E. Transposons, Genome Size, and Evolutionary Insights in Animals. *Cytogenet. Genome Res.* **2015**, *147*, 217–239. [CrossRef]
21. Kapusta, A.; Suh, A.; Feschotte, C. Dynamics of genome size evolution in birds and mammals. *Proc. Natl. Acad. Sci. USA* **2017**, *114*, E1460–E1469. [CrossRef] [PubMed]
22. Sessegolo, C.; Burlet, N.; Haudry, A. Strong phylogenetic inertia on genome size and transposable element content among 26 species of flies. *Biol. Lett.* **2016**, *12*, 20160407. [CrossRef] [PubMed]
23. Bao, W.; Kojima, K.K.; Kohany, O. Repbase Update, a database of repetitive elements in eukaryotic genomes. *Mob. DNA* **2015**, *6*, 11. [CrossRef] [PubMed]
24. Llorens, C.; Futami, R.; Covelli, L.; Dominguez-Escriba, L.; Viu, J.M.; Tamarit, D.; Aguilar-Rodriguez, J.; Vicente-Ripolles, M.; Fuster, G.; Bernet, G.P.; et al. The gypsy database (gydb) of mobile genetic elements: Release 2.0. *Nucleic Acids Res.* **2011**, *39*, D70–D74. [CrossRef]
25. Lowe, T.M.; Eddy, S.R.; Avni, D.; Biberman, Y.; Meyuhas, O. tRNAscan-SE: A Program for Improved Detection of Transfer RNA Genes in Genomic Sequence. *Nucleic Acids Res.* **1997**, *25*, 955–964. [CrossRef]
26. Xu, Z.; Wang, H. LTR_FINDER: An efficient tool for the prediction of full-length LTR retrotransposons. *Nucleic Acids Res.* **2007**, *35*, W265–W268. [CrossRef] [PubMed]

27. Ellinghaus, D.; Kurtz, S.; Willhoeft, U. LTRharvest, an efficient and flexible software for de novo detection of LTR retrotransposons. *BMC Bioinform.* **2008**, *9*, 18. [CrossRef] [PubMed]
28. Steinbiss, S.; Willhoeft, U.; Gremme, G.; Kurtz, S. Fine-grained annotation and classification of de novo predicted LTR retrotransposons. *Nucleic Acids Res.* **2009**, *37*, 7002–7013. [CrossRef]
29. Edgar, R.C. Search and clustering orders of magnitude faster than BLAST. *Bioinformatics* **2010**, *26*, 2460–2461. [CrossRef] [PubMed]
30. Larkin, M.; Blackshields, G.; Brown, N.; Chenna, R.; Mcgettigan, P.; Mc William, H.; Valentin, F.; Wallace, I.; Wilm, A.; López, R.; et al. Clustal W and Clustal X version 2.0. *Bioinformatics* **2007**, *23*, 2947–2948. [CrossRef]
31. Smit, A.F.A.; Hubley, R.; Green, P. RepeatMasker Open-4.0 2013–2015. Available online: http://www.repeatmasker.org (accessed on 24 April 2017).
32. Hoede, C.; Arnoux, S.; Moisset, M.; Chaumier, T.; Inizan, O.; Jamilloux, V.; Quesneville, H. PASTEC: An Automatic Transposable Element Classification Tool. *PLoS ONE* **2014**, *9*, 91929. [CrossRef]
33. Abrusán, G.; Grundmann, N.; Demester, L.; Makalowski, W. TEclass—A tool for automated classification of unknown eukaryotic transposable elements. *Bioinformatics* **2009**, *25*, 1329–1330. [CrossRef]
34. Bailly-Bechet, M.; Haudry, A.; Lerat, E. "One code to find them all": A perl tool to conveniently parse repeatmasker output files. *Mob. DNA* **2014**, *5*, 13. [CrossRef]
35. Waterhouse, R.M.; Seppey, M.; Simão, F.; Manni, M.; Ioannidis, P.; Klioutchnikov, G.; Kriventseva, E.V.; Zdobnov, E.M. BUSCO Applications from Quality Assessments to Gene Prediction and Phylogenomics. *Mol. Biol. Evol.* **2017**, *35*, 543–548. [CrossRef]
36. Notredame, C.; Higgins, D.G.; Heringa, J. T-coffee: A novel method for fast and accurate multiple sequence alignment. *J. Mol. Biol.* **2000**, *302*, 205–217. [CrossRef]
37. Wernersson, R. RevTrans: Multiple alignment of coding DNA from aligned amino acid sequences. *Nucleic Acids Res.* **2003**, *31*, 3537–3539. [CrossRef]
38. Kumar, S.; Stecher, G.; Li, M.; Knyaz, C.; Tamura, K. MEGA X: Molecular Evolutionary Genetics Analysis across Computing Platforms. *Mol. Biol. Evol.* **2018**, *35*, 1547–1549. [CrossRef]
39. Misof, B.; Liu, S.; Meusemann, K.; Peters, R.S.; Donath, A.; Mayer, C.; Frandsen, P.B.; Ware, J.; Flouri, T.; Beutel, R.G.; et al. Phylogenomics resolves the timing and pattern of insect evolution. *Science* **2014**, *346*, 763–767. [CrossRef] [PubMed]
40. Yang, Z. PAML 4: Phylogenetic Analysis by Maximum Likelihood. *Mol. Biol. Evol.* **2007**, *24*, 1586–1591. [CrossRef] [PubMed]
41. Sanderson, M.J. r8s: Inferring absolute rates of molecular evolution and divergence times in the absence of a molecular clock. *Bioinformatics* **2003**, *19*, 301–302. [CrossRef]
42. Rehm, P.; Borner, J.; Meusemann, K.; Von Reumont, B.M.; Simon, S.; Hadrys, H.; Misof, B.; Burmester, T. Dating the arthropod tree based on large-scale transcriptome data. *Mol. Phylo. Evol.* **2011**, *61*, 880–887. [CrossRef]
43. Revell, L.J. Phytools: An r package for phylogenetic comparative biology (and other things). *Methods Ecol. Evol.* **2012**, *3*, 217–223. [CrossRef]
44. Pennell, M.W.; Eastman, J.M.; Slater, G.J.; Brown, J.; Uyeda, J.C.; Fitzjohn, R.G.; Alfaro, M.E.; Harmon, L.J. geiger v2.0: An expanded suite of methods for fitting macroevolutionary models to phylogenetic trees. *Bioinformatics* **2014**, *30*, 2216–2218. [CrossRef] [PubMed]
45. Felsenstein, J. Maximum-likelihood estimation of evolutionary trees from continuous characters. *Am. J. Hum. Genet.* **1973**, *25*, 471–492.
46. Butler, M.A.; King, A.A. Phylogenetic Comparative Analysis: A Modeling Approach for Adaptive Evolution. *Am. Nat.* **2004**, *164*, 683–695. [CrossRef] [PubMed]
47. Harmon, L.J.; Losos, J.B.; Davies, T.J.; Gillespie, R.G.; Gittleman, J.L.; Jennings, W.B.; Kozak, K.H.; McPeek, M.; Moreno-Roark, F.; Near, T.J.; et al. Early bursts of body size and shape evolution are rare in comparative data. *Evolution* **2010**, *64*, 2385–2396. [CrossRef] [PubMed]
48. Paradis, E.; Claude, J.; Strimmer, K. APE: Analyses of Phylogenetics and Evolution in R language. *Bioinformatics* **2004**, *20*, 289–290. [CrossRef]
49. Holm, S. A simple sequentially rejective multiple test procedure. *Scand. J. Statist.* **1979**, *6*, 65–70.
50. Pagel, M. Detecting correlated evolution on phylogenies: A general method for the comparative analysis of discrete characters. *Proc. R. Soc. Lond. B.* **1994**, *255*, 37–45.

51. Szitenberg, A.; Cha, S.; Opperman, C.H.; Bird, D.M.; Blaxter, M.L.; Lunt, D.H. Genetic Drift, Not Life History or RNAi, Determine Long-Term Evolution of Transposable Elements. *Genome Biol. Evol.* **2016**, *8*, 2964–2978. [CrossRef]
52. Szitenberg, A.; John, M.; Blaxter, M.L.; Lunt, D.H. ReproPhylo: An Environment for Reproducible Phylogenomics. *PLoS Comput. Biol.* **2015**, *11*, e1004447. [CrossRef] [PubMed]
53. Rabosky, D.L.; Donnellan, S.C.; Grundler, M.; Lovette, I.J.; West, C.; James, S.A.; Davey, R.P.; Dicks, J.; Roberts, I.N. Analysis and Visualization of Complex Macroevolutionary Dynamics: An Example from Australian Scincid Lizards. *Syst. Biol.* **2014**, *63*, 610–627. [CrossRef] [PubMed]
54. Rabosky, D.L.; Goldberg, E.E. Model Inadequacy and Mistaken Inferences of Trait-Dependent Speciation. *Syst. Biol.* **2015**, *64*, 340–355. [CrossRef]
55. Rabosky, D.L.; Santini, F.; Eastman, J.; Smith, S.A.; Sidlauskas, B.; Chang, J.; Alfaro, M.E. Rates of speciation and morphological evolution are correlated across the largest vertebrate radiation. *Nat. Commun.* **2013**, *4*, 1958. [CrossRef]
56. Rabosky, D.L.; Gründler, M.; Anderson, C.; Title, P.; Shi, J.J.; Brown, J.W.; Huang, H.; Larson, J.G. BAMMtools: An R package for the analysis of evolutionary dynamics on phylogenetic trees. *Methods Ecol. Evol.* **2014**, *5*, 701–707. [CrossRef]
57. Altschul, S.F.; Gish, W.; Miller, W.; Myers, E.W.; Lipman, D.J. Basic local alignment search tool. *J. Mol. Biol.* **1990**, *215*, 403–410. [CrossRef]
58. Slater, G.S.C.; Birney, E. Automated generation of heuristics for biological sequence comparison. *BMC Bioinform.* **2005**, *6*, 31. [CrossRef]
59. Price, M.N.; Dehal, P.S.; Arkin, A.P. FastTree: Computing Large Minimum Evolution Trees with Profiles instead of a Distance Matrix. *Mol. Biol. Evol.* **2009**, *26*, 1641–1650. [CrossRef]
60. Thomas, G.W.C.; Dohmen, E.; Hughes, D.S.T.; Murali, S.C.; Poelchau, M.; Glastad, K.; Anstead, C.A.; Ayoub, N.A.; Batterham, P.; Bellair, M.; et al. The genomic basis of arthropod diversity. *bioRxiv* **2018**, 382945.
61. Pearson, W.R. Rapid and sensitive sequence comparison with FASTP and FASTA. *Meth. Enzymol.* **1990**, *183*, 63–98.
62. Misra, S.; Crosby, M.; Mungall, C.J.; Matthews, B.B.; Campbell, K.S.; Hradecky, P.; Huang, Y.; Kaminker, J.S.; Millburn, G.H.; Prochnik, S.; et al. Annotation of the Drosophila melanogaster euchromatic genome: A systematic review. *Genome Biol.* **2002**, *3*, 83. [CrossRef]
63. Quesneville, H.; Bergman, C.M.; Andrieu, O.; Autard, D.; Nouaud, D.; Ashburner, M.; Anxolabéhère, D.; Stormo, G. Combined Evidence Annotation of Transposable Elements in Genome Sequences. *PLoS Comput. Biol.* **2005**, *1*, e22. [CrossRef]
64. Xu, H.-E.; Zhang, H.-H.; Xia, T.; Han, M.-J.; Shen, Y.-H.; Zhang, Z. BmTEdb: A collective database of transposable elements in the silkworm genome. *Database* **2013**, *2013*, 55. [CrossRef]
65. Obbard, D.J.; Gordon, K.H.; Buck, A.H.; Jiggins, F.M. The evolution of rnai as a defence against viruses and transposable elements. *Philos. Trans. R. Soc. Lond. B Biol. Sci.* **2009**, *364*, 99–115. [CrossRef]
66. Zhao, X.; Xiong, J.; Mao, F.; Sheng, Y.; Chen, X.; Feng, L.; Dui, W.; Yang, W.; Kapusta, A.; Feschotte, C.; et al. RNAi-dependent Polycomb repression controls transposable elements in Tetrahymena. *Genes Dev.* **2019**, *33*, 348–364. [CrossRef]
67. Revell, L.J.; Harmon, L.J.; Collar, D.C. Phylogenetic Signal, Evolutionary Process, and Rate. *Syst. Biol.* **2008**, *57*, 591–601. [CrossRef] [PubMed]
68. Kidwell, M.G. Transposable elements and the evolution of genome size in eukaryotes. *Genetica* **2002**, *115*, 49–63. [CrossRef]
69. Felsenstein, J. Phylogenies and the Comparative Method. *Am. Nat.* **1985**, *125*, 1–15. [CrossRef]
70. Daniels, S.B.; Peterson, K.R.; Strausbaugh, L.D.; Kidwell, M.G.; Chovnick, A. Evidence for Horizontal Transmission of the P Transposable Element between Drosophila Species. *Genetics* **1990**, *124*, 339–355.
71. Bingham, P.M.; Kidwell, M.G.; Rubin, G.M. The molecular basis of P-M hybrid dysgenesis: The role of the P element, a P-strain-specific transposon family. *Cell* **1982**, *29*, 995–1004. [CrossRef]
72. Gilbert, C.; Feschotte, C. Horizontal acquisition of transposable elements and viral sequences: Patterns and consequences. *Curr. Opin. Genet. Dev.* **2018**, *49*, 15–24. [CrossRef] [PubMed]
73. Peccoud, J.; Loiseau, V.; Cordaux, R.; Gilbert, C. Massive horizontal transfer of transposable elements in insects. *Proc. Natl. Acad. Sci. USA* **2017**, *114*, 4721–4726. [CrossRef] [PubMed]

74. Blumenstiel, J.P. Evolutionary dynamics of transposable elements in a small RNA world. *Trends Genet.* **2011**, *27*, 23–31. [CrossRef] [PubMed]
75. Reiss, D.; Mialdea, G.; Miele, V.; De Vienne, D.M.; Peccoud, J.; Gilbert, C.; Duret, L.; Charlat, S. Global survey of mobile DNA horizontal transfer in arthropods reveals Lepidoptera as a prime hotspot. *PLoS Genet.* **2019**, *15*, e1007965. [CrossRef] [PubMed]
76. Silva, J.C.; Loreto, E.L.; Clark, J.B. Factors that affect the horizontal transfer of transposable elements. *Curr. Issues Mol. Biol.* **2004**, *6*, 57–71. [PubMed]
77. Goerner-Potvin, P.; Bourque, G. Computational tools to unmask transposable elements. *Nat. Rev. Genet.* **2018**, *19*, 688–704. [CrossRef]
78. Pasyukova, E.G.; Nuzhdin, S.V.; Morozova, T.V.; Mackay, T.F.C. Accumulation of Transposable Elements in the Genome of Drosophila melanogaster is Associated with a Decrease in Fitness. *J. Hered.* **2004**, *95*, 284–290. [CrossRef]
79. Charlesworth, B. Effective population size and patterns of molecular evolution and variation. *Nat. Rev. Genet.* **2009**, *10*, 195–205. [CrossRef]
80. Bewick, A.J.; Vogel, K.J.; Moore, A.J.; Schmitz, R.J. Evolution of DNA methylation across insects. *Mol. Biol. Evol.* **2017**, *34*, 654–665. [CrossRef] [PubMed]
81. Lewis, S.H.; Quarles, K.A.; Yang, Y.; Tanguy, M.; Frezal, L.; Smith, S.A.; Sharma, P.P.; Cordaux, R.; Gilbert, C.; Giraud, I.; et al. Pan-arthropod analysis reveals somatic pirnas as an ancestral defence against transposable elements. *Nat. Ecol. Evol.* **2018**, *2*, 174–181. [CrossRef] [PubMed]
82. Lu, J.; Clark, A.G. Population dynamics of PIWI-interacting RNAs (piRNAs) and their targets in Drosophila. *Genome Res.* **2009**, *20*, 212–227. [CrossRef]
83. El Baidouri, M.; Cooke, R.; Gao, D.; Lasserre, E.; Llauro, C.; Mirouze, M.; Picault, N.; Panaud, O.; Carpentier, M.-C.; Jackson, S. Widespread and frequent horizontal transfers of transposable elements in plants. *Genome Res.* **2014**, *24*, 831–838. [CrossRef]
84. Drezen, J.-M.; Josse, T.; Bézier, A.; Gauthier, J.; Huguet, E.; Herniou, E.A. Impact of Lateral Transfers on the Genomes of Lepidoptera. *Genes* **2017**, *8*, 315. [CrossRef]
85. Gilbert, C.; Peccoud, J.; Chateigner, A.; Moumen, B.; Cordaux, R.; Herniou, E.A. Continuous Influx of Genetic Material from Host to Virus Populations. *PLoS Genet.* **2016**, *12*, e1005838. [CrossRef]
86. Luo, S.; Lu, J. Silencing of Transposable Elements by piRNAs in Drosophila: An Evolutionary Perspective. *Genom. Proteom. Bioinform.* **2017**, *15*, 164–176. [CrossRef]
87. Yamashiro, H.; Siomi, M.C. Piwi-interacting RNA in drosophila: Biogenesis, transposon regulation, and beyond. *Chem. Rev.* **2018**, *118*, 4404–4421. [CrossRef]
88. Bartolomé, C.; Bello, X.; Maside, X. Widespread evidence for horizontal transfer of transposable elements across Drosophila genomes. *Genome Biol.* **2009**, *10*, R22. [CrossRef]
89. Nigumann, P.; Redik, K.; Mätlik, K.; Speek, M. Many Human Genes Are Transcribed from the Antisense Promoter of L1 Retrotransposon. *Genomics* **2002**, *79*, 628–634. [CrossRef]
90. Van De Lagemaat, L.N.; Landry, J.-R.; Mager, D.L.; Medstrand, P. Transposable elements in mammals promote regulatory variation and diversification of genes with specialized functions. *Trends Genet.* **2003**, *19*, 530–536. [CrossRef]
91. Chen, B.; Shilova, V.Y.; Zatsepina, O.G.; Evgen'Ev, M.B.; Feder, M.E. Location of P element insertions in the proximal promoter region of Hsp70A is consequential for gene expression and correlated with fecundity in Drosophila melanogaster. *Cell Stress Chaperones* **2008**, *13*, 11–17. [CrossRef]
92. Bejerano, G.; Lowe, C.B.; Ahituv, N.; King, B.; Siepel, A.; Salama, S.R.; Rubin, E.M.; Kent, W.J.; Haussler, D. A distal enhancer and an ultraconserved exon are derived from a novel retroposon. *Nature* **2006**, *441*, 87–90. [CrossRef]
93. Smith, A.M.; Sanchez, M.-J.; Follows, G.A.; Kinston, S.; Donaldson, I.J.; Green, A.R.; Gottgens, B. A novel mode of enhancer evolution: The Tal1 stem cell enhancer recruited a MIR element to specifically boost its activity. *Genome Res.* **2008**, *18*, 1422–1432. [CrossRef] [PubMed]
94. Franchini, L.F.; Lopez-Leal, R.; Nasif, S.; Beati, P.; Gelman, D.M.; Low, M.J.; de Souza, F.J.; Rubinstein, M. Convergent evolution of two mammalian neuronal enhancers by sequential exaptation of unrelated retroposons. *Proc. Natl. Acad. Sci. USA* **2011**, *108*, 15270–15275. [CrossRef]
95. Bundock, P.; Hooykaas, P. An Arabidopsis hAT-like transposase is essential for plant development. *Nature* **2005**, *436*, 282–284. [CrossRef]

96. Kapitonov, V.V.; Jurka, J. RAG1 Core and V(D)J Recombination Signal Sequences Were Derived from Transib Transposons. *PLoS Biol.* **2005**, *3*, e181. [CrossRef] [PubMed]
97. Jangam, D.; Feschotte, C.; Betrán, E. Transposable element domestication as an adaptation to evolutionary conflicts. *Trends Genet.* **2017**, *33*, 817–831. [CrossRef]
98. De Souza, F.S.; Franchini, L.F.; Rubinstein, M. Exaptation of Transposable Elements into Novel Cis-Regulatory Elements: Is the Evidence Always Strong? *Mol. Biol. Evol.* **2013**, *30*, 1239–1251. [CrossRef]
99. Joly-Lopez, Z.; Bureau, T. Exaptation of transposable element coding sequences. *Curr. Opin. Genet. Dev.* **2018**, *49*, 34–42. [CrossRef]
100. Chuong, E.B.; Elde, N.C.; Feschotte, C. Regulatory activities of transposable elements: From conflicts to benefits. *Nat. Rev. Genet.* **2017**, *18*, 71–86. [CrossRef]
101. Aminetzach, Y.T.; MacPherson, J.M.; Petrov, D.A. Pesticide Resistance via Transposition-Mediated Adaptive Gene Truncation in Drosophila. *Science* **2005**, *309*, 764–767. [CrossRef]
102. Hof, A.E.V.; Campagne, P.; Rigden, D.J.; Yung, C.J.; Lingley, J.; Quail, M.A.; Hall, N.; Darby, A.C.; Saccheri, I.J. The industrial melanism mutation in British peppered moths is a transposable element. *Nature* **2016**, *534*, 102–105. [CrossRef] [PubMed]

© 2019 by the authors. Licensee MDPI, Basel, Switzerland. This article is an open access article distributed under the terms and conditions of the Creative Commons Attribution (CC BY) license (http://creativecommons.org/licenses/by/4.0/).

MDPI
St. Alban-Anlage 66
4052 Basel
Switzerland
Tel. +41 61 683 77 34
Fax +41 61 302 89 18
www.mdpi.com

Genes Editorial Office
E-mail: genes@mdpi.com
www.mdpi.com/journal/genes

www.ingramcontent.com/pod-product-compliance
Lightning Source LLC
LaVergne TN
LVHW071946080526
838202LV00064B/6689